014092596

KU-405-577

WITHDRAWN
FROM STOCK

Process Systems Engineering

Edited by
Michael C. Georgiadis,
Eustathios S. Kikkinides, and
Efstratios N. Pistikopoulos

Related Titles

Haber, R., Bars, R., Schmitz, U.

Predictive Control in Process Engineering

2008
ISBN: 978-3-527-31492-8

Dimian, A. C., Bildea, C. S.

Chemical Process Design

Computer-Aided Case Studies

2008
ISBN: 978-3-527-31403-4

Engell, S. (ed.)

Logistic Optimization of Chemical Production Processes

2008
ISBN: 978-3-527-30830-9

Ingham, J., Dunn, I. J., Heinzle, E., Prenosil, J. E., Snape, J. B.

Chemical Engineering Dynamics

An Introduction to Modelling and Computer Simulation

2007
ISBN: 978-3-527-31678-6

Keil, F. J. (ed.)

Modeling of Process Intensification

2007
ISBN: 978-3-527-31143-9

Bröckel, U., Meier, W., Wagner, G. (eds.)

Product Design and Engineering

Best Practices

2007
ISBN: 978-3-527-31529-1

Puigjaner, L., Heyen, G. (eds.)

Computer Aided Process and Product Engineering

2006
ISBN: 978-3-527-30804-0

Agachi, P. S., Nagy, Z. K., Cristea, M. V., Imre-Lucaci, A.

Model Based Control

Case Studies in Process Engineering

2007
ISBN: 978-3-527-31545-1

Sundmacher, K., Kienle, A., Seidel-Morgenstern, A. (eds.)

Integrated Chemical Processes

Synthesis, Operation, Analysis, and Control

2005
ISBN: 978-3-527-30831-6

Process Systems Engineering

Volume 5: Energy Systems Engineering

Edited by
Michael C. Georgiadis, Eustathios S. Kikkinides, and
Efstratios N. Pistikopoulos

WILEY-VCH Verlag GmbH & Co. KGaA

The Editors

Prof. Michael C. Georgiadis
Department of Engineering Informatics
    and Telecommunications
University of Western Macedonia
Vermiou and Lygeris Str.
Kozani, 50100
Greece
and
Centre for Process Systems Engineering
Department of Chemical Engineering
Imperial College London
Roderick Hill Building
London SW7 2AZ
United Kingdom

Prof. Eustathios S. Kikkinides
Department of Engineering and
    Management of Energy Resources
University of Western Macedonia
Sialvera and Bakola Str.
Kozani, 50100
Greece

Prof. Efstratios N. Pistikopoulos
Centre for Process Systems Engineering
Department of Chemical Engineering
Imperial College London
Roderic Hill Building
London SW7 2AZ
United Kingdom

All books published by Wiley-VCH are
carefully produced. Nevertheless, authors,
editors, and publisher do not warrant the
information contained in these books,
including this book, to be free of errors.
Readers are advised to keep in mind that
statements, data, illustrations, procedural
details or other items may inadvertently be
inaccurate.

Library of Congress Card No.:
applied for

British Library Cataloguing-in-Publication Data
A catalogue record for this book is available
from the British Library.

**Bibliographic information published by
the Deutsche Nationalbibliothek**
Die Deutsche Nationalbibliothek lists this
publication in the Deutsche
Nationalbibliografie; detailed bibliographic
data are available on the Internet at
<http://dnb.d-nb.de>.

© 2008 WILEY-VCH Verlag GmbH & Co.
KGaA, Weinheim

All rights reserved (including those of
translation into other languages). No part of
this book may be reproduced in any form – by
photoprinting, microfilm, or any other means
– nor transmitted or translated into a
machine language without written
permission from the publishers. Registered
names, trademarks, etc. used in this book,
even when not specifically marked as such,
are not to be considered unprotected by law.

Composition VTEX Ltd., Vilnius
Printing Strauss GmbH, Mörlenbach
Bookbinding Litges & Dopf GmbH,
Heppenheim

Printed in the Federal Republic of Germany
Printed on acid-free paper

ISBN: 978-3-527-31694-6

Contents

Preface *XI*
List of Contributors *XV*

1 **Polygeneration Systems Engineering** *1*
P. Liu, E.N. Pistikopoulos and Z. Li *1*
1.1 Introduction to Polygeneration Energy Systems *1*
1.1.1 Background *1*
1.1.2 Development Plans *4*
1.1.2.1 Vision 21 Plan *4*
1.1.2.2 China's Plan for Polygeneration Energy Systems *7*
1.2 Model for Strategic Investment Planning *8*
1.2.1 Mathematical Formulation *8*
1.2.2 Investment Planning for a Polygeneration Plant – a Case Study *15*
1.3 Model for Configuration Design *18*
1.3.1 Introduction *19*
1.3.2 Superstructure Representation *20*
1.3.3 Mathematical Model *23*
1.3.3.1 Physical Representation of the Process *24*
1.3.3.2 Economic Section of the Model *29*
1.3.4 A Polygeneration Plant for Electricity and Methanol – a Case
Study *31*
1.3.4.1 Scenario Settings and Model Results *31*
1.3.4.2 Discussion and Sensitivity Analysis *34*
1.4 Concluding Remarks *36*

2 **Multiobjective Design and Optimization of Urban Energy
Systems** *39*
F. Maréchal, C. Weber and D. Favrat *39*
2.1 Introduction *39*
2.2 Structuring Phase *42*
2.2.1 Definition of the List of Available Energy Sources *42*
2.2.2 Energy Consumption Profiles Including Power and Temperature
Levels *42*

Process Systems Engineering: Vol. 5 Energy Systems Engineering
Edited by Michael C. Georgiadis, Eustathios S. Kikkinides and Efstratios N. Pistikopoulos
Copyright © 2008 WILEY-VCH Verlag GmbH & Co. KGaA, Weinheim
ISBN: 978-3-527-31694-6

2.2.3   Routing Algorithm   *42*
2.2.4   Technology Database   *43*
2.3   Optimization Phase   *44*
2.4   The Postprocessing Phase   *44*
2.5   Resolution Strategy of the Optimization Phase   *45*
2.6   Mathematical Formulation of the Optimization Phase   *47*
2.6.1   The Master Optimization   *47*
2.6.1.1   Objective Function and Decision Variables   *47*
2.6.1.2   Constraints and Parameters   *47*
2.6.2   The Slave Optimization   *48*
2.6.2.1   Objective Function and Decision Variables   *48*
2.6.2.2   Constraints of the Slave Optimization   *48*
2.6.2.3   First Principle Energy Balances   *49*
2.6.2.4   Heat Cascade Constraints   *52*
2.6.2.5   Electricity Balances   *53*
2.6.2.6   Mass Balances   *55*
2.6.2.7   Connections of Nodes and Implementation/Location of the
         Technologies   *55*
2.6.2.8   Costing and CO_2 Emission Functions   *57*
2.6.2.9   Considering Multiobjective Aspects in the Slave Optimization   *60*
2.6.2.10   Empirical Constraints   *60*
2.6.3   Data Processing Routine   *61*
2.7   Test Case Application   *63*
2.7.1   Structuring of the Information   *63*
2.7.1.1   List of Available Energy Sources   *64*
2.7.1.2   Energy Consumption Profiles Including Power and Temperature
         Levels   *64*
2.7.1.3   Spatial Constraints and Possible Connections   *69*
2.7.1.4   Decentralized Technologies   *69*
2.8   Optimization   *70*
2.8.1   Decision Variables   *70*
2.8.1.1   Conditions Defined for the Optimization   *70*
2.8.1.2   Resolution Phase Analysis   *70*
2.8.1.3   Pareto Optimal Frontier   *72*
2.8.1.4   Space Restrictions   *79*
2.8.1.5   Inhomogeneous Requirements   *79*
2.8.2   Final Configuration Choice   *81*
2.9   Conclusions   *81*

3   Hydrogen-Based Energy Systems: The Storage Challenge   85
       E.S. Kikkinides   85
3.1   Introduction   *85*
3.2   Underground Hydrogen Storage   *87*
3.2.1   Process Operation and Energy Issues   *89*
3.3   Hydrogen Storage by Compression   *90*

3.3.1   Process Operation and Energy Issues   *91*
3.4   Hydrogen Storage by Liquefaction   *93*
3.4.1   Process Operation and Energy Issues   *94*
3.5   "Solid" Hydrogen Storage   *98*
3.5.1   Adsorbents   *99*
3.5.2   Metal and Chemical Hydrides   *101*
3.5.3   Process Operation and Energy Issues   *102*
3.5.3.1 Material Balance   *103*
3.5.3.2 Energy Balance   *103*
3.5.3.3 Momentum Balance   *103*
3.5.3.4 Adsorption Kinetics and Equilibrium   *104*
3.5.3.5 Absorption Kinetics and Equilibrium   *104*
3.5.3.6 Boundary and Initial Conditions   *106*
3.5.4   Solution of the Model   *108*
3.5.5   Determination of Optimal Adsorption Properties in a
         Carbon-Based Adsorbent   *108*
3.5.6   Effect of the Heat Conduction Coefficient   *111*
3.5.7   Optimal Design and Control of Hydrogen Storage in a
         Metal-Hydride Bed   *112*
3.5.8   Modeling the Heat Exchange Process   *113*
3.6   Dynamic Optimization Framework   *114*
3.6.1   Process Constraints   *114*
3.6.2   Solution Approach   *115*
3.6.3   Optimization Results for Controlling the Cooling Medium Flow
         Rate   *117*
3.7   Conclusions   *118*

4   **Hydrogen Energy Systems**   *125*
         L. Zheng, L. Chang, D. Gao, P. Liu and E.N. Pistikopoulos   125
4.1   Introduction of Hydrogen Production Technologies and their
         Integration into Polygeneration Energy Systems   *126*
4.1.1   Introduction of Hydrogen Production Technologies   *126*
4.1.2   Integration of Hydrogen Production into Polygeneration Energy
         Systems   *127*
4.1.2.1 Integration into a Hydrogen-Electricity Polygeneration System   *127*
4.1.2.2 Economic Analysis of Hydrogen-Electricity Polygeneration
         Systems   *130*
4.2   Methodology of Hydrogen Infrastructure Strategic Planning
         Model   *138*
4.2.1   Hydrogen Infrastructure Pathway Options   *138*
4.2.2   Model Overview   *141*
4.3   Case Study Application   *145*
4.3.1   Hydrogen Energy System Taking Methanol as an Agent   *145*

4.3.2   A Case Study of China Hydrogen Infrastructure Strategic
         Planning   *146*
4.4   Conclusions   *156*

5   **Integrated Optimization of Oil and Gas Production**   *159*
         M.C. Georgiadis and E.N. Pistikopoulos   159
5.1   Introduction   *159*
5.2   General Problem Statement   *163*
5.3   Mathematical Model   *163*
5.4   Separable Programming for Optimization of Production
         Operations   *170*
5.4.1   Well Oil Rate Upper and Lower Bounds and Piecewise Linear
         Approximation   *172*
5.5   Solution Procedure   *177*
         Remarks   *178*
         Remark   *179*
5.6   Accuracy and Robustness of the Formulation   *179*
5.7   Examples   *183*
5.7.1   Naturally Flowing Wells   *183*
5.7.2   Gas-Lift Well Example   *186*
5.7.3   Treelike Structure Pipeline Network   *188*
5.8   Conclusions   *189*

6   **Wind Turbines Modeling and Control**   *195*
         K. Kouramas and E.N. Pistikopoulos   195
6.1   Introduction   *195*
6.2   Wind Turbine System Modeling   *196*
6.2.1   Modeling of the Rotor Aerodynamics   *197*
6.2.2   Drive Train Model   *199*
6.2.3   Modeling the Induction Generator   *200*
6.2.3.1 The Fifth-Order Model   *201*
6.2.3.2 Reduced-Order Model   *202*
6.3   Dynamic Performance Simulation   *203*
6.3.1   Wind Turbine Dynamic Behavior   *203*
6.3.2   Simulations   *205*
6.4   Power Control   *207*
6.5   Conclusions   *213*

7   **Stochastic Optimization of Investment Planning Problems in the
         Electric Power Industry**   *215*
         D. Kuhn, P. Parpas and B. Rustem   215
7.1   Introduction   *215*
7.2   Stochastic Programming Concepts   *217*
7.2.1   Modeling Framework   *218*

7.3   A Capacity Expansion Model   *219*
7.4   Probabilistic Model   *223*
7.5   Decomposition Algorithms   *226*
7.5.1   Nested Benders Decomposition   *227*

8   Integrated Design of CO₂ Capture Processes from Natural Gas   *231*
        F.E. Pereira, E. Keskes, A. Galindo, G. Jackson and C.S. Adjiman   *231*
8.1   Introduction   *231*
8.1.1   Choice of Separation Technique   *232*
8.1.2   Choice of Solvent for Physical Absorption   *232*
8.1.3   A Basic Physical Absorption Process Flowsheet   *233*
8.2   Development of a Process Model   *234*
8.2.1   SAFT–VR Model for $CO_2/CH_4/n$-Alkane Mixtures   *235*
8.3   Optimization Methodology   *238*
8.3.1   Formulation of the Design Problem   *240*
8.3.2   Process Optimization   *241*
8.4   Optimal Designs   *242*
8.5   Conclusions   *246*

9   Energy Efficiency and Carbon Footprint Reduction   *249*
        J.J. Klemeš, I. Bulatov and S.J. Perry   *249*
9.1   Introduction   *249*
9.2   Overview of Energy-Saving Techniques   *251*
9.3   Screening and Scoping: Energy Audit, Benchmarking, and Good
        Housekeeping   *253*
9.4   Energy-Saving Analysis: Balancing and Flowsheeting
        Simulation   *255*
9.5   Integrated Approach: Heat Integration   *257*
9.6   Energy Efficiency Optimization   *265*
9.6.1   Energy Management and Control   *265*
9.6.2   Energy Management Systems (EMS) and Programs   *266*
9.7   Emerging Energy-Saving Technologies   *267*
9.7.1   Methodology – the Description of the Algorithm   *268*
9.7.2   EMINENT Tool Description   *272*
9.8   Emissions and Carbon Footprint   *274*
9.8.1   Technologies and Their CFPs   *277*
9.8.1.1 Micro-CHP   *277*
9.8.1.2 Reciprocating Engines   *277*
9.8.1.3 Stirling Engines   *277*
9.8.1.4 Fuel Cells   *278*
9.8.2   Biomass   *279*
9.8.3   Solar   *280*
9.8.4   Wind   *283*
9.8.5   Overall CFP Performance of the Technologies for Energy Use   *283*

9.9   A Selection of Industrial and Residential Sector Applications:
         Petrochemical, Hospital Complex, Food and Drink   284
9.9.1   Petrochemicals – Fluid Catalytic Cracking   284
9.9.2   Hospital Complex   289
9.9.3   Food and Drink   291
9.10   Conclusions   297

10   **Optimization of Structure and Operating Parameters of a Sequence
         of Distillation Columns for Thermal Separation of Hydrocarbon
         Mixtures**   *301*
         M. Markowski and K. Urbaniec   *301*
10.1   Introduction   *301*
10.2   Methodology of Thermal Integration of a System Comprising a
         Sequence of Distillation Columns and a Refrigeration
         Subsystem   *302*
10.3   Mathematical Model of an Ideal Distillation Column   *305*
10.4   Thermal Integration Procedure for a System Employing
         Conventional Distillation Columns   *308*
10.4.1   Theoretical Model   *308*
10.4.2   Thermal Integration   *309*
10.4.3   Example   *311*
10.5   Thermal Integration Procedure for a System Employing
         Heat-Integrated Distillation Columns   *315*
10.5.1   Adjustment of the Theoretical Model   *315*
10.5.2   Thermal Integration   *317*
10.5.3   Example   *317*
10.6   Conclusions   *321*

         Index   *323*

Preface

Energy is one of the most critical international issues at the moment and most likely to be so for the years to come. As part of the energy debate, it is becoming gradually accepted that current energy systems, networks encompassing everything from primary energy sources to final energy services, are becoming unsustainable. Driven primarily by concerns over urban air quality, global warming caused by greenhouse gas emissions, and dependence on depleting fossil fuel reserves, a transition to alternative energy systems is receiving serious attention. Such a tradition will certainly involve meeting the growing energy demand of the future with greater efficiency as well as using more renewable energy sources (such as wind, solar, biomass, etc.). While many technical options exist for developing a future sustainable and less environmentally damaging energy supply, they are often treated separately driven by their own technical communities and political groups (Pistikopoulos [1]).

Energy systems engineering provides a methodological scientific framework to arrive at realistic integrated solutions to the complex energy problems by adopting a holistic systems-based approach. This book demonstrates the potential of an energy-systems engineering-based approach to systematically quantify different options at different levels of complexity (technology, plant, energy supply chain, megasystem) through a number of real-life applications.

In Chapter 1, Pistikopoulos and coworkers present a polygeneration energy systems approach, which combines power generation and chemical fuel synthesis in a single plant, thus providing a promising alternative pathway toward achieving sustainable and flexible economic development. A mixed-integer programming formulation is proposed in constructing long-term decision models which are suitable for investment planning and design of polygeneration infrastructure systems. More specifically, two models are presented: one for investment planning of a polygeneration plant, and the other for the configuration design – both models are then applied to a case study involving a polygeneration plant producing methanol and electricity.

The group of Marechal and coworkers in Chapter 2 proposes a systematic methodology for designing urban energy conversion systems. The methodology allows the integration of polygeneration energy conversion technologies as well as the design of heat and cold distribution networks. It leads to the optimal selection

Process Systems Engineering: Vol. 5 Energy Systems Engineering
Edited by Michael C. Georgiadis, Eustathios S. Kikkinides and Efstratios N. Pistikopoulos
Copyright © 2008 WILEY-VCH Verlag GmbH & Co. KGaA, Weinheim
ISBN: 978-3-527-31694-6

and combination of energy conversion technologies, their geographical implementation, the optimal supply and return temperatures of the distribution networks, the connection of buildings, and the use of local renewable resources that can be used for heat pumping. The methodology integrates thermodynamic considerations, mathematical programming techniques, conceptual design principles, and various alternative technologies, geographical locations and buildings.

Kikkinides in Chapter 3 presents a comprehensive review of important issues and challenges associated with hydrogen storage for the case of stationary and mobile applications. Different methods of storage have been analyzed in terms of process performance and heat management. The results of this analysis demonstrate that each method has its own advantages and disadvantages and the final selection depends on a variety of characteristics such as type of storage application, hydrogen supply requirements, etc. In the near term, compressed hydrogen will compete with liquefied hydrogen as the dominant storage method for fuel-cell vehicles at the demonstration level. Metal or chemical hydrides are expected to offer significant advantages when the current research and development efforts succeed in commercializing the required technology. It is illustrated that in metal hydride storage tanks enhanced heat transfer allows for faster tank filling while at the same time it prevents the formation of large temperature gradients in the storage tank. Finally, it is concluded that for this reason innovative design strategies for both heat transfer configurations and/or material microstructure should be developed to improve process performance.

The group of Li and coworkers, in collaboration with Liu and Pistikopoulos, presents in Chapter 4 a generic model for the optimal long-range planning and design of future hydrogen supply chains for fuel cell vehicles. The model relies on mixed-integer optimization techniques to provide optimal integrated investment strategies across a variety of supply chain decision-making stages. Key high-level decisions addressed by the model are the optimal selection of the primary energy feedstocks, allocation of conversion technologies to either central or distributed production sites, design of the distribution technology network, and selection of refueling technologies. At the strategic planning level, capacity expansions as well as technology shutdowns are captured to explicitly address the dynamics of the infrastructure and the timing of the investment. Low-level operational decisions addressed include the estimation of primary energy feedstock requirements and production, distribution, and refueling rates. Realizing that both financial and ecological concerns are driving the interest in hydrogen, formal multiobjective optimization techniques are used to establish the optimal tradeoff between the Net Present value of the investment and green house emissions. The applicability of the proposed models is illustrated by using a large-scale case study from China. It is shown how the model can identify optimal supply chain designs, capacity expansion policies, and investment strategies for a given geographical region.

Georgiadis and Pistikopoulos in Chapter 5 present a formulation for the optimization of oil and gas production operations. The formulation simultaneously optimizes well rates and gas lift allocation, is able to handle flow interactions among wells and can be applied to difficult situations where some wells are too weak to flow to the manifold or require a certain amount of gas lift to flow. The accuracy, efficiency, and robustness of the formulation have been established by comparisons with an exact optimization formulation that was solved using an SQP method in a number of examples. The algorithm is applicable both to tree-like structure pipeline networks and to pipeline networks with loops. Due to the ability of the algorithm to account for flow interactions it always leads to superior operating policies compared to the heuristic rules typically applied in practice. The proposed optimization method can be used for real-time production control since all the variables required for the construction of a well model can be measured and the discrete data can be directly incorporated in the formulation.

In Chapter 6, Kouramas and Pistikopoulos present a model-based approach for power control of wind turbines including the design of a novel explicit multi-parametric model predictive controller. A wind turbine with its detailed nonlinear, approximating nonlinear and approximating ARX model is proposed. Simulation analysis is used to discuss the wind turbine's performance in the presence of wind speed variations as well as blade pitch angle variations. The latter was shown that can be manipulated to provide power regulation in the presence of wind speed disturbances. Finally, the steps for the design of an explicit model predictive controller are illustrated and the resulting controller is evaluated against the actual nonlinear model of the system.

The work of Kuhn, Parpas, and Rustem in Chapter 7 is motivated by the need to develop systematic methods for taking optimal decisions on whether to invest in new power system infrastructure. In this context, the timely expansion of generation and transmit ion capacities is crucial for the reliability of a power system and its ability to provide uninterrupted service under changing market conditions. Assuming a probabilistic model for future electricity demand, fuel prices, equipment failures, and electricity spot prices, a capacity expansion problem is formulated that minimizes the sum of the costs for upgrading the local power system and the costs for operating the upgraded system over an extended planning horizon. The arising optimization problem represents a two-stage stochastic program with binary first-stage decisions. Solution of this problem relies on a specialized algorithm, which constitutes a symbiosis of a regulated decomposition method and a branch-and-bound scheme.

Adjiman and coworkers in Chapter 8 present a process for the capture of CO_2 from natural gas streams through physical absorption into n-alkane solvent. A methodology for the optimization of the solvent blend and operating conditions is also introduced in the context of three different process flowsheets. The model indicates that this process is commercially viable, since profit through gas sales far outweighs capital and operating costs. Overall CO_2 inlet concentrations was studied. The process can be extended to investigate nonequilibrium strategies, or to the screening of solvents outside the n-alkane family. The work clearly demonstrates the benefits of employing a complex equation of state within a process modeling and optimization framework.

Klemes and coworkers in Chapter 9 present initially a critical overview of the energy-efficiency problem. Energy-saving techniques that are currently available, relatively simple but effective screening and scoping approaches are reviewed, including energy auditing, benchmarking, and good housekeeping. Energy-saving techniques, such as balancing and flowsheeting simulation, and a heat integration approach are also discussed in details. Then the increasingly significant issue of emissions and carbon footprint assessment and minimization is introduced. It is shown that even renewable energy sources make some contribution to the overall carbon footprint, which is frequently not accounted for in assessment studies. Various renewable technologies applicable in the domestic sector are assessed in a view of reducing the release of CO_2. A number of industrial applications are also presented including an energy-efficiency retrofit study from a large energy consuming petrochemicals industry, followed by a pulp and paper plant energy minimizations study and several examples from the food industry. Case studies related to the domestic energy sector are demonstrated as well. These cover a hospital and a total site comprising a sugar plant and a nearby town.

In the final chapter, Markowski and Urbaniec present a systematic methodology for the design of energy-efficient gas separation systems comprising a sequence of conventional distillation columns and a refrigeration subsystem. Tools and concepts derived from the well-known Pinch Technology are employed to address this problem. Assuming an ideal column and its profile, thermal integration of the underlying conventional systems is first considered. Extensions to flowsheets including heat-integrated columns are also presented. The efficiency of the proposed methodology is illustrated to flowsheets involving separation of complex hydrocarbon mixtures. The energy-saving potential of including heat-integrated distillation columns is revealed. Furthermore, extension to systems comprising also a heat-exchanger network and a refrigeration subsystem are illustrated, leading to significant reductions in the compressor shaftwork required for the gas separation process.

This collection represents a set of stand-alone works that capture recent research trends in energy systems engineering – the development and application of techniques, methodologies, algorithms, and tools for optimizing various energy systems. We hope that by the end of the book, the reader will have developed a commanding comprehension of the main aspects of integrated energy systems and energy supply chains, the ability to critically access the key characteristics and elements related to the design and operation of energy systems and the capacity to implement the new technology in practice.

We are extremely grateful to the authors for their outstanding contributions and for their patience, which have led to a final product that exceeded our expectations.

References

1 PISTIKOPOULOS, E.N., Energy Systems Engineering – an integrated approach for the energy systems of the future. *Presented in the European Congress of Chemical Engineering – 6, Copenhagen, Denmark, 16–21 September* **2007**.

Michael Georgiadis   June 2008
Eustathios Kikkinides
Efstratios Pistikopoulos

List of Contributors

Claire Adjiman
Imperial College London
Centre for Process Systems Engineering
Department of Chemical Engineering
Roderic Hill Building
South Kensington Campus
London SW7 2AZ
United Kingdom

Igor Bulatov
University of Manchester
School of Chemical Engineering and
Analytical Science
Centre of Process Integration
PO BOX 88
Manchester M60 1QD
United Kingdom

Le Chang
Tsinghua University
Tsinghua-BP Clean Energy Center
Department of Thermal Engineering
Shunde Building
Beijing 100084
China

Daniel Favrat
Industrial Energy Systems Laboratory
Ecole Polytechnique Federale de
Lausanne
Station 9
1015 Lausanne
Switzerland

Amparo Galindo
Imperial College London
Centre for Process Systems Engineering
Department of Chemical Engineering
Roderic Hill Building
South Kensington Campus
London SW7 2AZ
United Kingdom

Dan Gao
Tsinghua University
Tsinghua-BP Clean Energy Center
Department of Thermal Engineering
Shunde Building
Beijing 100084
China

Michael Georgiadis
University of Western Macedonia
Department of Engineering Informatics
and Telecommunications
Karamanli and Lygeris Street
Kozani 50100
Greece

and

Imperial College London
Centre for Process Systems Engineering
Department of Chemical Engineering
Roderic Hill Building
South Kensington Campus
London SW7 2AZ
United Kingdom

Process Systems Engineering: Vol. 5 Energy Systems Engineering
Edited by Michael C. Georgiadis, Eustathios S. Kikkinides and Efstratios N. Pistikopoulos
Copyright © 2008 WILEY-VCH Verlag GmbH & Co. KGaA, Weinheim
ISBN: 978-3-527-31694-6

George Jackson
Imperial College London
Centre for Process Systems Engineering
Department of Chemical Engineering
Roderic Hill Building
South Kensington Campus
London SW7 2AZ
United Kingdom

Emmanuel Keskes
Imperial College London
Centre for Process Systems Engineering
Department of Chemical Engineering
Roderic Hill Building
South Kensington Campus
London SW7 2AZ
United Kingdom

Eustathios Kikkinides
University of Western Macedonia
Department of Engineering and
Management of Energy Resources
Bakola and Sialvera Street
Kozani, 50100
Greece

Jiří Jaromír Klemeš
University of Manchester
School of Chemical Engineering and
Analytical Science
Centre of Process Integration
PO BOX 88
Manchester M60 1QD
United Kingdom

Kostas Kouramas
Imperial College London
Centre for Process Systems Engineering
Department of Chemical Engineering
Roderic Hill Building
South Kensington Campus
London SW7 2AZ
United Kingdom

Daniel Kuhn
Imperial College London
Department of Computing
180 Queen's Gate
South Kensington Campus
London SW7 2AZ
United Kingdom

Zheng Li
Tsinghua University
Tsinghua–BP Clean Energy Center
Department of Thermal Engineering
Shunde Building
Beijing 100084
China

Pei Liu
Imperial College London
Centre for Process Systems Engineering
Department of Chemical Engineering
Roderic Hill Building
South Kensington Campus
London SW7 2AZ
United Kingdom

Francois Marechall
Industrial Energy Systems Laboratory
Ecole Polytechnique Federale de
Lausanne
Station 9
1015 Lausanne
Switzerland

Mariusz Markowski
Warsaw University of Technology
CERED Centre of Excellence
Plock Campus
Jachowicza 2/4, 09-402 Plock
Poland

Panos Parpas
Imperial College London
Department of Computing
180 Queen's Gate
South Kensington Campus
London SW7 2AZ
United Kingdom

Frances Pereira
Imperial College London
Centre for Process Systems Engineering
Department of Chemical Engineering
Roderic Hill Building
South Kensington Campus
London SW7 2AZ
United Kingdom

Simon John Perry
University of Manchester
School of Chemical Engineering and
analytical science
Centre of Process Integration
PO BOX 88
Manchester M60 1QD
United Kingdom

Efstratios N. Pistikopoulos
Imperial College London
Centre for Process Systems Engineering
Department of Chemical Engineering
Roderic Hill Building
South Kensington Campus
London SW7 2AZ
United Kingdom

Berc Rustem
Imperial College London
Department of Computing
180 Queen's Gate
South Kensington Campus
London SW7 2AZ
United Kingdom

Krzysztof Urbaniec
Warsaw University of Technology
CERED Centre of Excellence
Plock Campus
Jachowicza 2/4, 09-402 Plock
Poland

Celine Weber
Industrial Energy Systems Laboratory
Ecole Polytechnique Federale de
Lausanne
Station 9
1015 Lausanne
Switzerland

1
Polygeneration Systems Engineering
Pei Liu, Efstratios N. Pistikopoulos, and Zheng Li

In this chapter we introduce polygeneration energy systems in the context of future energy systems, and modeling and optimization issues involved in planning and configuration design of polygeneration processes. Two models are presented: one is for investment planning of a polygeneration plant, and the other is for the configuration design. Applications of both models are realized in case studies for a polygeneration plant producing methanol and electricity. The algorithm of mixed-integer programming (MIP) is employed in both models.

1.1
Introduction to Polygeneration Energy Systems

1.1.1
Background

Global energy consumption has been rising since 1970s, and according to the projection of Department of Energy (DOE) of the United States, it will keep on rising for quite a long period in the future [1], see Fig. 1.1. However, the global greenhouse gas (GHG) emissions have to be restricted to a certain level since most countries around the world (excluding the United States) had signed the Kyoto Protocol by 2005, which requires its participating countries to reduce their GHG emissions to below emission levels in 1990 by 2012 [1], see Fig. 1.2.

A severe and lasting global energy problem is the shortage of liquid fuels. Worldwide proved oil reserves amount to 1293 billion barrels by 2006, and the daily consumption in 2003 was 80 million barrels [1]. If this consumption rate were not to increase, all global oil reserves would be depleted in about 44 years. Moreover, 57% of the oil reserves are found in the Middle East, the most politically unstable region around the world. Thus, countries that depend heavily on oil importation need to seek diversification of liquid fuel suppliers to enhance national energy security.

One possible solution to the current acute energy and environmental problems is to utilize more advanced, innovative, and efficient primary energy technologies. Power generation is the largest primary energy consumer, accounting for 40% of

Process Systems Engineering: Vol. 5 Energy Systems Engineering
Edited by Michael C. Georgiadis, Eustathios S. Kikkinides and Efstratios N. Pistikopoulos
Copyright © 2008 WILEY-VCH Verlag GmbH & Co. KGaA, Weinheim
ISBN: 978-3-527-31694-6

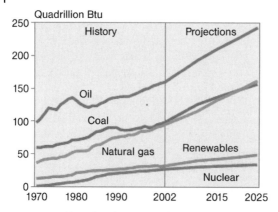

Fig. 1.1 World marketed energy use by energy type, 1970–2025.

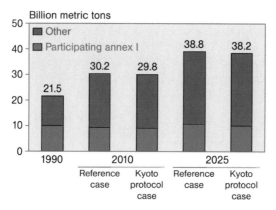

Fig. 1.2 World carbon dioxide emissions.

the primary energy and using all energy resources, like coal, natural gas, and oil. Consequently, it is a colossal source of GHG emissions, being the cause for the release of more than 7.7 billion tons of carbon dioxide annually; thus, power generation accounts for 37.5% of the total annual carbon dioxide emissions [2]. Innovations and improvements of power generation technologies for higher efficiency and lower emissions have never ceased to emerge and be licensed over the decades. The Integrated Gasification Combined Cycle (IGCC), which combines a gasifier with a gas turbine cycle and a steam turbine cycle, is one of the most promising alternatives.

Fortunately, oil is not the only energy source for the production of liquid fuels. They can also be synthesized from other fossil fuels, like coal, natural gas, and petroleum coke, as well as renewable energy sources such as biomass. Synthetic liquid fuels have the potential to substitute conventional, oil-based liquid fuels; for example, methanol and dimethyl ether (DME) can be used as gasoline and diesel oil, respectively.

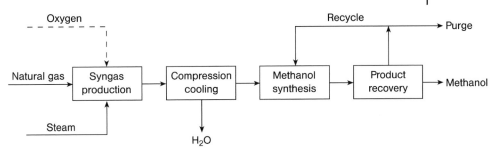

Fig. 1.3 Methanol synthesis from natural gas.

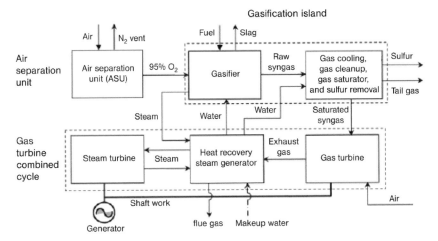

Fig. 1.4 Conceptional process illustration of an IGCC power plant.

The concept of polygeneration comes from the similarities between liquid fuel synthesis processes and combined cycle power generation processes. Both processes require synthesis gas (syngas), mainly consisting of carbon dioxide and hydrogen, as an intermediate product. Figure 1.3 shows a conventional process for methanol synthesis from natural gas [3]. Natural gas is first fed into a reactor together with adequate amount of oxygen and steam to produce syngas. The syngas is then cooled down to remove extra amount of steam and impurities like sulfurated hydrogen. The cooled syngas is compressed to certain pressure, fed into a synthesis reactor, and catalyzed to produce crude methanol. The effluent is condensed and distillated to produce the final product, either fuel-class or chemical-class methanol. Unconverted gas is recycled to the synthesis reactor. Figure 1.4 shows a typical conceptional structure of an IGCC power plant [4]. It usually uses coal as main fuel, but can also use petroleum coke or biomass. Fuel is fed into a gasifier first, where it is gasified to produce syngas. The syngas goes through a series of cleaning modules to remove solid particles and acid gases. The clean syngas is then fed into the combustion chamber of a gas turbine, where it combusts with a large amount of air, producing hot flue gas. This hot gas expands in the gas turbine first, and then goes through a heat recovery steam generator (HRSG), producing high, medium, and low pressure steam to drive corresponding pressured steam turbines. The similarities existing between liquid fuel synthesis processes and IGCC processes indicate a possibility of co-producing electricity, synthetic fuels, heat, and other chemicals in one process, with higher conversion efficiency that will result in lower polluting emission levels.

A polygeneration energy system can improve profit margins and market penetration, decrease the overall capital investment cost, reduce GHG emissions, increase feedstock flexibility, and alleviate the current grave dependence on crude oil and all refinery fuels. An exemplary polygeneration energy system for integrated production of methanol and electricity is shown in Fig. 1.5, in which coal or other carbon-based fuels are fed to a gasifier, where they react with oxygen to produce syngas; part of it is fed to a chemical synthesis plant to produce methanol, which can be either sold, stored in the plant for peak-time power generation, or transported to other power plants for peak-time power generation. The flue gas from the chemical synthesis plant, together with the other part of the fresh syngas flow, undergoes combustion in a power generation plant to generate electricity [5].

Polygeneration energy systems have many advantages over conventional stand-alone power or chemical plants. The production cost for methanol can be reduced by 40% in a polygeneration plant coproducing methanol, heat, and electricity. For a quad-generation plant coproducing syngas, methanol, heat, and power, the reduction over conventional plants is 46% for syngas production cost, 38% for capital investment, 31% for operating cost per energy unit, and 22.6% for GHG emission [6]. In a polygeneration plant coproducing DME and electricity, the DME production cost will be 6 to 6.5 $/GJ, thus comparable with conventional fuel prices [7].

Overall, polygeneration energy systems have many advantages over conventional stand-alone power generation or chemical synthesis technologies, like higher energy conversion efficiency, lower emissions, and lower capital investment. All these advantages make polygeneration a very competitive technology and attract many countries to make research and development plans. Attracted by these advantages, several countries have made plans for research and developing polygeneration energy systems. A review of plans for polygeneration energy systems published by the U.S. government and Chinese government is presented next.

1.1.2
Development Plans

1.1.2.1 **Vision 21 Plan**
The U.S. Department of Energy (DOE) published its Vision 21 program in 1999, which was an approach to producing energy that addresses pollution control as an integral part of high-efficiency energy production, the so-called *21st Century Energy Plant* [8].

Two main goals of the Vision 21 plan are shown in Table 1.1. Its primary purpose is to decrease emissions from fossil fuels. According to its standard, air pollutants such as sulfur dioxide, nitrogen oxides, and mercury will be reduced to around zero levels. Carbon dioxide emissions will also be reduced significantly because of

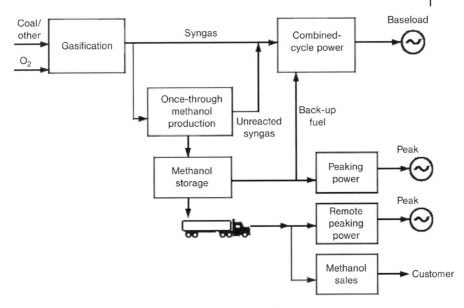

Fig. 1.5 A polygeneration plant co-producing methanol and electricity.

Table 1.1 Vision 21 plant targets.

	Reference plant	2010	2020
Air emissions			
SO_2	98%	99%	>99%
NO_x	0.15 lb/10^6 Btu	0.05 lb/10^6 Btu	<0.01 lb/10^6 Btu
Mercury	\	90% removal	95% removal
By product utilization	30%	50%	Near 100%
Plant efficiency (HHV)	40%	45–50%	50–60%

the enhanced efficiency. The plant could even reduce GHG emissions to the near zero level when carbon dioxide capture and sequestration (CCS) technologies are applied. The secondary purpose of the Vision 21 plan is to maximize the plant efficiency and use as much of the energy in the fuel as possible. This purpose can be realized by combining several advanced energy utilization processes together, such as combining an IGCC and a chemical synthesis process to produce electricity and high-value chemical fuels, like methanol and DME. A series of research projects about these advanced energy utilization technologies were developed as subprojects of the Vision 21 plan. Figure 1.6 shows a conceptual structure of a Vision 21 plant based on different modules [9].

Vision 21 plants are based on many technologies in diverse fields, as shown in Fig. 1.7. Modules of the plant are based on enabling technologies, of which some are already demonstrated (e.g., gasification and advanced combustion), while others, such as gas separation, need further improvement. Supporting technologies

Vision 21 technology modules

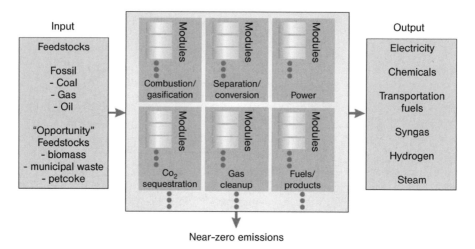

Fig. 1.6 Vision 21 technology module.

Vision 21 technology portfolio

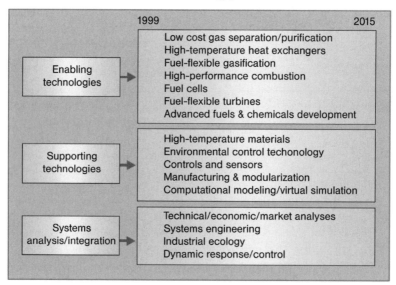

Fig. 1.7 Vision 21 technology portfolio.

are the bases for components and support systems of Vision 21 plants. Most of the supporting technologies listed in Fig. 1.7 are still under development. Systems analysis and integration are essential technologies for process configuration and optimization, such that maximum potential of processes can be utilized.

Another feature of the Vision 21 plan is that configurations of energy plants are according to specific local conditions, such as the market demand of products and

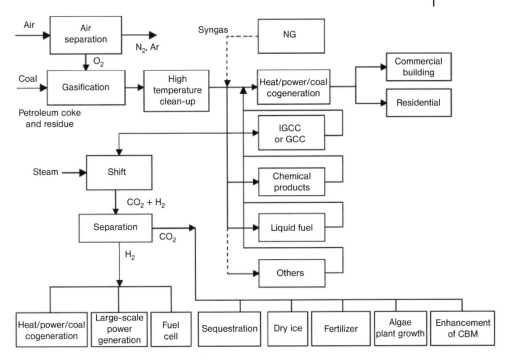

Fig. 1.8 Polygeneration energy systems in China.

availability of fuels. A plant can be designed to produce certain products to meet the nearby market needs, to utilize the most approachable and cheapest fuels, as well as to cooperate with existing plants at sites.

1.1.2.2 China's Plan for Polygeneration Energy Systems

China's oil and natural gas reserve cannot meet its increasing demand of energy. The present coal-based energy system in China will not be changed for a long period, since the coal reserve is abundant compared with that of oil and natural gas. However, because of its fast economic, especially the dramatic increase of automobile development, China needs more and more liquid fuels. This demand has already exceeded its oil production ability. On the other hand, China also faces severe environmental problems of air pollution and GHG emissions, mainly caused by conventional power plants [10].

Polygeneration energy systems can both meet China's demands for liquid fuels and solve the environmental problems. A mid-term plan for polygeneration will focus on using coal and other solid fuel (e.g., petroleum coke and heavy oil residues) as main feedstocks. Products include electricity, liquid fuels, such as methanol, DME, Fischer-Tropsch (F-T) fuel, as well as heat for commercial and residential use. Hydrogen will be hopefully included into its long-term plan, see Fig. 1.8.

China's mid-term plan for polygeneration energy systems focuses on the following key technologies and system research:

- *Gasification.* Texaco and Shell's entrained bed gasifiers with a capacity greater than 2000 tpd of coal are in favor, especially Texaco's coal–water slurry gasification technology, which has already entered China's power market successfully.
- *Gas turbine.* New generation of gas turbines with an efficiency higher than 60%, initial temperature over 1400 °C, pressure ratio from 20 to 30 is of interest. Specific gas turbine research that uses low heat value syngas as its main fuel is employed during China's Tenth Five-Year Plan period and 863 Program, in order to make gas turbines suitable to cooperate with coal gasifiers.
- *Syngas purification.* High-temperature (500–600 °C) purification technologies are of great interest and believed to be a very effective direction to reduce the whole plant's capital cost, because the efficiency of the whole cycle is increased significantly since most of the apparent heat in syngas can be utilized at such a high temperature.
- *Syngas reactor.* Mature synthesis reaction technologies, such as F-T oil synthesis, Liquid Phase Methanol (LPMeOH) synthesis, two-step method for DME, mainly focusing on their stand-alone conversion rate, need to be modified to be suitable for polygeneration energy systems, which combine the electricity and chemical production and require a total economical and environmental optimization.
- *Gas separation technologies.* Membrane separation technologies have the potential to reduce the large amount energy consumed in air and tail gas separation processes using a conventional method.
- *Carbon dioxide treatment.* Carbon dioxide produced from polygeneration energy systems can be buried in deep sea or injected into oil or gas fields to enhance its production.

1.2
Model for Strategic Investment Planning

A general model for facilitating strategic decisions in the development progress of polygeneration energy systems was built using the algorithm of MIP. The purpose of this model is to help make plans for the future development of polygeneration energy systems in specific regions or countries. It includes all alternative technologies, feedstocks, and products and selects the most profitable combination by maximizing the net present value (NPV) over a future planning time horizon. Its mathematical formulation is presented first, followed by an application to a case study for a polygeneration plant producing electricity and methanol.

1.2.1
Mathematical Formulation

The general model for strategic investment planning uses the MIP algorithm to incorporate all alternative technologies, feedstocks, and products, from which the most profitable combination is selected by maximizing the NPV over a future planning time horizon. The model is presented by first presenting its main assumptions and features, then listing definitions of variables together with their physical meanings, followed by the mathematical equations, and last, linearization of non-linear equations.

The model relies on exhaustive enumeration of possible energy production alternatives, connecting the elements of each set to the permissible elements of the next set and then activating the eligible groups of pertinent equality and inequality constraints; for every combination, it then calculates economic quantities within the time interval considered. Then, it summarizes the economic results in all time intervals and gets a NPV value. The procedure is continued for the next combination and the NPV is evaluated again (the largest is stored). Then another round of iteration begins until all the element combinations are checked. When all possibilities have been evaluated, the optimal result (max NPV) is obtained.

Some unique features of polygeneration energy systems were explicitly considered before applying the MIP algorithm to model the strategic investment decisions associated with their development. More specifically, the model must be able to accommodate the following:

- A medium-term or long-term future planning horizon, as construction periods of polygeneration energy systems are around 5 years, whilst operation time can be as long as 30 years.
- Multiple and diverse primary energy feedstocks and production technologies. Diversity of feedstocks and alternative technologies is a key feature of polygeneration energy systems. The model has collected large sets of economic and technical data of feedstocks and alternative technologies to represent this feature.
- Transition from one production technology to another and capacity expansions over time. As more advanced technologies, and cheaper and more environmental friendly feedstocks may become available in the future, transitions of technologies may lead to more profit.
- Economics of scale of large-scale production technologies or so-called size effect. Capacity expansions are essential to the capital investment of polygeneration energy systems in the progress of development. Capital investment and installation costs decrease when the operation capacity expanses.

The model uses a superstructure to represent all possible alternatives for a polygeneration plant, shown in Fig. 1.9. The superstructure acts as the overriding model, capturing all the possible alternatives and interactions between the various production components. From this superstructure the optimization algorithm then searches for the best combinations by eliminating the existence of units and links between them. The superstructure starts with a set of primary energy resources:

$$F = \{\text{Fuel 1, Fuel 2, Fuel 3}, \ldots\}, \tag{1.1}$$

which can be used as feedstocks for polygeneration energy systems using any technologies in set A:

$$A = \{\text{Technology 1, Technology 2, Technology 3}, \ldots\}. \tag{1.2}$$

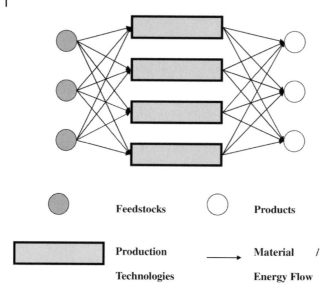

Fig. 1.9 Superstructure representation of a polygeneration plant.

Each of these production technologies is defined such that they can perform conversion of primary energy feedstocks into final products:

$$P = \{\text{Product 1, Product 2, Product 3,} \ldots\} \tag{1.3}$$

The primary objective of the model is to support the optimal strategic investment planning over a long-term future horizon, t in T. The model achieves this by making optimal decisions in terms of four levels:

- Level 1: Strategic process design, including selection of primary feedstocks, selection of conversion technologies, and assignment of feedstocks to technologies.
- Level 2: Capacity expansion planning, including estimation of when to expand which technologies and how much the exact quantity should be.
- Level 3: Production planning, including estimation of how much of each primary energy feedstocks the selected technologies require and what the rates of products production are.
- Level 4: Performance index assessment, including computation of financial objective such as NPV.

Using MIP modeling techniques, constrains can be formulated representing these various decisions. In addition to the constraints describing the physical phenomena, an objective function also has to be formulated for the financial performance criteria. This objective function drives the optimization in search of the best investment strategy. Since a long-term future investment horizon is of interest, the NPV is chosen as the financial performance measure, thereby capturing both the capital expenditure and operating cost requirements of processes as well as the time value of the investment over the future horizon.

Model assumptions are summarized below:

- Existing capacity of all technologies before the planning time horizon is not considered, such that capacity of all technologies is 0 at the beginning of the planning horizon.
- Capacity expansion and decrease cannot exist in the same time interval.
- Production of chemical products cannot exceed market demands, whilst production of electricity and other possible products and byproducts does not have such a constraint and all these products are sold to the market regardless of quantity.
- Variable operating cost of polygeneration plants includes only purchase of feedstocks, which overwhelm other components in real operations.

Model formulations are shown below. Nomenclature of model variables can be found in Table 1.2.

The objective function is the NPV, which is a summation of discounted net cash flows over all time intervals:

$$\text{NPV} = \sum_t \frac{NetCashFlow(t) * Years(t)}{(1 + DiscountRate)^{n(t)}} \qquad (1.4)$$

Production and economic constraints consist of both equalities and inequalities. The total installed capacity for a specific technology is an accumulation of capacity expansions:

$$F(a, t) = FE(a, t), \quad t = t1 \qquad (1.5)$$

$$F(a, t) = F(a, t - 1) + FE(a, t) - FD(a, t), \quad t > t1 \qquad (1.6)$$

Capacity expansion of a technology during a time interval cannot exceed its proper upper limit:

$$0 \leqslant FE(a, t) \leqslant Y(a, t) * UpperLimit \qquad (1.7)$$

Capacity decrease of a technology during a time interval cannot exceed its proper upper limit:

$$0 \leqslant FD(a, t) \leqslant \big(1 - Y(a, t)\big) * UpperLimit \qquad (1.8)$$

Conversion from feedstocks to products can occur through a series of different technologies:

$$\sum_f Fuel(a, f, t) * ConversionRate(a, p) = Product(a, p, t) \qquad (1.9)$$

Production of chemical products cannot exceed the demand of the respective products:

Table 1.2 Nomenclature for model variables.

Symbol	Definition
Continuous variables	
$F(a, t)$	Capacity of technology a during time interval t
$FE(a, t)$	Amount of capacity expansion of technology a during time interval t
$FD(a, t)$	Amount of capacity decrease of technology a during time interval t
$Fuel(a, f, t)$	Consumption of feedstock f through technology a during time interval t
$Product(a, p, t)$	Production of product p through technology a during time interval t
$Invest(a, t)$	Capital investment in technology a during time interval t
$FixedCost(a, t)$	Fixed operating cost of technology a during time interval t
$VarCost(a, t)$	Variable operating cost of technology a during time interval t
$Income(a, t)$	Income of technology a during time interval t
$NetCashFlow(t)$	Net cash flow before tax during time interval t
NPV	Net present value over the whole planning horizon
Discrete variable	
$Y(a, t)$	1 if the capacity of technology a expands or remains unchanged during time interval t, 0 if it decreases
Parameters	
DiscountRate	Discount rate throughout the planning horizon
OperatingTimePerYear	Operating time per year for all technologies
$Years(t)$	Number of years within time interval t
$FuelSupply(f, t)$	Maximum supply of fuel f during time interval t
$Demand(p, t)$	Market demand for product p during time interval t
$FuelPrice(f, t)$	Price of feedstock f during time interval t
$ProductPrice(p, t)$	Price of product p during time interval t
$RefCapacity(a)$	Capacity of the reference plant for technology a
$RefInvest(a)$	Capital investment of the reference plant for technology a
$RefFixedCost(a)$	Fixed cost of the reference plant for technology a
$SizeFactor(a)$	Size factor of technology a
$ConversionRate(a, p)$	Conversion rate of feedstocks to product p through technology a, on a HHV (higher heating value) basis
$FuelTechnology(a, f)$	Binary, 1 if technology a can utilize feedstock f, 0 otherwise
$N(t)$	Number of years from beginning of the planning horizon to the beginning of time interval t
$Size\_factor(a)$	An indicator of size effect of technology a

$$\sum_a Product(a, p, t) \leqslant Demand(p, t) \qquad (1.10)$$

Consumption of feedstocks through a technology should also not exceed its installed capacity:

$$\sum_f Fuel(a, f, t) \leqslant F(a, t) * OperatingTimePerYear \qquad (1.11)$$

Furthermore, the consumption of feedstocks should not exceed available supplies:

$$\sum_f Fuel(a, f, t) \leqslant FuelSupply(f, t) \qquad (1.12)$$

Capital investment and fixed cost of a technology are calculated from a reference plant on an annual basis, considering the size effect:

$$Invest(a, t) = \frac{RefInvest(a)}{Years(t)} * \left(\frac{FE(a, t)}{RefCapacity}\right)^{SizeFactor(a)} \qquad (1.13)$$

$$FixedCost(a, t) = \frac{RefFixedCost(a)}{Years(t)} * \left(\frac{F(a, t)}{RefCapacity}\right)^{SizeFactor(a)} \qquad (1.14)$$

The variable operating cost is a summation of the cost of feedstocks:

$$VarCost(a, t) = \sum_f FuelPrice(f, t) * Fuel(a, f, t) \qquad (1.15)$$

The income of a technology occurs as a result of the sale of its products:

$$Income(a, t) = \sum_p ProductPrice(p, t) * Product(a, p, t) \qquad (1.16)$$

The net cash flow during a time interval is the algebraic summation of capital investment, fixed cost, variable cost, and income:

$$NetCashFlow(t) = \sum_a \left(Income(a, t) - \left(Invest(a, t)\right.\right.$$

$$\left.\left. + FixedCost(a, t) + VarCost(a, t)\right)\right) \qquad (1.17)$$

The model shown above is nonlinear because of Eqs. (1.13) and (1.14). Since most equations in the model are linear except Eqs. (1.13) and (1.14), it is desirable to linearize this model to avoid the nonlinearity, which requires demonstration of convexity and extensive computation time to obtain a global optimum. Linearization of nonlinear equations involving binary variables can be conducted using the method given by Williams [11]. Applying the linearization algorithm, Eqs. (1.13) and (1.4) become

$$Invest(a, t) = \frac{RefInvest(a)}{Years(t)} * \left(\frac{FE'(a, t)}{RefCapacity}\right)^{SizeFactor(a)} \qquad (1.18)$$

$$FE'(a, t) = \sum_{i=1}^{n} \lambda_{FE,i}(a, t) * FE_i(a, t) \qquad (1.19)$$

$$Invest(a, t) = \sum_{i=1}^{n} \lambda_{FE,i}(a, t) * Invest_i(a, t) \tag{1.20}$$

$$\sum_{i=1}^{n} \lambda_{FE,i} = 1 \tag{1.21}$$

$$\lambda_{FE,i} \geqslant 0, \quad i = 1, 2, \ldots, n \tag{1.22}$$

$$\sum_{i=1}^{n-1} \delta_{FE,i} = 1 \tag{1.23}$$

$$\lambda_{FE,1} - \delta_{FE,1} \leqslant 0 \tag{1.24}$$

$$\lambda_{FE,i} - \delta_{FE,i-1} - \delta_{FE,i} \leqslant 0, \quad i = 2, 3, \ldots, n-1 \tag{1.25}$$

$$\lambda_{FE,n} - \delta_{FE,n-1} \leqslant 0 \tag{1.26}$$

$$FixedCost(a, t) = \frac{RefFixedCost(a)}{Years(t)} * \left(\frac{F'(a, t)}{RefCapacity}\right)^{SizeFactor(a)} \tag{1.27}$$

$$F'(a, t) = \sum_{i=1}^{n} \lambda_{F,i}(a, t) * F_i(a, t) \tag{1.28}$$

$$FixedCost(a, t) = \sum_{i=1}^{n} \lambda_{F,i}(a, t) * FixedCost_i(a, t) \tag{1.29}$$

$$\sum_{i=1}^{n} \lambda_{F,i} = 1 \tag{1.30}$$

$$\lambda_{F,i} \geqslant 0, \quad i = 1, 2, \ldots, n \tag{1.31}$$

$$\sum_{i=1}^{n-1} \delta_{F,i} = 1 \tag{1.32}$$

$$\lambda_{F,1} - \delta_{F,1} \leqslant 0 \tag{1.33}$$

$$\lambda_{F,i} - \delta_{F,i-1} - \delta_{F,i} \leqslant 0, \quad i = 2, 3, \ldots, n-1 \tag{1.34}$$

$$\lambda_{F,n} - \delta_{F,n-1} \leqslant 0 \tag{1.35}$$

where $FE_i(a, t)$ and $FixedCost_i(a, t)$ are discrete values of $FE(a, t)$ and $FixedCost(a, t)$ at fixed points, $\lambda_{FE,i}$ and $\lambda_{F,i}$ are two sets of continuous variables, and $\delta_{FE,i}$ and $\delta_{F,i}$ are two sets of binary variables.

Table 1.3 Prices and supply capability of available fuels.

	Price (million $/EJ)					Supply capability (EJ/year)				
	t1	t2	t3	t4	t5	t1	t2	t3	t4	t5
Coal	1038	1038	1038	1038	1038	38	42	45	49	53
Domestic NG	2999	3152	3313	3482	3659	2.0	2.8	4.0	5.6	7.8
Imported NG	5805	6102	6413	6740	7084	\	\	\	\	\
Biomass	5188	4151	3113	2075	1038	\	\	\	\	\

Table 1.4 Prices and market demands for products.

	Price (million $/EJ)					Market demand (EJ/year)				
	t1	t2	t3	t4	t5	t1	t2	t3	t4	t5
Methanol	15 114	15 217	15 320	15 423	15 526	0.19	0.45	0.79	1.18	1.68
Electricity	21 689	21 ,011	21 ,011	21 011	21 011	\	\	\	\	\

After the linearization, the model can be solved using MIP algorithms and a global optimum can be guaranteed. Next, a case study for polygeneration plants producing electricity and methanol is presented to illustrate the application of the model.

1.2.2
Investment Planning for a Polygeneration Plant – a Case Study

A case study for the development of polygeneration plants in China during the period between 2010 and 2035 was conducted using the model, focusing on nationwide investment planning for polygeneration plants producing electricity and methanol.

In the case study, the time horizon is from 2010 to 2035 and is divided into five time intervals with 5 years in each, represented as t1 to t5. Available fuels include coal, domestic and imported natural gas (NG), and biomass. Prices and supply capability of the feels are shown in Table 1.3. Products are electricity and methanol, prices and market demands for which are shown in Table 1.4. A set of 12 processes has been selected to incorporate all possible alternative paths for transforming primary energy feedstocks to final products, including 6 polygeneration pathways and 6 conventional, stand-alone technologies. The processes are represented as {COAL-LPMeOHe-CC-P, COAL-LPMeOHm-CC-P, COAL-GPMeOH-CC-P, NG-SMRRMS-NONE-M, NG-ATROTMS-NONE-M, NG-ATRRMS-NONE-M, BIO-LPMeOHm-CC-P, BIO-LPMeOHe-CC-P, BIO-LPMeOHhg-CC-P, BIO-LPMeOH-SC-M, BIO-GPMeOH-SC-M, BIO-GPMeOHhg-SC-M}. Nomenclature for technologies consisting of the processes is shown in Table 1.5. Each process has a reference plant as calculation basis. Table 1.6 shows capacity, capital invest-

Table 1.5 Nomenclature for alternative technologies.

Symbol	Meaning
COAL	Using coal as feeding fuel
NG	Using natural gas as feeding fuel
BIO	Using biomass as feeding fuel
LPMEOHe	Liquid phase methanol synthesis designed for high electricity generation
LPMEOHm	Liquid phase methanol synthesis designed for high methanol production
LPMEOHhg	Liquid phase methanol synthesis with hot gas cleaning
GPMEOH	Gas phase methanol synthesis
GPMEOHhg	Gas phase methanol synthesis with hot gas cleaning
SMRRMS	Steam methane reforming and recycle methane synthesis
ATROTMS	Auto-thermal reforming and once-through methane synthesis
ATRRMS	Auto-thermal reforming and recycle methane synthesis
CC	Combined cycle of a gas turbine and a steam turbine
NONE	No electricity generation
P	Polygeneration of electricity and methanol
M	Stand-alone methanol production

Table 1.6 Key parameters of reference processes.

Process	Capacity (GW fuel)	Investment cost (million $)	Fixed operation cost (million $/year)	Conversion rate to electricity	Conversion rate to methanol
COAL-LPMEOHe-CC-P	1.29	628	35.3	0.156	0.247
COAL-LPMEOHm-CC-P	1.29	594	39.9	−0.002	0.482
COAL-GPMEOH-CC-P	1.29	496	31.9	−0.066	0.517
NG-SMRRMS-NONE-M	0.744	429	23.6	−0.007	0.635
NG-ATROTMS-NONE-M	0.705	369	20.3	−0.007	0.670
NG-ATRRMS-NONE-M	0.716	326	17.9	−0.007	0.660
BIO-LPMEOHm-CC-P	0.428	279	11.2	0.124	0.376
BIO-LPMEOHe-CC-P	0.428	288	11.5	0.244	0.265
BIO-LPMEOHhg-CC-P	0.428	323	12.9	0.144	0.403
BIO-LPMEOH-SC-M	0.432	256	10.3	0	0.570
BIO-GPMEOH-SC-M	0.428	322	12.9	0.035	0.515
BIO-GPMEOHhg-SC-M	0.432	271	10.8	−0.040	0.589

ment cost, fixed operation cost, and conversion rates from fuel to products of these reference plants [12–14].

The optimized development procedure is shown in Fig. 1.10. Two technologies emerge as optimal throughout the planning horizon considered. The process of coal-based liquid phase methanol synthesis integrated with a combine cycle (de-

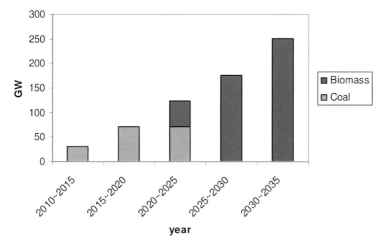

Fig. 1.10 Installed capacity over the time horizon.

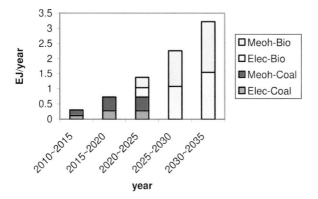

Fig. 1.11 Production rates over the time horizon, by product.

noted by Coal in Fig. 1.10) is optimal in the beginning. Another process, biomass-based liquid phase methanol synthesis integrated with a combine cycle (denoted by Biomass in Fig. 1.10), is found to be optimal afterward. Clearly, polygeneration processes exhibit overwhelming advantages over conventional stand-alone processes, none of which are selected by the model. On the supply side, none of natural gas-based processes are selected due to the comparatively high price of natural gas. The reason for the substitution of coal-based processes by biomass-based ones lies in two main aspects: higher conversion efficiency and the optimistic yet plausible assumption that the price of biomass will gradually drop to a competitive level of coal in the near future. On the production side, both processes tend to produce the largest amount of electricity, due to the relatively higher price of electricity than methanol (on an energy basis), see Fig. 1.11.

Cash flow by category over the time horizon for each process is shown in Fig. 1.12. Total income keeps on increasing over the time horizon because of the

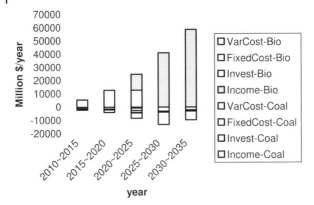

Fig. 1.12 Cash flow over the time horizon.

increasing market demand for methanol and consequent expansions of production capacity. The total costs get to the maximum during the time interval 2025 to 2030, and decrease afterward, due to the decrease in the biomass price.

In conclusion, the case study illustrates following characters of polygeneration energy systems:

- Polygeneration processes are in general superior to conventional stand-alone processes.
- Biomass-based polygeneration processes can be quite cost effective if the price of biomass drops to the same level as that of coal.
- Polygeneration processes producing more electricity are more profitable under scenarios with much higher price of electricity than methanol.
- Natural gas-based processes do not show any advantages because of inherent lower conversion rates of stand-alone processes and high price of natural gas. These processes become competitive only in circumstances where the price of natural gas is much lower and the ratio between electricity price and methanol price is lower as well.
- Fuel price has a much stronger influence on the net present value than other economic factors.

The model illustrated in this section is for mainly strategic planning for the development of polygeneration plants. Once the strategic developing direction is made, the next stage is specific process design in unit and equipment level. A model developed for this purpose is presented in the following section.

1.3
Model for Configuration Design

A mixed-integer nonlinear programming (MINLP) model is developed for the design optimization of polygeneration energy systems. A suitable superstructure is introduced, based on partitioning a general polygeneration energy system into four major blocks, for each of which alternative available technologies and types of equipment are considered. A detailed case study, involving a coal-based polygeneration plant producing electricity and methanol, is presented to demonstrate the key features and applicability of the proposed approach.

1.3.1
Introduction

Due to the high degree of integration and coupling between the power generation and chemical synthesis parts, determining the optimal configuration and design of a polygeneration energy system is quite a challenging task. Different process designs have been reported in the literature. Ma et al. proposed a group of sequential and parallel process designs for a coal-based polygeneration plant producing electricity and methanol. By comparing energy efficiency and economic characteristics, they concluded that the sequential design with a once-through methanol synthesis unit exhibits the best overall performance [15, 16]. Liu et al. tested the dynamic operating behavior of the processes designed by Ma et al. operating with changing power loads, concluding that a parallel process design has better performance under certain part-lord operating conditions [17]. Liu et al. proposed a novel process design producing electricity and DME from natural gas with the purpose of seeking a better way to transport natural gas from West China to East China [18]. Chen et al. compared the energy and exergy efficiencies between polygeneration plants producing electricity and DME and stand-alone DME plants, concluding that the energy saving ratio in a polygeneration plant can be as high as 16.6% [19]. Besides general processes producing electricity and chemical fuels, there are also other forms of polygeneration process designs with specific purposes, like exploring the potential of coal gas generated in coke ovens in iron and steel industry, and combining an ammonia process with a coal-based power generation process for a higher energy utilization rate [20, 21].

The reported works have some requirements for process design in common, like selection of technologies, sizing of equipment, process integration, prediction of technical performance, and prediction of economic characteristics. However, these processes are usually designed according to specific requirements or conditions, and few of them intend to incorporate all aspects arising in process design problems. It is therefore necessary to provide a general methodology for the process design of polygeneration energy systems which covers all design requirements and can be applied according to specific conditions. In this work, we present a superstructure representation of a polygeneration process and a comprehensive mixed-integer optimization model which systematically attempts to address this challenge.

This section is structured as follows. The superstructure representation is described next. Then the detailed mathematical model is illustrated, followed by a detailed presentation of a case study, modeling a polygeneration plant for the production of methanol and electricity.

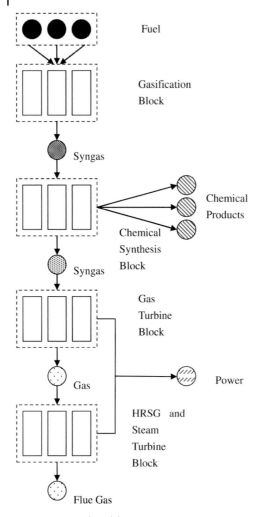

Fuel

Gasification
Block

Syngas

Chemical
Products

Chemical
Synthesis
Block

Syngas

Gas
Turbine
Block

Gas

Power

HRSG and
Steam
Turbine
Block

Flue Gas

Fig. 1.13 General model superstructure.

1.3.2
Superstructure Representation

A general superstructure representation of a polygeneration plant is shown in Fig. 1.13, consisting of four blocks: gasification, chemical synthesis, gas turbine, and heat recovery steam generator (HRSG) and steam turbine. The superstructure acts as the overriding model, capturing all the possible alternatives and intersections between process components. For each block, several alternative technologies and types of equipment are available for selection. All combinations of these technologies and types of equipment form the design space of the plant. The optimal process design is the best combination of these components, obtained by eliminating the existence of units and links between them.

To further illustrate the model superstructure and its utilization in modeling, a four-block superstructure of a coal-based polygeneration process producing electricity and methanol is discussed in detail below. Figure 1.14 shows the superstructure and all alternative technologies and types of equipment for each block.

The function of the gasification block is to prepare clean syngas for downstream utilization by gasifying feedstocks, usually coal, in a high-temperature, high-pressure, and reductive atmosphere. The crude syngas consists mainly of hydrogen, carbon monoxide, carbon dioxide, hydrogen sulfide (H_2S), carbonyl sulfide (COS), unconverted carbon, and ash. The hot crude syngas can either be quenched by cold water or cooled through a series of radiative and convective heat exchangers where heat can be recovered and used for power generation. Once it is cooled down, slag is removed and fine solid particles of unburned carbon are separated and recycled. After that, the syngas goes through a cleanup process to remove acid components which are extremely hazardous to downstream units and catalysts. Depending on the temperature of the syngas entering the cleanup process, two types of cleanup technologies are available: cold gas cleanup (CGCU) and hot gas cleanup (HGCU). In the model superstructure, the gasification block is further divided into two subblocks, representing the cooling part and the cleanup part. Technologies and types of equipment for the gasification block are denoted by

- Q: Quench
- LRC: Low-temperature radiative and convective cooling
- HRC: High-temperature radiative and convective cooling
- CC: Cold syngas cleanup
- HC: Hot syngas cleanup

Syngas leaving the gasification block enters the methanol synthesis block. There are two kinds of commercially matured methanol synthesis technologies. According to the phase of the synthesis reaction, they are known as gas phase methanol synthesis (GPMeOH) and liquid phase methanol synthesis (LPMeOH). In a GPMeOH reactor, reactants are in gas phase and react with each other on the surface of solid catalysts. In a LPMeOH reactor, gaseous reactants resolve in inert oil with solid catalyst particles being suspended in.

The methanol synthesis progress mainly consists of three reactions (shown in Eqs. (1.36) to (1.37)), and only two of them are independent:

$$CO + 2H_2 \rightarrow CH_3OH \tag{1.36}$$

$$CO_2 + 3H_2 \rightarrow CH_3OH + H_2O \tag{1.37}$$

$$CO + H_2O \rightarrow CO_2 + H_2 \tag{1.38}$$

Besides the main reactor, some accessory units are necessary to ensure an optimal performance for the reactor. First of all, since the synthesis reactions are

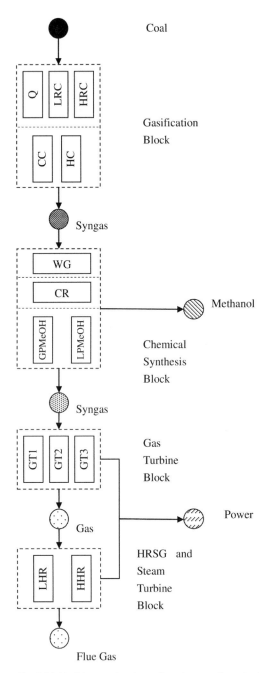

Fig. 1.14 Model superstructure of a polygeneration plant producing electricity and methanol.

highly exothermic, heat released in the synthesis reaction should be either recovered for power generation or absorbed by cooling water to obtain an isothermal atmosphere. For the ease of controlling the reaction heat, GPMeOH has an upper limit for the carbon monoxide content in reactants and needs a water gas shift reactor before it to adjust the composition of the feeding syngas. However, LPMeOH reactors do not have such a constraint. Therefore, a water gas shift reactor always exists before a GPMeOH reactor. Secondly, to obtain a maximum conversion rate, both GPMeOH and LPMeOH reactors have an optimal value of approximately 5% for the carbon dioxide volume fraction, making the catalyst staying at the most active level. With this requirement, a carbon dioxide removal unit is usually necessary before the reactor. Technologies and types of equipment for the methanol synthesis block are denoted by

- WG: Water gas shift reactor
- CR: Carbon dioxide removal unit
- GPMeOH: Gas phase methanol synthesis
- LPMeOH: Liquid phase methanol synthesis

Fluegas leaving the methanol synthesis block enters the gas turbine block for power generation. This block consists of a combustion chamber where fuel burns with pressurized air to produce pressurized hot gas, an air compressor that compresses air into the combustion chamber, and a turbine that transforms the thermal energy of the hot gas to mechanical work. Technologies and types of equipment for this block are given by

- GT1: Gas turbine whose first-stage inlet temperature is 1703 K
- GT2: Gas turbine whose first-stage inlet temperature is 1589 K
- GT3: Gas turbine whose first-stage inlet temperature is 1473 K

The exhausted gas leaving the gas turbine block enters HRSG where its heat is recovered to generate steam for the steam turbine, where the thermal energy in the steam is transformed into mechanical work. Technologies and types of equipment for this block are given by

- LHR: Low heat recovery technology with an exhaust gas temperature of 450 K
- HHR: High heat recovery technology with an exhaust gas temperature of 400 K

The four-block superstructure of the process has been introduced above. Its mathematical realization is discussed next.

1.3.3
Mathematical Model

The mathematical model consists of a physical representation of the process, including selection of technologies and types of equipment, sizing of equipment, and mass and energy balance, and an economic part regarding investment costs, operation and maintenance (O&M) costs, expenses on fuels, and income from sales of products. The objective function, annual profit of the plant, is also included in the economic part of the model. Before going into further details, general rules and nomenclature of the model are listed below:

- All logical choices representing selection of technologies and types of equipment are denoted by binary variables y with different subscripts representing corresponding situations. A value of 1 means selection whilst zero otherwise.
- All symbols with a "0" in the subscript are model parameters, having the same physical meaning with corresponding variables of the same format without "0."
- Naming of technologies and types of equipment follow those in Fig. 1.14.

Next, the physical representation of the process is introduced block by block first, followed by the illustration of the economic part of the model.

1.3.3.1 Physical Representation of the Process
Gasification Block Mass composition, temperature, and pressure of the fuel stream fed to the gasification block are given by

$$z(ie) = \sum_{ft} z_0(ft, ie) * y_f(ft) \tag{1.39}$$

$$T_f = \sum_{ft} T_{f,0}(ft) * y_f(ft) \tag{1.40}$$

$$P_f = \sum_{ft} P_{f,0}(ft) * y_f(ft) \tag{1.41}$$

$$\sum_{ft} y_f(ft) = 1 \tag{1.42}$$

where ft is the set of available fuel feedstocks, ie is the set of elements consisting a fuel feedstock, z is the mass fraction of element ie in a fuel feedstock, T_f is the temperature of a fuel feedstock, P_f is the pressure of a fuel feedstock, and y_f is a set of binary variables representing the selection of fuel feedstocks.

Physical properties of other feeding streams, like water or steam, and oxygen or air, can be expressed in a similar way, omitted here for conciseness.

Key operation parameters of the gasifier are calculated as follows:

$$R_{waterfuel} = \sum_{gft} R_{waterfuel,0} * y_{gas}(gft) \tag{1.43}$$

$$R_{O_2fuel} = \sum_{gft} R_{O_2fuel,0} * y_{gas}(gft) \tag{1.44}$$

$$C_{gas}(rs) = \sum_{gft} C_{gas,0}(rs, gft) * y_{gas}(gft) \tag{1.45}$$

$$T_{gas} = \sum_{gft} T_{gas,0} * y_{gas}(gft) \tag{1.46}$$

$$P_{gas} = \sum_{gft} P_{gas,0} * y_{gas}(gft) \tag{1.47}$$

$$\sum_{gft} y_{gas}(gft) = 1 \tag{1.48}$$

where $R_{waterfuel}$ is the ratio between the mass flowrates of the water/steam stream and the fuel stream, R_{O2fuel} is the ratio between the mass flowrates of the oxygen/air stream and the fuel stream, C_{gas} is the ratio of mole fractions between certain components in crude syngas, rs is a set of the mole ratios for the crude syngas, T_{gas} and P_{gas} are gasification temperature and pressure, respectively, and y_{gas} is a set of binary variables representing the selection of gasification technologies.

Using these parameters, mass relations between the feeding steams of the gasifier can be set up:

$$ma_{gfwater} = ma_f * R_{waterfuel} \tag{1.49}$$

$$ma_{gfO_2} = ma_f * R_{O_2fuel} \tag{1.50}$$

where $ma_{gfwater}$, ma_{gfO2}, and ma_f represent the mass flowrate of the water/steam stream, oxygen/air stream, and fuel stream, respectively.

After that, mass balance between the feedstocks of the gasifier and the crude syngas is built on an elementary basis:

$$f\left(z(ie), ma_f, ma_{gfwater}, ma_{gfO_2}, ma_{rawsg}\right) = 0 \tag{1.51}$$

where ma_{rawsg} represents the mass flowrate of the crude syngas.

The mole flowrate and the mass flowrate of the crude syngas can be connected to each other through its mole composition and the molecular weight of its components:

$$f(ma_{rawsg}, mo_{rawsg}, \vec{x}_{rawsg}) = 0 \tag{1.52}$$

where mo_{rawsg} is the mole flowrate of the crude syngas, and $\vec{x}_{rawsg}$ is a vector of mole composition of the crude syngas.

Equation (1.51) is on an elementary basis, while Eq. (1.52) is on a component basis. Considering the fact that more types of components exist in the crude syngas than elements in the feedstocks, ratios of mole fractions between certain components in the crude syngas, which is typical character of a certain type of gasification technology, are added below as complements to Eqs. (1.51) and (1.52).

$$f(\vec{x}_{rawsg}, C_{gas}) = 0 \tag{1.53}$$

Now that the mole composition, the temperature, and the pressure of the crude syngas are all here, its specific enthalpy and total enthalpy can be expressed by

$$h_{\text{rawsg}} = h(T_{\text{gas}}, p_{\text{gas}}, \vec{x}_{\text{rawsg}}) \tag{1.54}$$

$$H_{\text{rawsg}} = mo_{\text{rawsg}} * h_{\text{rawsg}} \tag{1.55}$$

where h_{rawsg} and H_{rawsg} are specific enthalpy and total enthalpy, respectively.

So far, selection of gasification technologies and mass balance for the gasifier has been built. The sizing of the gasifier is given by

$$ma_f - F_{\text{gas}} \leqslant 0 \tag{1.56}$$

where F_{gas} is the capacity of the gasifier for fuel processing.

Selection of technologies for the gasifier cooler, sizing of the cooler, and physical properties of the syngas leaving the cooler were modeled using the same method as for the gasifier, omitted here for conciseness.

Heat recovered in the gasifier cooler is given by

$$\Delta H_{\text{cooler}} = H_{\text{coolsg}} - H_{\text{rawsg}} \tag{1.57}$$

The recovered heat is used for power generation. The amount of power generation from this stream of heat depends on the temperature and pressure of the working fluid carrying it and the working process in the HRSG and steam turbine block. Instead of going into extensive technical details of heat transfer and fluid engineering, which is not the focus of this model, all the influential factors involved in generating power from the recovered heat are incorporated in a single parameter η_{cooler}, defined as the ratio of the power generated by the recovered heat to the total amount of the recovered heat. Thus the power generated indirectly from the gasifier cooler is given by

$$W_{cooler} = \Delta H_{cooler} * \eta_{cooler} \tag{1.58}$$

Note that some technologies are not compatible. For example, a hot gas cleanup unit can never be used after a quench cooler; instead, it requires a cooler using high-temperature radiative and convective technology, denoted by

$$y_{\text{cleanup}}(HC) - y_{\text{cooler}}(HRC) \leqslant 0 \tag{1.59}$$

Calculations of physical properties, mole composition, mass flowrate, and enthalpy for streams in the other blocks follow a similar way as stated above, thus omitted for conciseness. Only formulations representing unique characteristics of blocks or streams are depicted below.

Chemical Synthesis Block Leaving the cleanup unit, the clean syngas is split into two streams. One goes through an optional water gas shift reactor, and the other is bypassed, both joining together again after the water gas reactor. Mole composition of the stream going through the reactor is changed through the water gas shift reaction

$$CO + H_2O \rightarrow CO_2 + H_2 \tag{1.60}$$

while that of the bypassed stream remains unchanged. This is a means of adjusting the mole composition of the syngas according to the requirements of the methanol synthesis reactor. The degree of adjustment depends on the design parameter of split ratio R_{split} by the following equation:

$$f(R_{split}, ma_{clsg}, \vec{x}_{clsg}, \vec{x}_{wgsg}) = 0 \tag{1.61}$$

where ma_{clsg} is the mass flowrate of the clean syngas, $\vec{x}_{clsg}$ is the mole composition of the clean syngas, and $\vec{x}_{wgsg}$ is the mole composition of the combined syngas after the water gas shift reactor.

After the combination, the syngas goes through a carbon dioxide removal unit, where the fraction of carbon dioxide in the syngas is adjusted to an appropriate level for the best performance of the catalysts in the methanol synthesis reaction, given by

$$x_{sg}(CO_2) = x_{sg,0}(CO_2) \tag{1.62}$$

After the carbon dioxide removal, the syngas goes to the methanol synthesis reactor to produce methanol. Gas phase synthesis technology has a strict upper limit on the mole fraction of carbon monoxide in the syngas, given by

$$x_{sg}(H_2) - (2x_{sg}(CO) + 3x_{sg}(CO_2)) \geqslant (y_{meoh}(GPMEOH) - 1) * U \tag{1.63}$$

where y_{meoh} stands for the selection of synthesis technologies, and U is a very large positive number.

Parameters of the reactor, such as reaction degree and compositions of the synthesis product, depend on the selection of synthesis technologies, given by

$$R_{meoh}(pmeoh) = \sum_{meoh} R_{meoh,0}(pmeoh, meoh) * y(meoh) \tag{1.64}$$

Using these parameters, mass balance between the incoming syngas and product gas is given by

$$f(mo_{sg}, \vec{x}_{sg}, mo_{pg}, \vec{x}_{pg}, R_{meoh}) = 0 \tag{1.65}$$

where pg denotes product gas. One realization of (1.65) is shown below as an example:

$$mo_{sg} * (x_{sg}(H_2) - 2R_{meoh}(CO)x_{sg}(CO) - 3R_{meoh}(CO_2)x_{sg}(CO_2))$$

$$= mo_{pg} * x_{pg}(H_2) \tag{1.66}$$

Crude methanol produced in the synthesis reactor goes through a series of distillation columns to produce methanol as a final product, either of fuel degree or chemical degree. Mathematically, this process is formulated as splitting the crude product into two streams. One stream contains mainly methanol and a minor content of water, depending on the product degree, whilst the other stream includes all the other components in the crude methanol. The mass flowrate of the final product methanol, or its production rate, must meet its market demand, given by Eq. (1.82).

Gas Turbine Block The exhausted gas leaving the synthesis block, also known as fuel gas, goes to the gas turbine block, where it combusts in the combustion chamber with a large amount of compressed air to produce gas with sufficient high temperature and pressure. Mathematically, the combustion procedure is expressed as an oxidation reaction with excessive oxygen. Assuming that complete combustion takes place in the combustion chamber, all carbon monoxide and hydrogen in the fuel gas is converted to carbon dioxide and steam.

The selection of gasification technologies determines the temperature and pressure of the gas entering and leaving the gas turbine, denoted by T_1 and p_1, and T_4 and p_4, respectively. Through energy balance, the mass flowrate of the air flowing into the compressor of the gas turbine is a function with respect to T_1, the mass flowrate of the fuel gas, and its mole composition, given by

$$f(ma_{\text{air}}, ma_{\text{fg}}, \vec{x}_{\text{fg}}, T_1) = 0 \tag{1.67}$$

where ma_{fg} and ma_{air} are the mass flowrates of the fuel gas and the air to the compressor, respectively.

Now that the flowrate of the air flowing through the compressor and its physical properties at the inlet and outlet point of compressor are known, the compression work consumed by the compressor can be expressed as their function:

$$W_{\text{gc}} = f(ma_{\text{air}}, T_1, p_1, p_2) \tag{1.68}$$

The realization of Eq. (1.68) is shown as follows [22]:

$$W_{\text{gc}} = \frac{1}{\eta_{\text{isen}}} ma_{\text{air}} C_p T_1 \left(\left(\frac{p_2}{p_1} \right)^{\frac{\gamma-1}{\gamma}} - 1 \right) \tag{1.69}$$

where η_{isen} is the isentropic efficiency of the compressor, C_p is the specific heat capacity of the working fluid, and γ is the adiabatic coefficient of the working fluid.

The mechanical work generated by the gas turbine is a function of the mass flowrate of the gas flowing through the gas turbine, its composition, and its physical properties at the point before and after the turbine, denoted by

$$W_{\text{gt}} = f(ma_{\text{gas1}}, \vec{x}_{\text{gas1}}, T_1, p_1, T_4, p_4) \tag{1.70}$$

HRSG and Steam Turbine Block Gas leaving the gas turbine enters the HRSG and steam turbine block, where its heat is recovered in the HRSG and transformed to mechanical work in the steam turbine. An overall efficiency, denoted by η_{st}, is used to represent different technologies for HRSG and the steam turbine, given by

$$\eta_{st} = \sum_{st} \eta_{st,0}(hst) * y_{st}(hst) \tag{1.71}$$

where *hst* is a set of available technologies for HRSG and the steam turbine.
Work generated by the steam turbine is thus given by

$$W_{st} = H_{gas4} * \eta_{st} \tag{1.72}$$

So far, all streams of the mechanical work consumed and generated in the process have been presented, based on which the net work generated by the process is given by

$$W = W_{gt} + W_{st} + W_{cooler} + W_{meoh} - W_{ASU} - W_{comp} \tag{1.73}$$

where W_{ASU} is the work consumption in the air separation unit (if there is one) which provides oxygen for the gasifier. It is a function of the mass flowrate of the oxygen steam to the gasifier [23].

The mechanical work is transformed to electricity through a generator, and the electricity generation is given by

$$E = W * \eta_G \tag{1.74}$$

where η_G represents the efficiency of the generator. The electricity generation should meet its market demand, given by Eq. (1.82).

1.3.3.2 Economic Section of the Model

The objective function of the model is the annual profit of the polygeneration plant over lifetime, given by

$$Profit = Income - CostEquip - CostFuel \tag{1.75}$$

where *Income* is what obtained from sale of products, *CostEquip* is the total costs of equipment, including depreciated investment costs and annual O&M costs, and *CostFuel* is the annual expense on purchase of fuels. The income from the sale of products is given by

$$Income = \sum_{p} PriceP(p) * ProRate(p) * OpTime \tag{1.76}$$

where *p* is the set of products, *PriceP* is the product price, *ProRate* is the production rate of a certain product per unit time, and *OpTime* is the annual operation time of the plant. Total costs of equipment include annual depreciated investment cost, fixed O&M cost, and variable O&M cost, represented by *Inv*, *OMFix*, and *OMVar*, respectively, given by

$$CostEquip = \sum_{e} Inv(e) + OMFix(e) + OMVar(e) \tag{1.77}$$

where *e* is the set for equipment. Assuming there are *ei* kinds of technologies or types of equipment *e* available for a certain block or unit, the investment costs are expressed as

$$Inv(e) = \sum_{ei} UInv0(e, ei) * F(e, ei) \tag{1.78}$$

$$0 \leqslant F(e, ei) \leqslant y(e, ei) * UL \tag{1.79}$$

$$\sum_{ei} y(e, ei) = 1 \tag{1.80}$$

where *UInv0* is a set of parameters representing the investment cost per unit processing capacity, *F* is the operation capacity of equipment *e* using technology or of type *ei*, and *UL* is the upper limit for the operation capacity. Equation (1.79) ensures that if a technology or type of equipment is not selected, its corresponding capacity is zero, whilst if it is selected, the operation capacity can take any value between zero and the upper limit. Equation (1.80) makes sure that one and only one kind of technology or type is selected for a piece of equipment. Equations for calculating the fixed and variable O&M costs are similar, omitted here for conciseness.

Expense on purchase of fuels is expressed by its price *PriceF* and consumption rate *FuelRate* as follows:

$$CostFuel = \sum_{f} PriceF(f) * FuelRate(f) * OpTime \tag{1.81}$$

where *f* is the set for fuels.

Assuming that there is an upper bound and lower bound for market demands, the production rate should meet the following constraints:

$$LDemand(p) \leqslant ProRate(p) \leqslant UDemand(p) \tag{1.82}$$

where *LDemand* and *UDemand* are the lower bound and upper bound of market demands, respectively.

The model is illustrated above in two major parts, a physical part as a representation of the process in terms of mass and energy balance and selection of technologies and types of equipment, and an economic part incorporating calculations of costs and profits. A practical application of the model is shown in the next section in a case study for a polygeneration plant producing electricity and methanol.

Table 1.7 Ultimate analysis of Illinois #6 coal (wt.%, dry).

C	H	O	N	S	Ash
71.72	5.06	7.75	1.41	2.82	11.24

Table 1.8 Conversion rates of methanol synthesis.

Technology	CO to methanol	CO_2 to methanol
Gas phase	0.446	0.199
Liquid phase	0.128	0.0075

1.3.4
A Polygeneration Plant for Electricity and Methanol – a Case Study

Using the model presented above, a case study was conducted for designing the configuration of a polygeneration plant producing electricity and methanol.

1.3.4.1 Scenario Settings and Model Results

All the four blocks use technologies and types of equipment listed in the model superstructure (shown in Fig. 1.14) as available selections. Eleven chemical components are involved in the model for calculations of mass and energy balance and chemical reactions, including O_2, N_2, H_2, CO, CO_2, H_2O, CH_4, H_2S, SO_2, COS, and CH_3OH.

In the gasification block, Texaco gasification technology is applied to the gasifier, which uses dry pulverized coal, pure oxygen, and steam from power generation sector as main feedstocks. The gasification temperature and pressure is 1371 °C and 42 bar, respectively. Parameters of the gasification and power generation units are from NETL's report of the Texaco IGCC case study [24].

Major technical parameters used in the model are shown in Tables 1.7–1.9. Table 1.7 shows the ultimate analysis of Illinois #6 coal, the main feedstock of the gasifier. Table 1.8 shows conversion rates of gas phase and liquid phase methanol synthesis technologies. Table 1.9 shows the temperature and pressure or pressure losses in the blocks and units, a positive number denoting absolute pressure and a negative number denoting a pressure loss.

In the case study, it is assumed that the market demand for methanol is from 400 to 700 ton/day, and the electricity demand is from 100 to 300 MW. The annual operating time of the plant is set to 6500 h.

Economic parameters of prices and investment costs for all units are shown in Table 1.10. Investment costs are depreciated over a 30-year time horizon. It is assumed that fixed O&M costs are 10% of the depreciated annual investment cost, and variable O&M costs are proportional to the utilization ratio of units.

Table 1.9 Temperature and pressure loss for each technology.

Unit	Technology	Temperature (K)	Pressure/pressure loss (bar)
Syngas cooler	Quench	491	−1
	Low temperature radiative and convective	477	−3.3
	High temperature radiative and convective	813	−3.3
Syngas cleanup unit	Cold cleanup	320	−4.6
	Hot cleanup	840	−5.6
Water gas shift reactor	Water gas shift	473	−1
Methanol synthesis	Gas phase	523	−5.5
	Liquid phase	523	−5.5
Gas turbine	Gas turbine technology 1	1703	19
	Gas turbine technology 2	1589	18
	Gas turbine technology 3	1473	17
HRSG and steam turbines	High heat recovery	400	1.05
	Low heat recovery	450	1.05

The model realization for the case study is highly nonlinear and nonconvex. It was implemented in GAMS (General Algebraic Modeling System) using the DI-COPT solver, with 299 continuous variables, 15 binary variables, and 313 equations, including 107 nonlinear equations and 20 inequality constraints.

Model results show that a maximum annual profit of 140.6 million dollars can be achieved by generating 300 MW electricity and 700 tons of methanol per day, and consuming 2991 tons of coal per day. Annual income from sale of electricity is 117 million dollars, and annual income from sale of methanol is 64.5 million dollars. Annual expense on purchase of fuel is 28.4 million dollars. Depreciated investment costs on equipment and O&M costs are 12.6 million dollars per year.

The plant uses low-temperature radiative and convective technology for cooling of crude syngas, followed by a low-temperature cleanup unit. The methanol synthesis part uses gas phase synthesis technology with a water gas shift reactor before it. The power generation unit uses gas turbine technology which has the highest first stage temperature and pressure, together with the technology of high heat recovery for the HRSG and the steam turbine.

For comparison, Table 1.11 shows the production rates, annual profit, and economic efficiency of other applicable combinations of available technologies. Economic efficiency is defined as the ratio of annual income to expense. Each configuration of the process is denoted by a combination of symbols denoting technologies listed in Fig. 1.14. For example, "LRC-CC-WG-GPMeOH-GT1-HHR" represents the optimized configuration generated by the model.

From the model results, it is clear that the most profitable configuration for the plant is the combination of gas phase methanol synthesis technology and highly efficient power generation technology. Because of its high conversion rate, gas phase

Table 1.10 Economic parameters.

Parameter	Value
Coal price ($/ton)	35
Methanol ($/ton)	340
Electricity price ($/kWh)	0.06
Investment cost for the gasifier ($/((kg/s coal)*y))	28 500
Investment cost for the cooler, quench ($/((kg/s syngas)*y))	3000
Investment cost for the cooler, low temperature radiative and connective ($/((kg/s syngas)*y))	45 000
Investment cost for the cooler, high temperature radiative and connective ($/((kg/s syngas)*y))	30 000
Investment cost for the cleanup unit, low temperature ($/((kg/s syngas)*y))	20 000
Investment cost for the cleanup unit, high temperature ($/((kg/s syngas)*y))	40 000
Investment cost for the water gas shift reactor ($/((kg/s syngas)*y))	5000
Investment cost for the CO_2 removal unit ($/((kg/s syngas)*y))	5000
Investment cost for the methanol synthesis unit, gas phase ($/((kg/s syngas)*y))	15 000
Investment cost for the methanol synthesis unit, liquid phase ($/((kg/s syngas)*y))	20 000
Investment cost for the gas turbine compressor ($/((kg/s air)*y))	2000
Investment cost for the gas turbine, technology 1 ($/((kg/s gas)*y))	3000
Investment cost for the gas turbine, technology 2 ($/((kg/s gas)*y))	2500
Investment cost for the gas turbine, technology 3 ($/((kg/s gas)*y))	2000
Investment cost for the HRSG and steam turbines, technology 1 ($/((kg/s gas)*y))	3000
Investment cost for the HRSG and steam turbines, technology 2 ($/((kg/s gas)*y))	2500

Table 1.11 Model results for technology combinations.

Technology combination	Power (MW)	Methanol (ton/day)	Coal (ton/day)	Profit (million dollar)	Economic efficiency
LRC-CC-WG-GPMeOH-GT1-HHR	300	700	2991	140.6	4.436
LRC-CC-WG-GPMeOH-GT1-LHR	300	700	3050	139.9	4.370
LRC-CC-WG-GPMeOH-GT2-HHR	300	700	3173	137.7	4.148
LRC-CC-WG-GPMeOH-GT3-HHR	300	700	3460	133.2	3.762
Q-CC-WG-GPMeOH-GT1-HHR	300	700	3618	136.9	4.068
LRC-CC-LPMeOH-GT1-HHR	300	474	2567	125.0	4.508
Q-CC-LPMeOH-GT1-HHR	300	588	3182	131.4	4.310
Q-CC-LPMeOH-GT2-LHR	300	673	3643	133.5	3.937
Q-CC-WG-LPMeOH-GT3-LHR	300	700	4113	129.6	3.500

methanol synthesis produces more methanol to meet the market demand, whilst the methanol production rates in most configurations using liquid phase methanol synthesis are below the maximum market demand. However, on the other hand, configurations using liquid phase methanol synthesis are in general more cost effective, due to its low operation pressure which leads to less consumption of compression work in the air separation unit.

From a technical aspect of view, gas phase methanol synthesis has some disadvantages. It has high investment cost, requires additional water gas shift reactor to adjust the composition of inlet syngas, and needs more compression work in the air separation to obtain its high operation pressure. However, it also has high methanol production rate, compensating all these disadvantages when the methanol price is high and more production of methanol is favorable.

In conclusion, gas phase methanol synthesis and highly efficient power generation technologies are the best for a polygeneration plant producing methanol and electricity under conditions set in this case study. Next, a sensitivity analysis is presented to show how changes of parameters and variables influence the model results.

1.3.4.2 Discussion and Sensitivity Analysis

Because of inherent characteristic, the design procedure for a process inevitably involves uncertainties. It is therefore imperative to investigate how and to what extent these uncertainties influence the model results. Here, sensitivity analysis is used to conduct this task by finding out the most important model input factors, i.e. parameters or variables with the highest contribution to the model outputs.

Mathematically, pseudoglobal sensitivity of model outputs with respect to model parameters is investigated to prioritize their significance. The general algorithm for conducting a pseudoglobal sensitivity analysis is shown below [25].

Assuming that A_1 to A_n are a set of model outputs, and k_1 to k_m are a set of uncertain model parameters, the model can be represented as

$$[A_1, A_2, \ldots, A_n] = f(k_1, k_2, \ldots, k_m) \tag{1.83}$$

Moreover, all these parameters are assumed to follow normal distribution, denoted as

$$kj_j \sim N(k_j^0, \sigma_{k_j}), \quad j = 1 \text{ to } m \tag{1.84}$$

where k_j^0 is the nominal value of k_j, and σ_{k_j} is its standard deviation.

The pseudoglobal sensitivity of a model result A_i with respect to an uncertain parameter k_j is given by

$$S_{k_j}^\sigma = \frac{\sigma_{k_j}}{\sigma_{A_i}} \frac{\partial A_i}{\partial k_j} \tag{1.85}$$

where S represents the pseudoglobal sensitivity, and σ_{A_i} is the standard deviation of A_i.

The exact value of σ_{A_i} is not easy to obtain, but an approximation can be obtained from the following equation:

$$\sigma_{A_i}^2 \cong \sum_j \sigma_{k_j}^2 \left(\frac{\partial A_i}{\partial k_j}\right)^2 \qquad (1.86)$$

This algorithm provides a practical way to compare the significance of model parameters. The model parameter with the largest absolute pseudoglobal sensitivity value has the most significant influence on the model outputs, and the sign of the pseudoglobal sensitivity shows whether this influence is beneficial or hazardous.

Applying this algorithm of pseudoglobal sensitivity analysis to our model, four model outputs were selected as A_i, including the power generation rate, methanol production rate, coal consumption rate, and annual profit of the plant. Uncertain model parameters are classified by their inherent character as either economic or technical. The economic category includes coal price, electricity price, methanol price, upper and lower bounds of market demands for products, investment costs, and O&M costs. The technical category consists of the inlet temperature at the first stage of gas turbine, energy efficiency of HRSG and the steam turbine, conversion ratio of methanol synthesis reaction, and the annual operation time of the process.

Without loss of generality, it was assumed that all the uncertain parameters took the values used for the case study as nominal values, and the standard deviation is 10% of the nominal value for all of them.

Values of pseudoglobal sensitivities for the model outputs with respect to uncertain parameters are shown in Table 1.12. The power generation and methanol production rates appear to be only sensitive to the market demands, which also have considerable effects on the annual profit. It is therefore always desirable to extend the market for more production. In the aspect of coal consumption rate, application of more advanced gas turbine technology has the largest potential to reduce the coal consumption and fuel expense.

The situation of the annual profit is a little more complex since it is influenced by several factors. The strongest influence comes from the annual operation time, which is a measure of the overall technical level of equipment manufacture and process integration. The second and third are the electricity price and demand, showing the heavy dependence of the plant's economic performance upon the power generation sector. Besides this, the price of methanol and its market demand also show certain influences on the profit. Coal price, investment cost, O&M costs, and conversion ratio of methanol synthesis have negative values on the profit, of which coal price being the largest.

Interestingly, increasing the conversion ratio of the methanol synthesis does not lead to less coal consumption but the other way round. This is because the conversion ratio is already sufficiently high; further increase will only make the flue-gas leaving the synthesis reactor to have lower flowrate and heating value. More coal consumption is therefore required to generate more syngas to keep the power generation. At the point in the case study, this extra amount of coal consumption

Table 1.12 Pseudoglobal sensitivities.

Uncertain parameters	Power (MW)	Methanol (ton/day)	Coal (ton/day)	Profit (million dollar)
Annual operation time	0	0	0	0.656
Market demand for electricity, upper bound	1	0	0.628	0.363
Market demand for electricity, lower bound	0	0	0	0
Market demand for methanol, upper bound	0	1	0.190	0.239
Market demand for electricity, lower bound	0	0	0	0
Coal price	0	0	0	−0.121
Electricity price	0	0	0	0.501
Methanol price	0	0	0	0.276
Investment cost	0	0	0	−0.045
O&M costs	0	0	0	−0.009
Temperature at first stage of gas turbine	0	0	−0.738	0.190
Energy efficiency of HRSG and steam turbines	0	0	−0.157	0.035
Conversion ratio of methanol synthesis	0	0	0.031	−0.005

cannot be compensated by the enhanced efficiency of methanol synthesis; thus the overall coal consumption increases.

The sensitivity analysis provides prioritized procedures for the development of polygeneration energy systems. First of all, it is crucial to increase the technical levels in every aspect, from single unit to the whole process integration. Then, market demands and product prices are also very important to the economic behavior of a polygeneration plant.

1.4
Concluding Remarks

Two mathematical models of different purposes and functions for planning and design of polygeneration energy systems are presented in this chapter. The model for strategic investment planning is suitable to be used in the very first stage of development planning for polygeneration plants, facilitating to make strategic decisions like which types of products to produce. The model for configuration design is developed for the second stage of planning for a specific plant, choosing the optimal combination of technologies and types of equipment according to particular conditions. Case studies incorporating applications of the models are provided after both models, illustrating how these models can be integrated into real industrial projects.

Acknowledgements

Pei Liu would like to thank Kwoks' Foundation for its scholarship. The authors would also like to acknowledge the financial support from BP.

References

1 DOE, International energy outlook **2006**. Available at www.eia.doe.gov/oiaf/ieo/index.html. Accessed 2006.

2 SIMS, R. E. H., ROGNER, H., GREGORY, K., Carbon emission and mitigation cost comparisons between fossil fuel, nuclear and renewable energy resources for electricity generation. *Energy Pol.* 31(13) (**2003**), pp. 1315–1326.

3 GRADASSI, M. J., WAYNE GREEN, N., Economics of natural gas conversion processes. *Fuel Process. Technol. Trends Nat. Gas Utilis.* 42(2) (**1995**), pp. 65–83.

4 FREY, H. C., ZHU, Y., Improved system integration for integrated gasification combined cycle (IGCC) systems. *Environ. Sci. Technol.* 40(5) (**2006**), pp. 1693–1699.

5 NATIONAL ENERGY TECHNOLOGY LABORATORY, Commercial-scale demonstration of the liquid phase methanol (LPMEOH) process. United States, **2003**.

6 NI, W., LI, Z., YUAN, X., National energy futures analysis and energy security perspectives in China: strategic thinking on the energy issue in the 10th Five-Year Plan. *Workshop on East Asia Energy Futures,* **2000**.

7 COCCO, D., PETTINAU, A., CAU, G., Energy and economic assessment of IGCC power plants integrated with DME synthesis processes. *Proc. Inst. Mech. Eng. A: J. Power Energy.* 220(2) (**2006**), pp. 95–102.

8 DEPARTMENT OF ENERGY, *Vision 21 Program Plan: Clean Energy Plants for the 21st Century,* Washington, DC, **1999**.

9 DEPARTMENT OF ENERGY, *Vision 21: Fossil Fuel Options for the Future,* 1st edn., National Academy Press, Washington, DC, **2000**.

10 LI, Z., NI, W., ZHENG, H., MA, L., Polygeneration energy system based on coal gasification. *Energy Sustain. Develop.* 7(4) (**2003**), pp. 57–62.

11 WILLIAMS, H. P., *Model Building in Mathematical Programming,* th ed., Wiley, New York, **1978**.

12 STRICKLAND, D., Wabash River IMPPCCT, integrated methanol and power production from clean coal technologies, *Quarterly Technical Progress Report* 07, **2001**.

13 LANGE, J. P., Perspectives for manufacturing methanol at fuel value. *Ind. Eng. Chem. Res.* 36(10) (**1997**), pp. 4282–4290.

14 HAMELINCK, C. N., FAAIJ, A. P. C., Future prospects for production of methanol and hydrogen from biomass *J. Power Sources.* 111(1) (**2002**), pp. 1–22.

15 MA, L., NI, W., LI, Z., REN, T., Analysis of the polygeneration system of methanol and electricity based on coal gasification (1). *Power Eng.* 24(3) (**2004**), pp. 451–456.

16 MA, L., NI, W., LI, Z., REN, T., Analysis of the polygeneration system of methanol and electricity based on coal gasification (2). *Power Eng.* 24(4) (**2004**), pp. 603–608.

17 LIU, P., GAO, J., LI, Z., Performance alteration of methanol/electricity polygeneration systems for various modes of operation. *Dongli Gongcheng/Power Eng.* 26(4) (**2006**), pp. 587–591.

18 LIU, G., LIU, Y., LI, Z., NI, W., XU, H., Design and analysis of a new dimethyl ether-electric power polygeneration system. *Dongli Gongcheng/Power Eng.* 26(2) (**2006**), pp. 295–299.

19 CHEN, B., JIN, H., GAO, L., Study of DME/power individual generation and polygeneration. *Kung Cheng Je Wu Li Hsueh Pao/J. Eng. Thermophys.* 27(5) (**2006**), pp. 721–724.

20 ZHANG, Y., NI, W., LI, Z., Research on superclean polygeneration energy system of iron and steel industry. *Proceedings of the Seventh IASTED International Conference on Power and Energy Systems,* Clearwater Beach, FL, USA, **2004**.

21 ZHANG, X., GAO, L., JIN, H., CAI, R., Design and analysis of coal based

ammonia-power polygeneration. *Dongli Gongcheng/Power Eng.* 26(2) **(2006)**, pp. 289–294.

22 AL-HAMDAN, Q. Z., EBAID, M. S. Y., Modeling and simulation of a gas turbine engine for power generation. *J. Eng. Gas Turb. Power – Trans. ASME* 128(2) **(2006)**, pp. 302–311.

23 MARTINEZ-FRIAS, J. M., ACEVES, S. M., SMITH, J. R., BRANDT, H., Thermody-namic analysis of zero-atmospheric emissions power plant. *J. Eng. Gas Turb. Power – Trans. ASME* 126(1) **(2004)**, pp. 2–8.

24 SHELTON, W., LYONS, J., *Texaco gasifier IGCC base cases*, DOE/NETL, **1998**.

25 SALTELLI, A., RATTO, M., TARANTOLA, S., CAMPOLONGO, F., Sensitivity analysis for chemical models. *Chem. Rev.* 105(7) **(2005)**, pp. 2811–2827.

2
Multiobjective Design and Optimization of Urban Energy Systems

François Maréchal, Céline Weber, and Daniel Favrat

Keywords

mixed-integer nonlinear programming (MINLP), multiobjective optimization problem, branch-and-bound algorithm, district heating systems, CO_2 emissions, energy conversion technologies, Pareto optimal frontier

2.1
Introduction

When considering the energy flows of a country like Switzerland, more than 50% of the energy consumption is devoted to the production of heating services at rather low temperature level (less than 100 °C). Furthermore, when considering that about 75% of the population lives in urban areas, the design of urban energy systems is one of the most important problem when addressing the problem of CO_2 mitigation (in Switzerland about 50% of the CO_2 emissions comes from the heating requirements of households).

District energy systems provide heating, hot water, cooling, and electricity to two or more buildings. They include the energy conversion technologies (preferably polygeneration units) together with the distribution networks. Unlike district heating systems, that provide only heating, district energy systems have not yet been much studied. District heating systems are often implemented when excess heat, from a geothermal source or from the combustion of waste for instance, is to be recycled. In the latter case, the system can also be designed to generate electricity. The majority of the literature on district heating systems concerns the optimization (mainly financial) of the operation strategies and/or the thermoeconomic optimization of the energy conversion technologies [1–7], as well as the energetic and/or exergetic performance of the complete district heating system [8, 9]. In nearly all of these papers, the distribution network of the analyzed district heating system already exists. Its design is only seldom mentioned, and if ever, then in a very simplified manner. One reason for the researchers not to be more interested in the design of distribution networks could be the belief that the design of the distribution network is anyway solved by politicians and urban planners, without involving any quantitative support [4], and that it is therefore useless to include the design of the

Process Systems Engineering: Vol. 5 Energy Systems Engineering
Edited by Michael C. Georgiadis, Eustathios S. Kikkinides and Efstratios N. Pistikopoulos
Copyright © 2008 WILEY-VCH Verlag GmbH & Co. KGaA, Weinheim
ISBN: 978-3-527-31694-6

distribution network when studying the thermoenvironomic optimization of district energy systems. It is, however, the belief of the authors that politicians and urban planners could be interested in using quantitative support, if they had the tools to do so. Söderman has studied the design of distributed energy systems [10] and developed a tool for decision makers. However the study doesn't take into account the temperature levels at which the energy services have to be delivered. Friedler et al. studied process synthesis and optimization for the chemical industry [11, 12], and it would be possible, considering some analogies and modifications, to adapt his method to district energy systems. However this work does not take into account any spatial constraints regarding the location of the technologies. Besides, the chemical processes that are designed are of the continuous type, whereas district energy systems are more related to a batch type operating mode (the energy requirements vary from one period to the other).

The method presented combines the design of the network together with the definition of the sizes and the operating strategies of the technologies that are best suited to meet the energy requirements of the district. Details about this work may be found in the thesis of Céline Weber [31]. The method takes into account the *spatial* and *temporal* aspects that are characteristic of district energy systems, as well as the temperature levels at which the energy services are requested. It addresses the following question: How shall a district energy system, a system that comprises the energy conversion technologies that transform the primary energy into the requested energy form (heating, cooling, electricity and hot water), and the distribution network from the plant to the customers, be designed, to minimize the overall costs and the CO_2 emissions while delivering the hourly energy services requested by the customers?

Because it combines spatial (location of the buildings) and temporal (consumption profiles) aspects, the design of a polygeneration district energy system is a complex problem. The number of possible configurations is very high since it concerns the identification of the energy conversion plant location, the different ways of connecting the buildings together with the definition of the size and the type of the energy conversion units. In addition, the performance of the system requires the definition of the optimal operating strategy that accounts for the consumption profiles of the different energy services (electricity, heating, cooling) that vary during the day, and from one day to another in a stochastic manner.

The method developed (Fig. 2.1) comprises a structuring phase in which all the relevant data regarding the district considered are gathered, an optimization phase in which the optimal district energy system is designed, and a postprocessing phase in which the total costs and CO_2 emissions of the system are computed. Due to the complexity of the analyzed problem, the optimization phase is decomposed in a master optimization and a slave optimization.

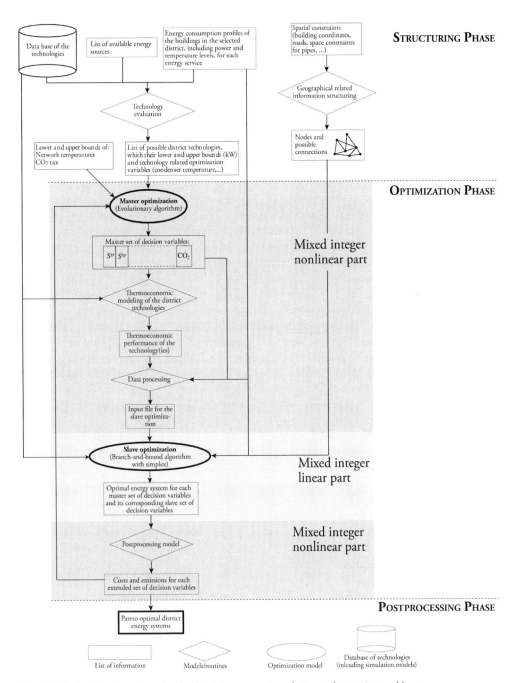

Fig. 2.1 Method developed to solve the district energy system design and operation problem.

2.2
Structuring Phase

Having defined the system boundaries, the structuring phase is the phase in which all the relevant information regarding the district for which an energy system needs to be designed, are gathered, structured, and put in the required form to be used in the optimization phase.

2.2.1
Definition of the List of Available Energy Sources

A comprehensive analysis of the available energy sources in the district will be realized. These include fossil fuels but also renewable energy sources such as wind, waste, sewage water, lakes, rivers, sun, geothermal energy, etc. For each energy source, the necessary parameters have to be gathered in order to assess the energy conversion integration. These concern both the quantity available and the quality such as the temperature level or the wind speed.

2.2.2
Energy Consumption Profiles Including Power and Temperature Levels

For each building, the energy consumption profiles for the different energy services need to be known. For the thermal energy requirements (heating, hot water, and cooling), it is important to know the power required as well as its required temperature level. Besides, for all energy services the *power level* is of utmost importance to design the size of the energy conversion technologies. Therefore, the profiles of the energy requirements, for *all* energy services, are best given in average hourly rates for each period (in kW), together with the duration of each period in hours. Considering the stochastic nature of the inhabitants' behavior and of the meteorological conditions, typical days representing the operating conditions to be considered for the design will be defined. In many cases, a six-period profile including a day and a night period for summer, winter and mid-season, can already be considered sufficient.

2.2.3
Routing Algorithm

The system has to fit into an existing geographical area. Therefore constraints such as buildings location, roads layout, space constraints in existing underground channels, etc., have to be taken into account when computing the possible layout of the network. Typically, a pipe will not be allowed to follow the shortest path between two buildings, but has to follow certain routes as shown in Fig. 2.2.

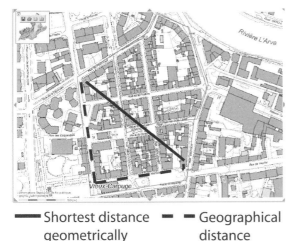

━━━ Shortest distance   ━ ━ Geographical
       geometrically   distance

Fig. 2.2 Difference between the geographical route that has to be followed to connect two buildings, and the shortest distance geometrically [13].

2.2.4
Technology Database

The *technology database* includes centralized and decentralized (backup) energy conversion technologies. Centralized technologies are the energy conversion technologies that *are connected* to the distribution networks and provide heating, hot water, cooling, and electricity to different customers in the district via the networks. Decentralized technologies are implemented directly in the building they serve, in case the building is not connected to the network or when the district energy system cannot meet the energy requirements of this building (for instance, if the temperature level at which the heat is required by a single given building is above the temperature level of the network, or if the building is located too far away from the rest of the buildings to justify a connection). The database includes the thermoeconomic simulation models of different energy conversion technologies. Thermoeconomic models means that the model will compute, as a function of the services required and the size of the equipment, the thermodynamic performances and the corresponding environmental impact or emissions, together with an estimate of the economical costs (including operating, maintenance, and investment costs).

Considering the list of available energy sources, together with the energy consumption profiles (including the temperature and power levels), the *Technology evaluation* tool analyzes the database to establish a list of possible technologies that can be used in the analyzed district. For instance, if the district is located in the vicinity of a lake or a river, heat pumps will be added in the list. Besides, if geothermal energy is available, a steam turbine or an organic Rankine cycle can be implemented. In contrast, considering wind mills in cities featuring no favourable wind conditions will not lead to a reasonable solution. Or else, if the maximum average hourly electricity rate amounts to 4000 kW for instance, no gas turbine combined cycle will be proposed.

2.3
Optimization Phase

The optimization phase aims at combining, in an optimal manner, the resources, and the energy conversion units in order to satisfy the requirements in the area. The optimization problem is defined in Fig. 2.1. It concerns the following decision variables:

1. Centralized and decentralized energy conversion technologies:

    - Type of the unit
    - Installed size
    - Location in the district
    - Unit specific operating conditions
    - Operation strategy (i.e., the level of usage at a given time)

2. District heating and cooling networks:

    - Existence and size of the pipes in the energy distribution networks (for heating and cooling)
    - Supply and return temperatures of the distribution networks in different periods
    - Flows in each pipe during each period.

A thermoeconomic model will compute the performances of the system. This is done by solving first the thermodynamic performances of the system and then estimating the corresponding investment costs.

A thermoeconomic multiobjective optimization problem will then be solved. The two objectives are the yearly costs (including investment, maintenance and operation) and the yearly CO_2 emissions (operation). The results of the multiobjective optimization problem define the Pareto optimal frontier of the system. In this curve, each point corresponds to one system configuration including the network design, the selected equipment, and the operation strategy for each of the units.

2.4
The Postprocessing Phase

The optimization phase results in a list of urban energy system configurations, the performances of which have to be evaluated so that the stakeholders will make their final choice. In the postprocessing phase, the performances of each configuration are evaluated and a multicriteria analysis will be realized. Using the calculated performance indicators, the decision makers will apply criteria such as

- the available budget,
- the maturity of the technology,
- the CO_2 emissions reduction targets,
- the degree of usage of renewable energy,
- the sensitivity analysis on some parameters,
- the acceptance of the population for given solutions,
- the implementation delays,
- the political program,
- ...

The multicriteria analysis will be based on the weight given to each of the performance indicators by the stakeholders. The success of this phase strongly relies on a close collaboration between all the decision makers, stakeholders, and engineers.

2.5
Resolution Strategy of the Optimization Phase

Due to the complexity of the problem, a specific optimization strategy is needed. The optimization problem involves both discrete and continuous variables and the models may be nonlinear. Therefore the problem to be solved is a mixed-integer nonlinear programming (MINLP) multiobjective optimization problem.

The optimization problem is decomposed into two parts, a nonlinear part, which is defined as a *multiobjective* master optimization problem solved by an evolutionary algorithm and a mixed-integer linear part, which is defined as a *single-objective* slave optimization problem solved by a branch-and-bound algorithm combined with the simplex method. It is therefore necessary to divide the variables set into two subsets in such a way that the network design can be solved as a linear problem. The decomposition strategy is mainly justified by the nonlinear nature of the problem and the number of decision variables, especially the discrete variables required to define the configuration of the network. The nonlinear nature of the problem makes the application of branch-and-bound strategies impractical while the combinatorial nature of the problem makes the use of evolutionary algorithm also impractical. In fact, the application of evolutionary algorithm would furthermore require a careful analysis of the degree of freedom of the problem and the definition of a very efficient nonlinear solving procedure. The branch-and-bound algorithm is typically used for solving network optimization problems like, for instance, the Traveling Salesman Problem or the Assignment Problem [14, 10], and shows very good convergence properties. It is therefore chosen to solve the network design and optimization. The branch-and-bound algorithm is less well suited when nonlinear issues like the efficiencies of given technologies for instance are implied. The purpose of the decomposition is therefore to combine the advantages of both optimization algorithms, i.e. the ability of the evolutionary algorithm to solve any type of optimization without requiring too drastic simplifications, and the speed of the branch-and-bound algorithm, together with the advantage of the decision variables separation.

The set of decision variables handled by the master optimization will be called hereafter the *master set* of decision variables, and the set of decision variables solved in the slave optimization will be called the *slave set* of decision variables. The combination of both sets of decision variables is the *extended set* of decision variables.

In addition to the mathematical nature of the decomposition, the decision variables set partitioning has been realized considering the physical meaning of the variables. The master set of decision variables is directly linked to the type and size of the technologies and given as input to the thermodynamic models in order to compute the performances of the technologies (efficiencies, mass flow rates, etc.). The results of the thermodynamic models, together with further decision variables of the master set (the network supply temperature and the temperature difference of the network in the decentralized heat pumps), the consumption profiles and the possible connections are used to compute the parameters of the slave optimization. The slave optimization then computes the slave set of decision variables that defines the optimal energy system configuration and its operation, as a function of the given master set of decision variables.

The multiobjective evolutionary optimizer used for this study was developed at the Industrial Energy Systems Laboratory at the Ecole Polytechnique Fédérale de Lausanne [15]. This optimizer uses the technique of the evolutionary algorithms to compute the tradeoffs between multiple objectives. In our case, two objectives have to be minimized: the costs (including operation and investment) and the CO_2 emissions of the system. In order to find the optimal configurations with the best performances in terms of CO_2 emissions and costs, the evolutionary algorithm creates a population of individuals (a set of master decision variables) by choosing randomly, for each individual, a set of values (genome) representing the decision variables. The "scores" or performances of each individual are then computed by solving the slave optimization problem. New individuals are then selected based on the scores of the existing individuals, using a set of combination operators such as mutation and crossover. After the evolution process is continued sufficiently, keeping the best individuals in the nondominated set (according to CO_2 emissions and costs), the optimal solutions can be found. This multiobjective strategy results in an estimation of the Pareto optimal frontier (hereafter Pareto curve) that represents the set of optimal points that can be considered to be optimal in terms of one or both of the two objectives. Each point of this curve corresponds to one configuration of the system and the optimal way of operating it on a yearly basis.

Table 2.1 Lower and upper bounds of the continuous decision variables optimized by the master optimizer.

Decision variable	Lower bound–upper bound
District energy conversion technologies	Minimum available size (market size) for the technology – Minimum between the maximum requirements and the maximum available market size
Specific technology related, optimization parameters (if any)	
Supply temperature for the heating network for each period	40–120 °C
Return temperature for the heating network for each period	30 °C – supply temperature of the respective period
Supply temperature for the cooling network for each period	6–20 °C
Return temperature for the cooling network for each period	Supply temperature of the respective period – 22 °C
CO_2 tax	0 CHF/t CO_2 – 50 CHF/t CO_2

2.6
Mathematical Formulation of the Optimization Phase

2.6.1
The Master Optimization

2.6.1.1 Objective Function and Decision Variables

The inputs of the master optimization are the list of decision variables with their lower and upper bounds. These decision variables include: binary variables for the choice of the district energy conversion technologies, continuous variables for the size of the district energy conversion technologies (S_{ct} and S_{ht}), various technology related parameters (like the condenser temperature for heat pumps for instance), the temperatures of the networks ($T_{\mathrm{hs},t}$, $T_{\mathrm{hr},t}$, $T_{\mathrm{cs},t}$ and $T_{\mathrm{cr},t}$) and the CO_2 tax. Regarding the temperatures, the method allows to optimize the supply and return temperatures of each network for each period independently. The lower and upper bounds of the *continuous* variables are given in Table 2.1.

In the problem formulation, the ranges of all the continuous decision variables have been normalized to vary between 0 and 1, in order to homogenize the ranges of different decision variables and therefore ensure a better covering up of the search space.

2.6.1.2 Constraints and Parameters

The constraints of the master optimization include the models of the district energy conversion technologies, the equations of the data processing routine, as well as some values needed to define the slave optimization problem.

The parameters include:

- the parameters used in the models of the district energy conversion technologies (exergetic efficiency of heat pumps or isentropic efficiencies of compressors for instance),
- the parameters used in the data processing routine (velocity of the water in the pipes, specific heat of water, etc.),
- the parameters of the slave optimization (fix and proportional costs of the pipes or individual backup energy conversion technologies, for instance).

Except for the thermodynamic parameters such as the specific heat of water, or else parameters such as the velocity of the water in the pipes, the values of the parameters change from district to district.

2.6.2
The Slave Optimization

2.6.2.1 Objective Function and Decision Variables

The objective function minimizes the annual operating costs and the annual investment costs, considering the possible income I^{el} of the excess electricity generated by the district energy conversion technologies. The annual operating costs include the costs for the natural gas, the grid electricity, the oil and the CO_2 emissions taxes. The annual investment costs include the costs for the pipes, circulation pumps, heat exchangers, and individual backup energy conversion technologies (air/water heat pumps, water/water heat pumps, boilers, and chillers). The slave optimization is therefore formulated as Eq. (2.1).

$$\min\left(C^{\text{gas}} + C^{\text{grid}} + C^{\text{oil}} + C^{\text{CO}_2} + C^{\text{pipes}} + C^{\text{pump}} + C^{\text{hex}} + C^{\text{aw}}\right.$$

$$\left. + C^{\text{ww}} + C^{\text{boiler}} + C^{\text{chiller}} - I^{\text{el}}\right) \quad [\text{CHF/year}] \tag{2.1}$$

2.6.2.2 Constraints of the Slave Optimization

The following types of constraints are considered in the slave optimization subproblem:

1. Energy balances at each node

   - First principle[1] thermal energy balances
   - Heat cascades
   - Electrical energy balances

2. Mass balances in the network
3. Connections of nodes and implementation/location of the technologies (district and backup technologies)
4. Performance functions (costing and CO_2 emission)
5. Empirical (engineer knowledge based) constraints

---

1) The term *First principle* refers to the thermodynamic principles.

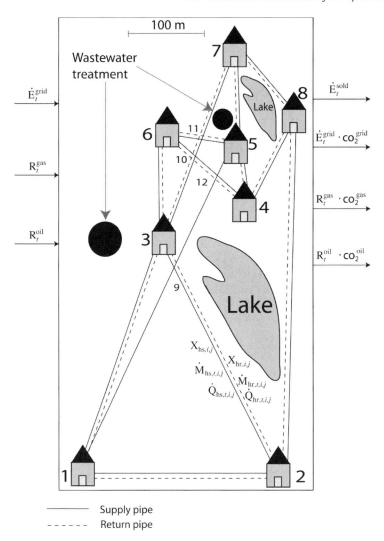

Fig. 2.3 Structure of the district (the plain lines between the buildings refer to the supply part of the distribution network and the dotted lines to the return part).

Figures 2.3 and 2.4 show, respectively, the superstructure of the network and the superstructure considered at each node as well as the decision variables considered in the slave optimization.

2.6.2.3 First Principle Energy Balances

1. Energy balances at the consumer's place (see also Fig. 2.4)

Energy balances are defined for heating and hot water requirements (indices H and HW in Eq. (2.2)), as well as for cooling requirements (index C in Eq. (2.3)). These energy balances link the energy required by a node, with

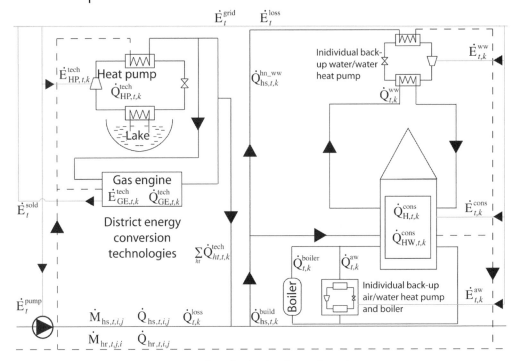

Fig. 2.4 Superstructure of a node (the plain lines refer to the supply part of the distribution network and the dotted lines to the return part).

the energy provided by the network $\dot{Q}^{build}_{hs,t,k}$, respectively $\dot{Q}^{build}_{cs,t,k}$, or the energy provided by decentralized technologies ($\dot{Q}^{ww}_{t,k}$, $\dot{Q}^{aw}_{t,k}$, or $\dot{Q}^{boiler}_{t,k}$ for heating, and $\dot{Q}^{chiller}_{t,k}$ for cooling). $\dot{Q}^{loss}_{t,k}$ represents all the losses that occur in the heating distribution network, and that have to be compensated by the district energy conversion technologies located at node k, if any (heat "losses" have been neglected for the cooling network).

Heating and hot water:

$$\dot{Q}^{cons}_{H,t,k} + \dot{Q}^{cons}_{HW,t,k} + \dot{Q}^{loss}_{t,k}$$

$$= \dot{Q}^{build}_{hs,t,k} + \dot{Q}^{aw}_{t,k} + \dot{Q}^{ww}_{t,k} + \dot{Q}^{boiler}_{t,k} \qquad \forall t, k \qquad (2.2)$$

Cooling:

$$\dot{Q}^{cons}_{C,t,k} = \dot{Q}^{build}_{cs,t,k} + \dot{Q}^{chiller}_{t,k} \qquad \forall t, k \qquad (2.3)$$

2. Energy balances between the consumers and the network (see also Fig. 2.3)

For each building connected to the network, the following balances relate the energy supplied by the networks (heating and cooling) to the building, with the mass flow rate of water flowing from the networks to the buildings. In the following equations, $T_{hs,t}$ and $T_{hr,t}$ are the supply, respectively return, temperatures of the heating network, and $T_{cs,t}$ and $T_{cr,t}$ the supply and return temperatures of the cooling network.

Heat supplied by the network to the building:

$$\dot{Q}_{hs,t,k}^{build} = \dot{M}_{hn,t,k}^{build} \cdot cp_{H_2O}^{liq} \cdot (T_{hs,t} - T_{hr,t}) \quad \forall t, k \tag{2.4}$$

Heat supplied by the heating network to the water/water heat pump in case individual backup water/water heat pumps locally increase the temperature level of the network:

$$\dot{Q}_{hs,t,k}^{hn\_ww} = \dot{M}_{hn,t,k}^{hn\_ww} \cdot cp_{H_2O}^{liq} \cdot (T_{hs,t} - T_{hr,t}) \quad \forall t, k \tag{2.5}$$

with

$$\dot{Q}_{hs,t,k}^{hn\_ww} = \left(1 - \frac{1}{COP_{t,k}^{ww}}\right) \cdot \dot{Q}_{t,k}^{ww} \quad \forall t, k \tag{2.6}$$

and

$$COP_{t,k}^{ww} = \eta^{hp} \cdot \frac{T_{Hs,t,k} + \Delta T}{T_{Hs,t,k} - T_{hr,t}} \quad \forall t, k \tag{2.7}$$

where η^{hp} is the exergetic efficiency of the heat pump, $T_{Hs,t,k}$ is the supply temperature of the hydronic circuit in the building, and $T_{hr,t}$ the return temperature of the heating network.

Cooling supplied by the network to the building:

$$\dot{Q}_{cs,t,k}^{build} = \dot{M}_{cn,t,k}^{build} \cdot cp_{H_2O}^{liq} \cdot (T_{cr,t} - T_{cs,t}) \quad \forall t, k \tag{2.8}$$

3. Energy balances at the plant node[2]

Heat provided by the district technology ht at node k to the heating network:

$$\dot{Q}_{ht,t,k}^{tech} = cp_{H_2O}^{liq} \cdot \dot{M}_{hn,t,k}^{tech} \cdot \Delta T_{ht} \quad \forall t, ht, k \tag{2.9}$$

Heat provided by the sum of all the technologies at node k to the heating network:

$$\sum_{ht} \dot{Q}_{ht,t,k}^{tech} = cp_{H_2O}^{liq} \cdot \dot{M}_{hn,t,k}^{tech} \cdot (T_{hs,t} - T_{hr,t}) \quad \forall t, k \tag{2.10}$$

[2] The term *plant* is used to designate a node where district energy conversion technologies are implemented. A given node can be simultaneously a plant node and a consumer node as shown in Fig. 2.4.

Cooling load provided by the technology ct at node k to the cooling network:

$$\dot{Q}^{tech}_{ct,t,k} = cp^{liq}_{H_2O} \cdot \dot{M}^{tech}_{cn,t,k} \cdot \Delta T_{ct} \quad \forall t, ct, k \tag{2.11}$$

Cooling load provided by the technologies at node k to the cooling network:

$$\sum_{ct} \dot{Q}^{tech}_{ct,t,k} = cp^{liq}_{H_2O} \cdot \dot{M}^{tech}_{cn,t,k} \cdot (T_{cr,t} - T_{cs,t}) \quad \forall t, k \tag{2.12}$$

In Eqs. (2.9) and (2.11) ΔT refer to temperatures in the heat cascade.

4. Energy balances of a node

The following energy balances are defined at each node for the heating and cooling networks between the energy leaving the node on one side (left-hand side of the = sign), and the energy arriving at the node, consumed by the node, and provided by the node if a district energy conversion technology is located at that node, on the other side (right-hand side of the = sign).

Heating supply network hs:

$$\sum_{(k,j)} \dot{Q}_{hs,t,k,j} = \sum_{(i,k)} \dot{Q}_{hs,t,i,k} - \left(\dot{Q}^{build}_{hs,t,k} + \dot{Q}^{hn\_ww}_{hs,t,k}\right)$$

$$+ \dot{Q}^{tech}_{ht,t,k} \quad \forall t, k \tag{2.13}$$

Cooling supply network cs:

$$\sum_{(k,j)} \dot{Q}_{cs,t,k,j} = \sum_{(i,k)} \dot{Q}_{cs,t,i,k} - \dot{Q}^{build}_{cs,t,k} + \dot{Q}^{tech}_{ct,t,k} \quad \forall t, k \tag{2.14}$$

5. Heat losses

The heat losses for each period in the heating network[3] are computed by multiplying the specific heat losses computed as a function of the temperatures of the district network with the total length of the network:

$$\dot{Q}^{loss}_{t} = \sum_{(i,j)} Dist_{i,j} \cdot X_{hs,i,j} \cdot \dot{q}^{loss}_{t} \quad \forall t \tag{2.15}$$

In order to be able to integrate these heat losses to the energy balance given in Eq. (2.2), they are attributed as an additional energy requirement to the node hosting the district energy conversion technologies (therefore $\sum_{k} \dot{Q}^{loss}_{t,k} = \dot{Q}^{loss}_{t}$).

2.6.2.4 Heat Cascade Constraints

From the simulation models of the technologies and of the heat requirement profiles, the corrected temperatures are computed assuming a minimum approach

---

3) The heat losses in the cooling network have been neglected.

temperature in the heat exchanger. The corrected temperature is then ranked to allow for the calculation of the heat cascade's that define a set of linear constraints in the system. It includes:

Heat cascade between the district heating technologies and the heating network for each temperature interval:

$$\sum_{ht}\sum_{k} \dot{Q}^{tech}_{ht,t,k,d} + \dot{Q}^{tn}_{hs,t,(d-1)} - \dot{Q}^{tn}_{hs,t,d} \geqslant \dot{Q}^{net}_{hs,t,d} \quad \forall t, d \tag{2.16}$$

with $\dot{Q}^{tn}_{hs,t,k,(d-1)}$ the excess heat transferred from temperature interval $d - 1$ to interval d and $\dot{Q}^{net}_{hs,t,d}$ the heat required by the heating network in interval d.

Heat cascade between the heating network and the buildings for each temperature interval:

$$\dot{Q}^{net}_{hs,t,d} + \sum_{k} \dot{Q}^{nb}_{hs,t,k,(d-1)} - \sum_{k} \dot{Q}^{nb}_{hs,t,k,d}$$

$$\geqslant \sum_{k} (\dot{Q}^{req}_{H,t,k,d} + \dot{Q}^{req}_{HW,t,k,d}) \quad \forall t, d \tag{2.17}$$

with $\sum_{k} \dot{Q}^{nb}_{hs,t,k,(d-1)}$ the heat from the heating network transferred from interval $d - 1$ to interval d, and $\sum_{k} (\dot{Q}^{req}_{H,t,k,d} + \dot{Q}^{req}_{HW,t,k,d})$ the sum of all the heating and hot water requirements of the district in interval d.

The link between the heat cascade defined in Eq. (2.16) and the heat cascade defined in Eq. (2.17) is ensured by the balancing term $\dot{Q}^{net}_{hs,t,d}$ and its corresponding mass flow rate (given by Eq. (2.9) and/or (2.10)), which remains constant for all temperature intervals.

Heat cascade between the district cooling technologies and the cooling network for each temperature interval:

$$\sum_{ct}\sum_{k} \dot{Q}^{tech}_{ct,t,k,d} + \dot{Q}^{tn}_{cs,t,(d-1)} - \dot{Q}^{tn}_{cs,t,d} \geqslant \dot{Q}^{net}_{cs,t,d} \quad \forall t, d \tag{2.18}$$

Heat cascade between the cooling network and the buildings for each temperature interval:

$$\dot{Q}^{net}_{cs,t,d} + \sum_{k} \dot{Q}^{nb}_{cn,t,k,(d-1)} - \sum_{k} \dot{Q}^{nb}_{cn,t,k,d} \geqslant \sum_{k} \dot{Q}^{req}_{C,t,d,k} \quad \forall t, d \tag{2.19}$$

2.6.2.5 Electricity Balances

$$\sum_{k} \dot{E}^{cons}_{t,k} + \dot{E}^{pump}_{t} + \sum_{k} (\dot{E}^{aw}_{t,k} + \dot{E}^{ww}_{t,k} + \dot{E}^{chiller}_{t,k}) + \dot{E}^{loss}_{t} + \dot{E}^{sold}_{t}$$

$$= \dot{E}^{grid}_{t} + \sum_{ht} \dot{E}^{tech}_{ht,t,k} \quad \forall t \tag{2.20}$$

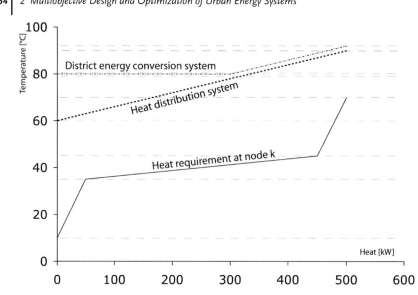

Fig. 2.5 Heat cascades at a given node: the first heat cascade concerns the energy conversion and the heat distribution system, the second concerns the heat distribution system and the heat requirement of the node.

Electricity generated by the district technologies:

$$\dot{E}_{ht,t,k}^{\text{tech}} = \frac{\dot{Q}_{ht,t,k}^{\text{tech}}}{\epsilon_{th}^{ht}} \cdot \epsilon_{el}^{ht} \quad \forall ht, t, k \tag{2.21}$$

If a district energy conversion technology requires electricity (such as a heat pump), its electric efficiency has been computed such that $\epsilon_{el}^{ht} < 0$. Therefore, $\dot{E}_{ht,t,k}^{\text{tech}}$ is also < 0.[4]

Electricity losses are computed considering the grid efficiency ϵ^{grid}:

$$\dot{E}_{t}^{\text{loss}} = \left(1 - \epsilon^{\text{grid}}\right) \cdot \sum_{ht} \dot{E}_{ht,t}^{\text{exp}} \quad \forall t \tag{2.22}$$

Electricity required for the decentralized water/water heat pump(s):

$$\dot{E}_{t,k}^{\text{ww}} = \frac{\dot{Q}_{t,k}^{\text{ww}}}{COP_{t,k}^{\text{ww}}} \quad \forall t, k \tag{2.23}$$

with $COP_{t,k}^{\text{ww}}$ given by Eq. (2.7)

---

4) For instance, if a district heat pump has a COP of 3, $\epsilon_{el}^{ht} = -0.33$ and $\epsilon_{th}^{ht} = 1$. Therefore, $\dot{E}_{ht,t,k}^{\text{tech}} = -0.33 \cdot \frac{\dot{Q}_{ht,t,k}^{\text{tech}}}{\epsilon_{th}^{ht}}$.

Electricity required for the decentralized air/water heat pump(s):

$$\dot{E}_{t,k}^{aw} = \frac{\dot{Q}_{t,k}^{aw}}{COP_{t,k}^{aw}} \quad \forall t, k \tag{2.24}$$

with

$$COP_{t,k}^{aw} = \eta^{hp} \cdot \frac{T_{Hs,t,k} + \Delta T}{T_{Hs,t,k} - T^{atm}} \quad \forall t, k \tag{2.25}$$

Electricity required for the individual backup chiller(s):

$$\dot{E}_{t,k}^{chiller} = \frac{\dot{Q}_{t,k}^{chiller}}{COP_{t,k}^{chiller} - 1} \quad \forall t, k \tag{2.26}$$

Electricity required by the network circulation pumps:

$$\dot{E}_{t}^{pump} = \frac{\left\{ \sum_{i,j} (\dot{M}_{hs,t,i,j} + \dot{M}_{hr,t,i,j} + \dot{M}_{cs,t,i,j} + \dot{M}_{cr,t,i,j}) \cdot Dist_{i,j} \right\} \cdot p_{t}^{drop}}{\rho} \quad \forall t \tag{2.27}$$

2.6.2.6 Mass Balances

Mass balances are defined at each node for both the supply and the return networks. They link the mass flow rate of water leaving the node (left-hand side of the $=$ sign in Eqs. (2.28) and (2.29)), with the mass flow rate of water arriving at the node, flowing through the building located at the node, and coming from the building at node k if a district energy conversion technology is located at that node (right-hand side of the $=$ sign).

Heating supply network:

$$\sum_{(k,j)} \dot{M}_{hs,t,k,j} = \sum_{(i,k)} \dot{M}_{hs,t,i,k} - \left(\dot{M}_{hn,t,k}^{build} + \dot{M}_{hn,t,k}^{hn\_ww} \right) + \dot{M}_{hn,t,k}^{tech} \quad \forall t, k \tag{2.28}$$

Cooling supply network:

$$\sum_{(k,j)} \dot{M}_{cs,t,k,j} = \sum_{(i,k)} \dot{M}_{cs,t,i,k} - \dot{M}_{cn,t,k}^{build} + \dot{M}_{cn,t,k}^{tech} \quad \forall t, k \tag{2.29}$$

(Symmetric equations are obviously defined for the return pipes calculation in each network.)

2.6.2.7 Connections of Nodes and Implementation/Location of the Technologies

The following constraints have been introduced for modeling the design of the network configuration (connections), and for defining the location of the centralized and decentralized conversion technologies:

1. Network configuration

Equations (2.30) and (2.31) define the existence of a connection between nodes i and j for networks hs and cs if energy (therefore water) flows from i to j ($\dot{M}_{hs}^{ub}$ and $\dot{M}_{cs}^{ub}$ are upper bounds computed in the data processing routine.):[5]

$$X_{hs,i,j} \cdot \dot{M}_{hs}^{ub} \geqslant \dot{M}_{hs,t,i,j} \quad \forall t, i, j \tag{2.30}$$

$$X_{cs,i,j} \cdot \dot{M}_{cs}^{ub} \geqslant \dot{M}_{cs,t,i,j} \quad \forall t, i, j \tag{2.31}$$

Equation (2.32) ensures that if a connection is implemented from node i to j in the heating supply network (hs) a connection is implemented from j to i in the heating return network (hr). The same is valid for the cooling network (between the supply cs and the return cr, Eq. (2.33)). Equation (2.34) ensures that the heating and the cooling network run parallel to each other.

$$X_{hs,i,j} = X_{hr,j,i} \quad \forall i, j \tag{2.32}$$

$$X_{cs,i,j} = X_{cr,j,i} \quad \forall i, j \tag{2.33}$$

$$X_{hs,i,j} = X_{cs,i,j} \quad \forall i, j \tag{2.34}$$

2. District energy conversion technologies

Equations (2.35) and (2.36) define the node where a district technology will be implemented and ensure that if the technology is operated at level $x_{ht,t}$, the heat generated by the district technology remains between the design load of the technology and the minimum technically acceptable part load κ_{ht}. Y_k^{tech} is a parameter defined in the structuring phase that defines the eligibility or not of a node to host a centralized technology. Analogous equations are defined for the district cooling technologies.

Heating technologies:

$$\dot{Q}_{ht,t,k}^{tech} \leqslant x_{ht,t} \cdot X_{ht,k} \cdot Y_k^{tech} \cdot S_{ht} \quad \forall ht, t, k \tag{2.35}$$

$$\dot{Q}_{ht,t,k}^{tech} \geqslant x_{ht,t} \cdot X_{ht,k} \cdot Y_k^{tech} \cdot S_{ht} \cdot \kappa_{ht} \quad \forall ht, t, k \tag{2.36}$$

$$\sum_k X_{ht,k} \leqslant 1 \quad \forall ht \tag{2.37}$$

3. Sizes of the decentralized energy conversion technologies

$$\dot{Q}_{t,k}^i \leqslant S_k^i \quad \forall t, k \tag{2.38}$$

$$X_k^i \cdot \dot{Q}_{H,k}^{max} \geqslant S_k^i \quad \forall k \in \text{Nodes} \tag{2.39}$$

---

5) Note that the binary variables are no indexed over the periods t since the final layout of the network is the same for all periods.

$$X_k^i \cdot \dot{Q}_{H,k}^{\min} \leqslant S_k^i \quad \forall k \in \text{Nodes} \tag{2.40}$$

It should be mentioned that in order to maintain the feasibility of the slave optimization, such constraints do not concern the backup boilers but only the other decentralized units. The backup oil burners will therefore be used to close the energy balance at each node and its size will be determined by the slave optimization. Obviously these units will also introduce the corresponding investment costs that will be considered in the objective function.

2.6.2.8 Costing and CO_2 Emission Functions
The costing functions implemented in the slave optimization include

1. the linearized investment costs of the pipes of the distribution networks, the circulation pumps, the heat exchangers between the networks and the buildings, and the individual backup energy conversion technologies,
2. the operating costs (including the maintenance and energy resources: natural gas, electricity from the grid and oil for the boilers),
3. the incomes from the electricity sales.

Annual investment costs estimation

1. *Pipes*
    In order to compute the investment costs for the pipes, the fixed costs are charged if a heating and/or cooling pipe is requested between two nodes ($X_{hs,i,j} = 1$ and/or $X_{cs,i,j} = 1$), but not yet existing ($Y_{hs,i,j}^{pipe} = 0$ and/or $Y_{cs,i,j}^{pipe} = 0$). If both heating and cooling pipes are requested, the fixed costs are charged only once.

$$C^{pipes} = \frac{Ms^{pipe}}{1306.6} \cdot (1 + Fm^{pipe}) \cdot An^{pipe} \cdot \sum_{(i,j)} Dist_{i,j}$$

$$\cdot \left(\left(X_{i,j} \cdot C_{fix}^{pipe} \right) + \left(\left(A_{hs,i,j} \cdot c_{prop}^{pipe} \right) \cdot \left(1 - Y_{hs,i,j}^{pipe} \right) \cdot X_{hs,i,j} \right) \right.$$

$$\left. + \left(\left(A_{cs,i,j} \cdot c_{prop}^{pipe} \right) \cdot \left(1 - Y_{cs,i,j}^{pipe} \right) \cdot X_{cs,i,j} \right) \right) \tag{2.41}$$

with

$$X_{i,j} \geqslant \left(X_{hs,i,j} \cdot \left(1 - Y_{hs,i,j}^{pipe} \right) \right) \quad \forall i, j \tag{2.42}$$

and:

$$X_{i,j} \geqslant \left(X_{cs,i,j} \cdot \left(1 - Y_{cs,i,j}^{pipe} \right) \right) \quad \forall i, j \tag{2.43}$$

Equations (2.42) and (2.43) ensure that the fixed costs for the pipes only accrue if a heating and/or cooling pipe is/are required and not already existing.

Moreover:

$$A_{hs,i,j} = \frac{\max_t \dot{M}_{hs,t,i,j}}{v \cdot \rho} \quad A_{cs,i,j} = \frac{\max_t \dot{M}_{cs,t,i,j}}{v \cdot \rho} \quad \forall i, j$$

2. *Pumps*

   Heating network:

$$C^{pump} = \frac{Ms^{pump}}{1306.6} \cdot \left(1 + Fm^{pump}\right) \cdot An^{pump}$$

$$\cdot \left(\left(X^{hn} \cdot C_{fix}^{pump} + S_{hn}^{pump} \cdot c_{prop}^{pump}\right)\right.$$

$$\left. + \left(X^{cn} \cdot C_{fix}^{pump} + S_{cn}^{pump} \cdot c_{prop}^{pump}\right)\right) \tag{2.44}$$

with

$$S_{hn}^{pump} = \frac{\max_t \left(\left(\dot{M}_{hs,t,i,j} + \dot{M}_{hr,t,j,i}\right) \cdot Dist_{i,j} \cdot p_t^{drop}\right)}{\rho} \quad \forall i, j$$

and

$$S_{hn}^{pump} \leqslant X^{hn} \cdot \left(\dot{M}_{hs}^{ub} + \dot{M}_{cs}^{ub}\right)$$

and symmetrically for the cooling network.

3. *Heat exchangers*

   The investment costs for the heat exchangers are estimated in the data processing routine and result in the following annual investment cost:

$$C^{hex} = \frac{Ms^{HEX}}{1306.6} \cdot \left(1 + Fm^{HEX}\right) \cdot An^{HEX}$$

$$\cdot \sum_k \left(X_{hn,k}^{HEX} \cdot C_{hs,k}^{HEX\_inv} + X_{cn,k}^{HEX} \cdot C_{cs,k}^{HEX\_inv}\right) \tag{2.45}$$

with

$$\dot{Q}_{hs,t,k}^{build} + \dot{Q}_{hs,t,k}^{hn\_ww} \leqslant X_{hn,k}^{HEX} \cdot \left(\dot{Q}_{H,t,k}^{cons} + \dot{Q}_{HW,t,k}^{cons}\right) \quad \forall t, k$$

and

$$\dot{Q}_{cs,t,k}^{build} \leqslant X_{cn,k}^{HEX} \cdot \dot{Q}_{C,t,k}^{cons} \quad \forall t, k$$

4. *Decentralized energy conversion technologies*

$$C^i = \frac{Ms^{aw}}{1306.6} \cdot \left(1 + Fm^i\right) \cdot An^i \cdot \sum_k \left(\left(X_k^i \cdot C_{fix}^i\right) + \left(S_k^i \cdot c_{prop}^i\right)\right) \tag{2.46}$$

In order to account for different expected lifetimes, annuity factors are considered for each equipment according to the following equation:

$$An^i = \frac{r^i \cdot (1 + r^i)^{N^i}}{(1 + r^i)^{N^i} - 1} \tag{2.47}$$

Operating costs The annual operating costs are obtained by summing the operating costs in each period and considering the duration of each period.

1. Grid costs

$$C^{\text{grid}} = \sum_t \dot{E}_t^{\text{grid}} \cdot D_t \cdot c^{\text{grid}} \tag{2.48}$$

2. Natural gas costs

$$C^{\text{gas}} = \sum_t R_t^{\text{gas}} \cdot D_t \cdot c^{\text{gas}} \tag{2.49}$$

in which

$$R_t^{\text{gas}} = \sum_{ht} \sum_k \frac{\dot{Q}_{ht,t,k}^{\text{tech}}}{\epsilon_{\text{th}}^{ht}} \tag{2.50}$$

and R_t^{gas} the average hourly gas consumption rate required by all the district energy conversion technologies operated with natural gas (no backup device is operated with natural gas), and D_t the duration of period t in hours.

3. Oil costs (only for individual backup boilers)

$$C^{\text{oil}} = \sum_t R_t^{\text{oil}} \cdot D_t \cdot c^{\text{oil}} \tag{2.51}$$

$$R_t^{\text{oil}} = \sum_k \frac{\dot{Q}_{t,k}^{\text{boiler}}}{\epsilon^{\text{boiler}}} \tag{2.52}$$

with R_t^{oil} the average hourly oil consumption rate required by the boilers, if any.

Incomes Incomes can be generated by the sale of superfluous electricity

$$I^{\text{el}} = \sum_t \dot{E}_t^{\text{sold}} \cdot D_t \cdot b^{\text{grid}} \tag{2.53}$$

where b^{grid} is the price paid by the grid when buying electricity.

CO₂ emissions The CO_2 emissions are computed considering the emissions related to the overall resources used in the system, i.e. the emissions of natural gas, oil, and electricity. When the net balance of electricity corresponds to an export of electricity, one may choose co_2^{subst} to consider the substitution based on the mean value of the grid using the electricity CO_2 mix or to consider the substitution of the CO_2 emissions of a new power plant to be build.

$$C^{CO_2} = \sum_t (R_t^{gas} \cdot co_2{}^{gas} + \dot{E}_t^{grid} \cdot co_2{}^{grid}$$

$$- \dot{E}_t^{sold} \cdot co_2{}^{subst} + R_t^{oil} \cdot co_2{}^{oil}) \cdot D_t \qquad (2.54)$$

2.6.2.9 Considering Multiobjective Aspects in the Slave Optimization

The optimization problem decomposition should be compatible with the multiobjective optimization strategy used at the master optimization level. For the slave optimization, we have therefore introduced a parametric programming dimension by introducing a weighting factor between the terms that represent the two objectives in the slave optimization. As the two objectives are on one hand the total cost and on the other hand the CO_2 emissions, a CO_2 tax is introduced as a parameter of the slave optimization. This allows us to weigh the importance of the CO_2 emissions in the slave optimization objective function. The value of the CO_2 tax is introduced as a decision variable in the master optimization, together with the sizes of the centralized technologies and the temperatures of the networks.

2.6.2.10 Empirical Constraints

Empirical constraints based on the experience of the engineer allow one to speed up the resolution time by reducing the search space, without excluding the optimal solution. These result from a systematic analysis of the problem statement. The slave optimization includes constraints based on heuristic rules referring to the implementation of the technologies (district technologies and backup technologies), and the pipe routing. Unlike cuts, which are managed directly by the solver, or valid inequalities (formulated by the programmer but redundant in the problem definition), the addition of empirical constraints changes the mathematical formulation of the problem by eliminating feasible integer solutions, however only suboptimal solutions.

Connection constraints Heating and/or cooling cannot, at a node k, be provided by two different routes.

$$\sum_i X_{hs,i,k} \leqslant 1 \quad \forall k \qquad (2.55)$$

This assumption is valid in the case of one centralized production unit. It will not be considered if more than one centralized production unit is allowed and considered in the master optimization problem.

Decentralized energy conversion technologies No more than one backup energy conversion technology can be implemented on a given node k to provide heating and hot water (for the cooling requirements there is anyway only one possible backup technology, so that a constraint such as (2.56) is not necessary):

$$X_k^{aw} + X_k^{ww} + X_k^{boiler} \leqslant 1 \quad \forall k \tag{2.56}$$

If the node is connected to the network, no air/water heat pump or boiler can be used at this same node:

$$X_k^{aw} + X_k^{boiler} + \sum_i X_{hs,i,k} \leqslant 1 \quad \forall k \tag{2.57}$$

Routing of the pipes From the analysis of the possible connections between nodes in the system, algorithms like the one used in [16] allow to identify the possible cycles. Considering the list of possible cycles in which a node is involved, the following constraints will be added in order to avoid the creation of cycles. It should be mentioned that these constraints could be modified if system reliability is considered.

$$\sum_{i \in cycle[c]} X_{i,next(i)} \leqslant card(cycle[c]) - 1; \tag{2.58}$$

2.6.3
Data Processing Routine

The evolutionary algorithm used in the master optimization builds a "genome" the performance of which is computed by solving the slave optimization problem. In order to set up the slave problem, a data preprocessing is realized.

1. Given the values of the decision variables in the master set, the data preprocessing routine first defines the heat cascade corrected temperature intervals, evaluates the heat exchanger surfaces and costs, and approximates the average specific heat and pressure losses. Applying cost estimation expressions, linearized equipment cost are also estimated for the decentralized technologies.
2. The empirical and feasibility constraints are defined. An example of feasibility constraint is, for instance, when the master optimizer defines a condenser temperature for a heat pump that is below the return temperature of the heating network, making the heat pump useless. In this case, the data processing routine does not pass the heat pump to the slave optimization, thus reducing its search space and avoiding the iterations that would be necessary in the slave optimization to find out that the heat pump is useless. The annual investment cost of the heat pump is nonetheless taken into account in the objective function of the master optimization. Hence the technology induces investment costs, but does not provide any useful energy, thus discouraging the master optimizer from choosing it. Such controls can be included for any type of district energy conversion technology, according to the specificities of the technology.

3. It defines the upper and lower bounds of the continuous decision variables of the slave optimization, that can be derived from the values given by the master optimization. Defining a more restrictive set of upper and lower bounds reduces the computation size of the search space and therefore drastically reduces the computations time.

The constraints of points 2 and 3 above allow one to reduce the computation time from hours to minutes. This step is therefore of prime importance when considering that the master optimization requires several thousands of iterations to reach convergence.

Once the slave optimization has converged, the results are again processed in the postprocessing phase, where the nonlinear performances of the system are computed. From the equipment sizes proposed by the master optimization, the investment of the centralized units is defined. From the results of the optimization, the flows in the district heating network are known. These are used to compute the size of the pipes and deduce their "nonlinear" cost. The size of the decentralized units is also resulting from the slave optimization and their cost is further computed considering nonlinear equations where appropriate. The cost of the heat exchangers is also re-evaluated and the annual operating cost of the system is computed together with the overall emissions.

It should be mentioned that the CO_2 emission tax is not considered when evaluating the annual operating cost, since it represents only a weighting factor for realizing the parametric multiobjective optimization in the salve optimization strategy.

In addition, the following general assumptions have been considered when building the model:

1. The differences in altitude have been neglected.
2. All heat exchangers are counterflow heat exchangers.
3. For the decentralized heat pumps, the condenser temperature has been chosen in function of the hot water requirements (regardless of whether the heat pump supplies heat or hot water requirements). This assumption penalizes especially low-temperature heating buildings and should therefore be refined in a more detailed approach.
4. If any excess electricity is generated by the district energy conversion technologies, it can be decided from case to case, whether this electricity shall be sold to the grid or not.
5. If any excess heat or cold is generated by the district energy conversion technologies, this energy is lost to the surroundings without any investment penalty.
6. Storage tanks are considered only for peak shaving and are supposed to be part of the equipment cost. This means that the mean heat demand can be used and includes compensation for heat load demand variations. It should be noted that in district heating systems, the volume of water in the pipes correspond to a significant buffer capacity.

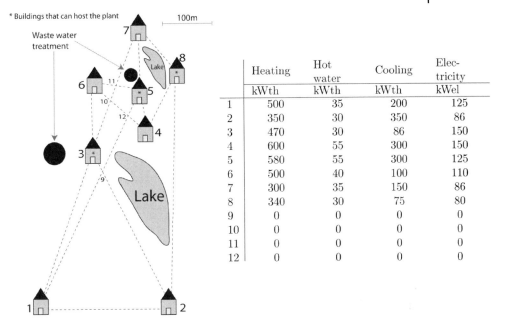

* Buildings that can host the plant

Waste water treatment

100m

	Heating	Hot water	Cooling	Electricity
	kWth	kWth	kWth	kWel
1	500	35	200	125
2	350	30	350	86
3	470	30	86	150
4	600	55	300	150
5	580	55	300	125
6	500	40	100	110
7	300	35	150	86
8	340	30	75	80
9	0	0	0	0
10	0	0	0	0
11	0	0	0	0
12	0	0	0	0

Fig. 2.6 District analyzed, with the maximum requirements in kW for each energy service and the buildings that can host the plant with the district energy conversion technologies.

7. Heat losses in the pipes are considered for the heating network but neglected for the cooling network, due to the relatively small temperature difference between the supply pipes and the surroundings as compared to the heating network.

2.7
Test Case Application

The method is applied to the test example presented in Fig. 2.6.

The goal of the study is to design the energy conversion system that will provide heating, hot water, cooling, and electricity to a district of eight buildings considering the minimization of the yearly CO_2 emissions and costs. The district to be developed (Fig. 2.6) is supposed to be new, so that no equipment is existing beforehand, except for the electricity grid. The dashed lines in Fig. 2.6 show the connections which are allowed between buildings and/or crossings (nodes).

2.7.1
Structuring of the Information

According to the method, the following information is gathered and structured:

Table 2.2 Periods considered in the test case with their duration and average atmospheric temperature.

Period	Duration (h/year)	T^{atm} (°C)
Summer day	1095	25
Summer night	1095	15
Mid-season day	2190	14
Mid-season night	2190	10
Winter day	1095	1
Winter night	1095	−5

2.7.1.1 List of Available Energy Sources

Regarding the energy sources, two types of renewable energy sources can be used, namely two lakes and two wastewater treatment facilities that can provide energy to heat pumps. Regarding the non-renewable sources, natural gas is the only one that is considered, since heating oil has been left out for centralized production. Other sources such as waste incineration or biomass have been assumed not to be available.

2.7.1.2 Energy Consumption Profiles Including Power and Temperature Levels

The energy consumption profiles have been developed. Six representative periods have been defined with their relative duration in hours/year, and average atmospheric temperature. These periods are defined in Table 2.2. As already mentioned, since the purpose of the method is the *design* a district energy system, and not the *controlling* of this system, it has been considered sufficient to divide the whole year into 6 periods only. This simplification has the disadvantage to exclude days with extreme weather events and could therefore favor undersized equipment. Increasing the number of representative periods however would lead to an increased number of variables and therefore an increase of the computation time required to solve the slave optimization problem. The analysis of the solutions with more detailed simulations will show the feasibility or not of the solutions proposed, for extreme conditions. In case the solutions are penalizing, extreme periods could be added to the period set as shown in [17]. The consumption profiles are presented in Figs. 2.8 to 2.10. The temperature levels for the heating requirements, corresponding to the supply temperature T_S and the return temperature T_R of the hydronic circuit in the building, have been set to $T_S = 35\,°C$ and $T_R = 32\,°C$ for the mid-season periods and $T_S = 58\,°C$ and $T_R = 48\,°C$ for the winter periods, according to Fig. 2.7. The water used for the hot water requirements (Fig. 2.9) has been assumed to enter the buildings at $10\,°C$ and to be heated to $60\,°C$. Finally for the cooling requirements (Fig. 2.10), the temperatures have been set for all the buildings to $T_S = 18\,°C$ and to $T_R = 21\,°C$. These latter values, which are higher than what is usually admitted ($T_S = 6\,°C$ and to $T_R = 12\,°C$) have been estimated based on what could be found in modern installations [18].

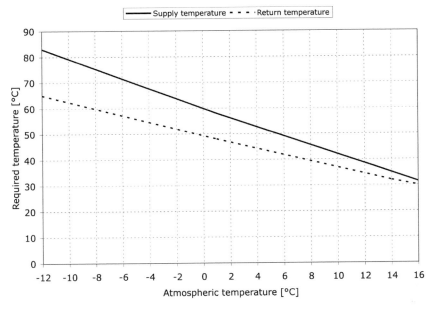

Fig. 2.7 Heating supply and return temperatures of the hydronic circuit of the buildings [19].

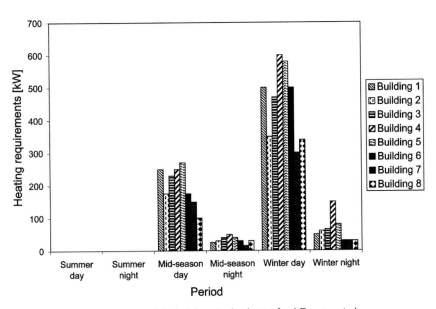

Fig. 2.8 Heating requirements of the buildings in the district for different periods.

Comparing the list of available energy sources together with the energy service requirements of the district including the temperature and power levels of these requirements, the following technologies are considered as reasonable and taken

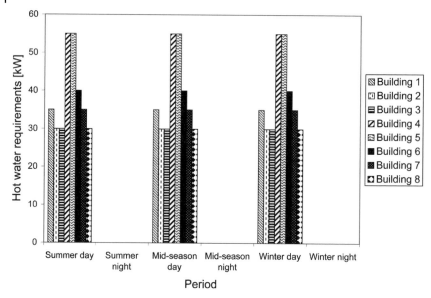

Fig. 2.9 Hot water requirements of the buildings in the district for different periods.

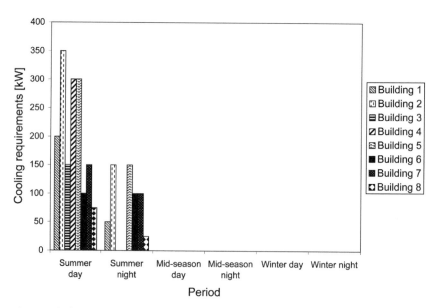

Fig. 2.10 Cooling requirements of the buildings in the district for different periods.

up in the list of possible district energy conversion technologies. The resulting energy conversion superstructure is presented in Fig. 2.12.

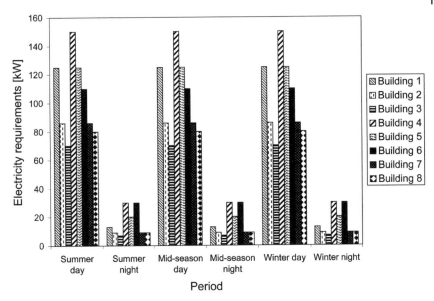

Fig. 2.11 Electricity requirements of the buildings in the district for different periods.

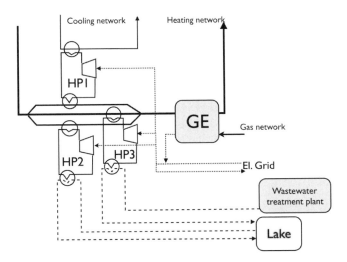

Fig. 2.12 Superstructure of the energy conversion system for the test case.

1. Heating and hot water:

    - A heat pump, connecting the return pipe of the cooling network on the evaporator side with the supply pipe of the heating network on the condenser side. This heat pump thus allows the energy transfer between the buildings requiring cooling and the buildings requiring heating. This heat pump will be abbreviated HP1 in the following, its lower bound is 1000 kWth and its upper bound 5000 kWth.

Table 2.3 Temperature level of the lake and the wastewater treatment facility.

Parameter	Summer	Mid-season	Winter
Temperature of the lakes (°C)	8	6	6
Temperature of the municipal wastewater treatment facility in summer (°C) [20]	22	18.5	16

- A second heat pump, valorizing energy from the lake(s) and having therefore its evaporator temperature determined by the temperature of the lake. A maximum temperature difference of 4 °C is assumed in the evaporator for the water coming from the lake, in order to avoid freezing in winter (the temperature of the water in the lake being 6 °C). This heat pump will be abbreviated HP2 in the following, its lower bound is 1000 kWth and its upper bound 5000 kWth.
- A third heat pump, converting energy recycled from the wastewater treatment facility(ies) and having therefore its evaporator temperature determined by the temperature of the wastewater treatment facility(ies): 16 °C in winter, 18.5 °C during the mid-seasons and 22 °C in summer. As mentioned in Section 2.6.3, a temperature difference of 10 °C is assumed in the evaporator for the water coming from the wastewater treatment facility. This heat pump will be abbreviated HP3 in the following, its lower bound is 1000 kWth and its upper bound 5000 kWth.

2. Cooling:

- The HP1 heat pump mentioned above, allowing heat pumping from the cooling return pipe.
- "Free" cooling using water from the lake(s) that is being circulated through the district network.

3. Combined heat and power:

- A gas engine (abbreviated GE), used in cogeneration mode, with a lower bound at 1000 kWel and an upper bound at 5000 kWel.

For the lake(s) and the wastewater treatment facility(ies), it is assumed that the resources are unlimited. In a more detailed analysis (for instance, if HP2 or HP3 type heat pumps appear to be promising technologies), the temperature difference of 4 °C, respectively 10 °C, assumed in the evaporator, should also be optimized. In this case however, the mass flow rate between the lake or the wastewater treatment facility(ies) and the heat pump, should also be considered. Besides, a limited amount of water could easily be taken into account by setting a constraint on the energy source. The temperature levels for the lake and the wastewater treatment facility are given in Table 2.3

Table 2.4 Parameters for the costing functions in the test case.

Parameter	Range	Value	Unit	Reference
C_{fix}^{pipe}	0–300 mm ø	753	CHF	[21]
c_{prop}^{pipe}	0–300 mm ø	7	CHF/m	[21]
C_{fix}^{boiler}	100–650 kWth	1000	CHF	[22]
c_{prop}^{boiler}	100–650 kWth	218	CHF/kWth	[22]
C_{fix}^{aw}	100–650 kWth	1000	CHF	[23]
c_{prop}^{aw}	100–650 kWth	1890	CHF/kWth	[23]
C_{fix}^{ww}	350–650 kWth	42833	CHF	[24]
c_{prop}^{ww}	350–650 kWth	126	CHF/kWth	[24]
$C_{fix}^{chiller}$	0–350	2500	CHF	Eval. after [24]
$c_{prop}^{chiller}$	0–350 kWth	2835	CHF/kWth	Eval. after [24]
r^i	$i = aw, boiler, chiller, hex, pipes, pump, ww$	0.08		[25]
$F_m{}^i$	$i = aw, boiler, chiller, hex, pipes, pump, ww$	0.06		[25]
$M_S{}^i$	year 2006 ($i = aw, boiler, chiller, ww$)	1306.6		[25]
$M_S{}^i$	year 1998 ($i = hex, pipes, pump$)	1306.6		[25]

Table 2.5 Efficiency parameters related to the individual back-up technologies, used in the test case.

Parameter	Value	Units	Reference
Exergetic efficiency of the individual back-up air/water heat pumps	0.34	(–)	[19]
Exergetic efficiency of the individual back-up water/water heat pumps	0.4	(–)	[19]
Exergetic efficiency of the individual back-up chillers	0.19	(–)	After [26]

2.7.1.3 Spatial Constraints and Possible Connections
Possible location of the district energy conversion technologies (Fig. 2.6): only buildings 3, 5, and 8 are possible candidates. Besides, the possible connections obtained from a routing algorithm, are given as dotted lines in Fig. 2.6. Other than this, no size related constraints are given.

2.7.1.4 Decentralized Technologies
As decentralized technologies, air/water and water/water heat pumps as well as boilers have been considered for heating and hot water while electric chillers have been considered for cooling. In order to compute the annual investment costs, the parameters given in Table 2.4 have been used. Besides, the efficiencies given in Table 2.5 have been used.

2.8
Optimization

2.8.1
Decision Variables

The master optimizer optimizes 17 decision variables:

1. Supply and return temperatures of the heating network in summer, and winter/mid-seasons: variables 1 to 4
2. Supply and return temperatures of the cooling network in summer (no cooling is assumed during the winter and the mid-seasons): variables 5 and 6
3. Investment or not in a HP1, HP2, and HP3 as well as gas engine GE: variables 7 to 10
4. Design size of heat pumps HP1, HP2, and HP3 as well as gas engine GE: variables 11 to 14
5. Condenser temperature of heat pumps HP1, HP2, and HP3: variables 15 to 17

Besides these decision variables, the master optimization also defines the CO_2 weighting factor. The condenser temperatures of the heat pumps do not necessarily equal the supply temperature of the heating network, as the three heat pumps and the gas engine can be put in series. In fact, in an optimally integrated system these energy conversion technologies *will* be in series, in order for the heat pumps to operate at better COPs. This will be computed in the heat cascade model.

All the parameters used for the test case are given in Tables 2.4 to 2.6.

2.8.1.1 Conditions Defined for the Optimization
To perform the optimization, it was decided that the excess electricity generated by the gas engine could be sold at a price of 0.09 CHF/kWel [27]. However, the avoided CO_2 when selling electricity to the grid, has not been accounted when evaluating the CO_2 emissions.

2.8.1.2 Resolution Phase Analysis
According to the values chosen by the master optimizer and the subsequent data processing performed by the data processor, the number of variables optimized by the slave optimizer ranges from 35 000 to 50 000 continuous variables and from 250 to 850 binary (0–1) variables. The large number of variables optimized by the slave optimizer is due to the number of indices for each type of variable. Let us analyze, for instance, the variable characterizing the energy streams between nodes in the network. For the slave optimizer, there are as many variables as there are possible combinations of connections (between two nodes), periods, temperature intervals[6]

---

6) The exact number of temperature intervals describing the whole system (the highest temperature being, for instance, the temperature of the heat from the gas engine and the lowest temperature the supply temperature of the cooling network) depends strongly on the set of variables defined by the master optimizer.

Table 2.6 Values of the thermodynamic and distributed energy related parameters, used in the optimization phase in general.

Parameter	Value	Units	Reference
CO_2 emissions of electricity from the grid $co_2{}^{grid}$ (European mix)	0.450	(kg/kWh)	[28]
CO_2 emissions for natural gas $co_2{}^{gas}$ (including transportation)	0.230	(kg/kWh)	[28]
CO_2 emissions for heating oil $co_2{}^{oil}$ (including transportation)	0.293	(kg/kWh)	[28]
Cost for electricity from the grid c^{grid}	0.13	(CHF/kWh)	[27]
Cost for natural gas c^{gas}	0.04	(CHF/kWh)	[27]
Cost for heating oil c^{oil}	0.0645	(CHF/kWh)	[27]
Efficiency of the circulation pumps	80%	(–)	
Friction coefficient in the pipes λ	0.02	(–)	[21]
Grid efficiency ϵ^{grid}	90%	(–)	
Heat transfer coefficient in the pipes K	0.203	(W/mK)	[21]
Isobaric specific heat of water $cp_{H_2O}^{liq}$	4.178	(kJ/(kg K))	
Overall heat transfer coefficient in heat exchangers	800	(W/(m^2 K))	[29]
Minimum ΔT in the heat-exchangers between the district networks and the hydronic circuit of the buildings	2	(°C)	
Selling price of electricity b^{grid}	0.09	(CHF/kWh)	[27]

and subnetworks[7]:

Number of possible connections in the test district: 21
Number of periods: 6
Number of temperature intervals: ~ 20
Number of subnetworks: 4

In other words, there are $21 \cdot 6 \cdot 20 \cdot 4 = 10\,080$ decision variables only for the energy streams. The same number of decision variables are related to the mass flows.

17 000 iterations of the master optimizer have been computed. Solving this problem took 4 days 8 h and 29 min of computation time on 4 Intel Pentium4 2'800 MHz computers (the resolution has been parallelized on 4 computers). The optimization time indicated here is of course strongly dependent on the number of iterations done by the master optimization[8] but also on the tolerance set in the MILP solver.

7) Supply and return networks for both heating and cooling.
8) The number of iterations is estimated empirically. The iterations are stopped when no improvement can be obtained from further iterations.

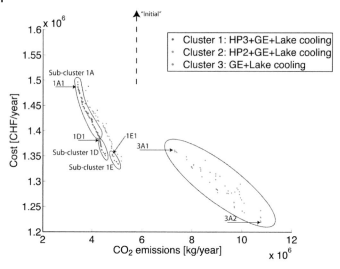

Fig. 2.13 Pareto optimal frontier.

2.8.1.3 Pareto Optimal Frontier

The results of the optimization are shown on the Pareto curve of Fig. 2.13. The most interesting configurations are summarized in Table 2.7. In this table, the indices a and b for the heat pump HP3 refer to the two heat pumps that are implemented in series at the wastewater treatment facility, due to the high temperature difference occurring at the evaporator side. When the temperature decrease of the heat source on the evaporator side exceeds $4\,°C$ indeed, two smaller heat pumps are implemented in series instead of one large heat pump. From the Pareto optimal frontier, three distinct groups of configuration types (hereafter clusters) can be recognized:

Cluster 1 "HP3+GE+Lake": The HP3 heat pump is implemented together with the gas engine, providing heating, hot water and electricity to all the buildings. The cooling is provided by the water from the lake supplied to all the buildings.

Cluster 2 "HP2+GE+Lake": The HP2 heat pump is implemented together with the gas engine, providing heating, hot water and electricity to all the buildings. The cooling is provided by the water from the lake to all the buildings.

Cluster 3 "GE+Lake": Only the gas engine is implemented to provide heating, hot water and electricity to some or all of the buildings. The cooling is provided by the water from the lake to all the buildings. No district heat pumps are implemented in this cluster.

The reference configuration, called "Initial," getting the electricity from the grid, heating and hot water from a boiler, and cooling from an electric chiller, does not appear in Fig. 2.13. Actually, such a configuration would emit 5779 tons-CO_2/year and cost 1.98 mio-CHF/year, thus overshooting the cost axis of the figure, and performing poorly in terms of CO_2 emissions.

Table 2.7 Main configurations for the test case.

Configuration	1A1	1A2	1B1	1D1	1E1	3A1	3A2
Annual CO_2 emissions ton-CO_2/year	3 407	4 256	3 582	4 350	4 824	7 303	10 769
Costs mio-CHF/year	1.49	1.39	1.46	1.37	1.35	1.36	1.22
Percentage Investment	53	46	53	46	44	27	22
Operational	47	54	47	54	56	73	78
Income I^{el} mio-CHF/year	0	0.18	0	0.22	0.26	0.44	1.2
Grid mio-CHF/year	0.28	0.26	0.14	0.26	0.13	0.02	0.02
Gas mio-CHF/year	0.42	0.58	0.54	0.6	0.76	1.26	1.86
HP3a Size MWth	1029	639	1040	634	634	0	0
Inv. CHF/year	174 816	123 469	176 134	122 901	122 901	0	0
T_{cond} °C	75	80	79	80	77	–	–
COP summer	3.06	3.03	3.05	3.03	3.04	–	–
COP mid-season	2.91	2.89	2.90	2.89	2.90	–	–
COP winter	2.82	2.79	2.80	2.79	2.81	–	–
HP3b Size MWth	1029	639	1040	634	634	0	0
Inv. CHF/year	174 816	123 469	176 134	122 901	122 901	0	0
T_{cond} °C	80	85	84	85	82	–	–
COP summer	3.10	3.08	3.09	3.08	3.09	–	–
COP mid-season	2.95	2.93	2.94	2.93	2.94	–	–
COP winter	2.86	2.83	2.84	2.83	2.84	–	–
GE Size MWel	1753	2532	1731	2540	2540	1978	3857
Inv. CHF/year	190 444	239 621	188 987	240 101	240 101	205 363	311 681
ϵ_{el} %	42.88	43.47	42.86	43.47	43.47	43.08	44.14
ϵ_{th} %	47.12	46.53	47.14	46.53	46.53	46.92	45.86

(continued on next page)

Table 2.7 (continued)

Configuration	1A1	1A2	1B1	1D1	1E1	3A1	3A2
Lake cooling kWth	1 625	1 625	1 625	1 625	1 625	1 625	1 625
Network Inv. CHF/year	192 121	187 630	199 014	189 293	190 908	176 933	186 590
Pumps Inv. CHF/year	5 624	5 493	5 813	5 542	5 661	4 324	5 683
$T_{hs,t}$ °C Summer	70.4	71.9	68.5	72.4	110.6	103.2	94.7
Mid-season and winter	77.0	77.6	75.9	77.6	77.5	98.0	103.0
$T_{hr,t}$ °C Summer	36.4	49.0	49.4	53.2	53.2	58.1	88.9
Mid-season and winter	52.6	50.1	51.8	51.5	51.5	51.7	53.5
$T_{hs,t} - T_{hr,t}$ °C Summer	34.0	22.9	19.1	19.2	57.4	45.1	5.8
Mid-season and winter	24.4	27.5	24.1	26.1	26.0	46.3	49.5
$T_{cs,t}$ °C At	10.2	10.0	11.0	10.0	10.0	10.5	10.2
$T_{cr,t}$ °C At	18.9	18.9	19.0	18.8	18.8	17.0	18.9

The optimal configuration in terms of CO_2 emissions (configuration 1A1 in Fig. 2.13 and Table 2.7) emits 3 407 tons-CO_2/year (59% of the "Initial" configuration) and costs 1.49 mio-CHF/year (75% of the "Initial" configuration). This configuration features a HP3 type district heat pump, together with a gas engine, a cooling network with water from the lake, and takes the electricity partly from the grid (summer periods as well as all the night periods) and partly from the gas engine (mid-season and winter day periods). All the buildings are connected to the networks, be it for cooling or for heating and hot water. The networks are shown in Fig. 2.14.

The optimal solution in terms of annual costs (configuration 3A2 in Fig. 2.13 and Table 2.7) emits 10 769 tons-CO_2/year and costs 1.22 mio-CHF/year. This solution features a gas engine that delivers heating and hot water to all the eight buildings, with lake cooling for all the buildings, and electricity from the grid during the summer night period (in all the other periods, the gas engine generates enough electricity for the district and even sells some to the grid). The heating and cooling networks are shown in Fig. 2.15. Due to the way of accounting for the CO_2 emissions, this configuration emits more CO_2 than the "Initial" configuration (186%). Since the avoided CO_2 of the grid electricity production is not accounted, such solutions will emit more CO_2 than the reference. When the avoided CO_2 is accounted, such solutions become better than the present "Initial" solution.

Figures 2.16 and 2.17 show for each period the composite curves for the district heating system for these two configurations (note that the scale for the X-axis changes according to the period). In the plots, the hot composite curve represents the heat provided by the district energy conversion technologies, and the cold composite curve the heat required by the district. Note that the periods during which the gas engine is not operated can be easily recognized in these figures, since no high temperature heat (up to 570 °C) is generated (for instance during period 1, the summer day period, of configuration 1A1). The least CO_2 emitting configuration features also the better energetic integration. For configuration 1A1, the entire heat is provided by the heat pump during periods 1 (summer day period) and 4 (mid-season night period). Due to the constraint that the heat pump cannot be operated less than 20% of the time (to avoid shutting the heat pump on and off for a few minutes only), the heat pump generates more heat than required during these two periods. In a more refined optimization, this constraint should be analyzed considering a higher number of period, including the heat storage aspects and considering more detailed models for the technologies.

It is noteworthy that in clusters 1 and 2, *all* the buildings are connected to both the heating and the cooling networks. In cluster 3, even if the heating network does not always connect all the buildings, the cooling network does provide cooling to all the buildings. Regarding the cooling network supply and return temperatures, one can see from Table 2.7 that they almost always correspond to the minimum, respectively maximum, allowed temperatures. For the supply temperature, 10 °C indeed corresponds to the temperature of the lake (8 °C) plus the minimum temperature difference of 2 °C, and for the return temperature, 19 °C correspond to the return temperature of the hydronic circuits in the buildings (21 °C) minus the

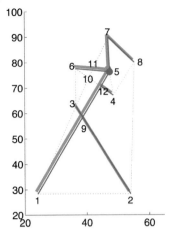

Fig. 2.14 Cooling network (blue), heating, and hot water network (red) for configuration 1A1 (see Fig. 2.13 and Table 2.7).

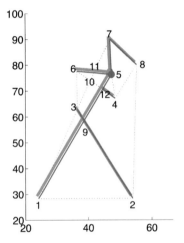

Fig. 2.15 Cooling network (blue), heating and hot water network (red), for configuration 3A2 (see Fig. 2.13 and Table 2.7).

minimum temperature difference of 2 °C. This fact is not surprising, since the heat "gains" (the opposite of the heat losses for heating networks) have been neglected and therefore advantage is taken of the maximum possible temperature difference between the supply and return pipes, in order to minimize the water that needs to be pumped. Finally, for all the configurations, an interaction between the district and the grid is always present, be it to buy or to sell electricity.

To be Pareto optimal, the configurations implementing district heat pumps need to have a heat pump using the wastewater treatment plant. The configurations of cluster 2, "HP2+GE+Lake," are never Pareto optimal. This is a consequence of the

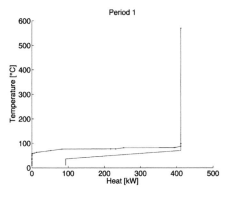

Composite curve of the summer day period

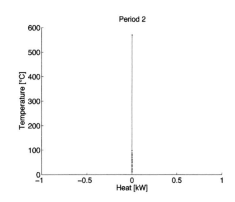

Composite curve of the summer night period

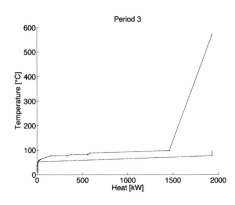

Composite curve of the mid-season day period

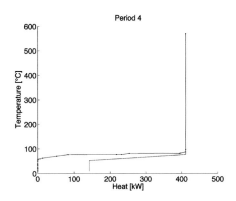

Composite curve of the mid-season night period

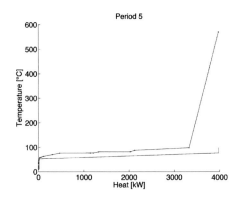

Composite curve of the winter day period

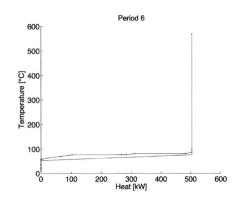

Composite curve of the winter night period

Fig. 2.16 Composite curves for the heating and hot water requirements of configuration 1A1 (see Fig. 2.13 and Table 2.7).

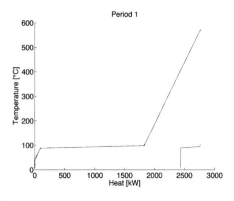

Composite curve of the summer day period

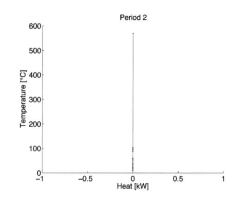

Composite curve of the summer night period

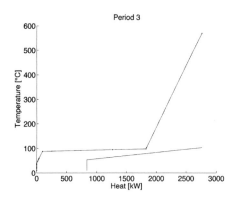

Composite curve of the mid-season day period

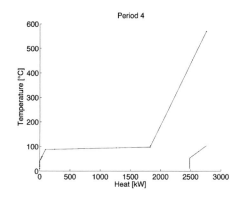

Composite curve of the mid-season night period

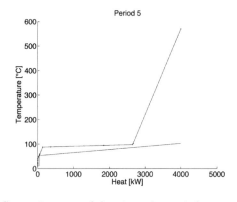

Composite curve of the winter day period

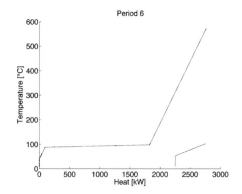

Composite curve of the winter night period

Fig. 2.17 Composite curves for the heating and hot water requirements for configuration 3A2 (see Fig. 2.13 and Table 2.7).

low temperature heat source being at a higher temperature level for HP3 (waste-water treatment facility) than for HP2 (lake), resulting in a better COP for HP3 compared to HP2. It might appear surprising that the difference of the results between clusters 1 and 2 is not larger. The main explanation lies in the fact that the COP difference for both types of heat pumps are less than 5%.

The district heat pump HP1 is never used. This is due to the tight constraints set on this technology. *District* cooling can indeed be provided either by this heat pump or by circulating water from the lake. For both options, pipes and a circulating pump need to be implemented, so that no difference exists between the two options at this level. However, HP1 generates an investment in the technology, which water from the lake does not require.[9] Besides, HP1 can only be used if in parallel with the cooling requirements, heating or hot water requirements are to be satisfied. In the case analyzed, this occurs only during the summer day period, for which cooling requirements and hot water requirements are to be satisfied simultaneously. Therefore if HP1 was implemented during the summer day period, another cooling technique should be sought for during the summer night period, making the overall resulting system for both periods less profitable than having only one cooling system.

2.8.1.4 Space Restrictions

Developing district energy systems in existing cities can be made difficult due to the lack of available space in existing underground channels. In order to demonstrate how the slave optimizer integrates the space restriction constraints and how this affects the result, a size restriction has been added to the network presented in Fig. 2.14. The resulting diameter between node 5, where the district technologies are implemented, and node 11, is 63 mm for the heating network and 112.4 mm for the cooling network. Figure 2.18 shows the new resulting network obtained when a maximum diameter of 10 mm is allowed between nodes 5 and 11 of the heating network. The slave optimizer has redesigned the configuration of Fig. 2.14 diverting the network over nodes 12 and 10. The size restriction on the pipe of the heating network also affects the cooling network (the shape of the cooling network is different between Figs. 2.18 and 2.14). This is explained by the fact that it is cheaper to have both networks running in parallel (only one gallery needs to be dug for both) than to have different routes for each network. For this new configuration, the investment costs amount to 193 787 CHF/year (compared to 192 121 CHF/year for configuration 1A1), and 5 738 CHF/year for the pump (compared to 5 624 CHF/year for configuration 1A1).

2.8.1.5 Inhomogeneous Requirements

For the test case analyzed in the present chapter, the buildings feature similar temperature levels for all the thermal energy requirements. It is therefore not a big surprise that all the buildings are connected to the heating (and cooling) networks in clusters 1 and 2. Figure 2.19 shows the resulting network for the same district, but

---

9) The costs for the heat exchanger between the lake and the network have been neglected.

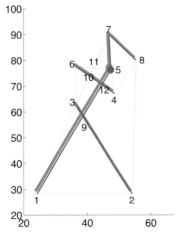

Fig. 2.18 Resulting network from the slave optimization, with the same input from the master optimization as for configuration 1A1 but with a size restriction imposed between buildings 5 and 11.

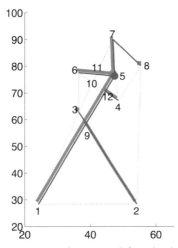

Fig. 2.19 Resulting network from the slave optimization, with the same input from the master optimization as for configuration 1A1 but with buildings featuring requirements with different temperature levels.

with less homogeneous temperature levels. To compute the network of Fig. 2.19, the slave optimizer got the same input from the master optimizer as for configuration 1A1, but with the following changes regarding the temperature levels of the heating and hot water requirements:

• The supply and return temperatures for the heating requirements in buildings 3 and 8 have been increased for periods 5 and 6 (winter periods) from $T_S = 58\,°\mathrm{C}$ to $T_S = 70\,°\mathrm{C}$ and from $T_R = 48\,°\mathrm{C}$ to $T_R = 55\,°\mathrm{C}$.

- In buildings 1 and 2, the temperature levels for the hot water requirements have been decreased from 60 °C to 40 °C, which would be enough for buildings having no showers for instance (office buildings).

From Fig. 2.19 one can see that since nothing changed for the temperature levels of the cooling requirements, the cooling network remains the same. However, the new heating network does not serve buildings 3 and 8. Due to their higher temperature levels, they now have their own boilers (and not heat pumps, due to the high temperature levels).

2.8.2
Final Configuration Choice

Once the Pareto optimal frontier has been computed, the most interesting options in terms of CO_2 emissions and cost are known. In order to choose which configuration will finally be implemented, the decision makers have to perform a multi-criteria analysis which will be specific to their situation. For instance if the priority is to decrease the CO_2 emissions, the least emitting configuration will be chosen. On the other hand, various "what if …" scenarios can also be tested. Or until what point would the centralized generation of heating and cooling still be an environmentally interesting solution, if the reverse individual heat pumps become popular? How much more are the consumers of a district willing to pay in order to get a more environmental friendly energy service provider? The answers to these questions will restrict the choice to a limited number of solutions which will then have to be refined with more detailed models, and why not a more accurate division of the year in periods.

2.9
Conclusions

A methodology for designing urban energy systems has been presented. It allows one to consider the integration of polygeneration energy conversion technologies as well as the design of heat and cold distribution network(s). Given a district with its buildings and energy consumption profiles, the method answers the following questions:

1. Which type of polygeneration energy conversion technologies are best suited for the district?
2. Where in the district shall these technologies be implemented (geographically)?
3. Is there an opportunity of combining several technologies in an integrated system?
4. What are the optimal supply and return temperatures of the distribution networks (heating and cooling), considering the requirements of the district and the technical limitations of the technologies?

5. How shall the buildings be connected?
6. What are the local renewable energy resources that can be valorized by heat pumping ?

The method integrates the following aspects:

1. Thermodynamic: joint consideration of different energy services, allowing for the implementation of efficient polygeneration energy conversion technologies, as well as the consideration of the *temperature levels* at which the thermal energy requirements have to be satisfied, thus allowing for an optimal system integration.
2. Mathematical programming and optimization: combination of different types of optimization algorithms, implementing each algorithm to optimize the variables it is best suited for. The optimization phase is therefore divided in a master and a slave optimization, following mathematical and hierarchical decomposition strategies. The use of a multiobjective optimization allows one to have a better understanding between the cost of energy and the CO_2 mitigation options.
3. Conceptual design: integration of time and space related constraints in a same problem.
4. Modularity: Possibility of designing energy systems for very different types of districts (different location, levels of requirements, building types, etc.) including very different types of technologies.
5. Multiscale: The method can be used to study small district problems with a limited number of buildings or larger problems considering district heating substations as nodes for large scale district heating systems.

An important result of the method to help answering the questions mentioned above is a Pareto optimal frontier showing the tradeoffs between the CO_2 emissions and the costs for different configurations of the analyzed district. It includes a list of configurations for which the key performances indicators are computed and among which the stakeholders will choose the best system configuration applying multicriteria analysis.

References

1 ROLFSMAN, B., Combined heat-and-power plants and district heating in a deregulated electricity market. *Appl. Energy* 78 (**2004**), pp. 37–52.

2 BENONYSSON, A., BOHM, B., RAVN, H. F., Operational optimization in a district heating system. *Energy Conv. Manag.* 5(36) (**1995**), pp. 297–314.

3 YAMAGUCHI, Y., SHIMODA, Y., MIZUNO, M., Development of district energy systems simulation model base on detailed energy demand model. *ibpsaNEWS* (1) (**2003**), pp. 27–34.

4 ARINGHIERI, R., MALUCELLI, F., Optimal operations management and network planning of a district system with a combined heat and power plant. *Ann. Oper. Res.* 120 (**2003**), pp. 173–199.

5 RYDSTRAND, M. C., WESTERMARK, M. O., BARLETT, M. A., An analysis of the efficiency and economy of humidified gas turbines in district heating applications. *Energy* 29 (**2004**), pp. 1945–1961.

6 VON SPAKOVSKY, M. R., CURTI, V., BATATO, M., The performance optimization of a gas turbine cogeneration/heat pump fa-

cility with thermal storage. *Transactions ASME* 117 (**1995**), pp. 2–9.

7 Hepbasli, A., Thermodynamic analysis of a ground-source heat pump system for district heating. *Int. J. Energy Res.* 29 (**2005**), pp. 671–687.

8 Ozgener, L., Hepbasli, A., Dincer, I., Performance investigation of two geothermal district heating systems for building applications: Energy analysis. *Energy Buildings* 38(4) (**2006**), pp. 286–292.

9 Comakli, K., Yüksel, B., Comakli, Ö., Evaluation of energy and exergy losses in district heating network. *Appl. Therm. Eng.* 24 (**2004**), pp. 1009–1017.

10 Söderman, J., Petterson, F., Structural and operational optimisation of distributed energy systems. *Appl. Thermal Eng.* 26 (**2005**), pp. 1400–1408.

11 Friedler, F., Tarjan, K., Huang, Y. W., Fan, L. T., Graph-theoretic approach to process synthesis: polynomial algorithm for maximal structure generation. *Comput. Chem. Eng.* (9) (**1993**), pp. 929–942.

12 Friedler, F., Varga, J. B., Fan, L. T., Decision-mapping: a tool for consistent and complete decisions in process synthesis. *Chem. Eng. Sci.* 50 (**1995**), pp. 1755–1768.

13 Weber, C., Maréchal, F., Favrat, D., Design and optimization of district energy systems. In: *10th International Symposium on District Heating and Cooling, Hannover. German Heat and Power Association (AGFW) and German Electricity Association (VDEW)*, **2006**.

14 Sierksma, G., Linear and Integer Programming – Theory and Practice, *Pure and Applied Mathematics*, vol. 245, Marcel Dekker, New York, **2002**.

15 Leyland, G. B., Multi-objective optimization applied to industrial energy problems. Ph.D. thesis n° 2572, Ecole Polytechnique Fédérale de Lausanne, **2002**.

16 Peng, Y., Méthodes d'optimisation et d'aide à la décision pour l'exploitation et l'extension de réseaux de chauffage à distance. PhD thesis, n° 1304, Ecole Polytechnique Fédérale de Lausanne, **1995**.

17 Weber, C., Maréchal, F., Favrat, D., Kraines, S., Optimization of an sofc-based

decentralized polygeneration system for providing energy services in an office-building in Tokyo. *Appl. Thermal Eng.* 26(13) (**2006**), pp. 1409–1419.

18 Zuccone, M., Personal communication. HSBC Private Bank, **2005**.

19 Girardin, L., Modélisation des besoins énergétiques de zones urbaines, projet sig-scane. Technical report, **2007**.

20 Bonnard & Gardel Ingénieurs Conseils, Technical report. Projet SIG-ScanE, **2007**.

21 Rodriguez, C., Méthodes d'analyse et de dimensionnement de réseaux – application aux réseaux de chauffage à distance. Lecture notes, Ecole Polytechnique Fédérale de Lausanne, **2003**.

22 Grandjean, Y., Personal communication. Conti & Associés Ingénieurs based on manufacturer data from Hoval, **2006**.

23 Zehnder, M., Personal communication. CTA.

24 Curti, V., Personal communication. Thermogamma based on manufacturer data from CIAT, **2007**.

25 Favrat, D., Maréchal, F., Modélisation et optimisation des systèmes énergétiques industriels. Lecture notes, Ecole Polytechnique Fédérale de Lausanne.

26 Curti, V., Thermogamma. Manufacturer data, **2005**.

27 Carron, J., Personal communication. Sinergy (Martigny, Switzerland), **2006**.

28 Swiss Centre for Life-Cycle Inventories. Ecoinvent, **2004**.

29 Curti, V., Modélisation et optimisation environomiques de systèmes de chauffage urbain alimentés par pompe à chaleur. PhD thesis, n° 1776, Ecole Polytechnique Fédérale de Lausanne, **1998**.

30 Vanderbei, R. J., *Linear Programming – Foundations and Extensions*, Second Edition. *International Series in Operations Research & Management Science*. Springer, Berlin, **2001**.

31 Weber, C. J., Multi-objective design and optimization of district energy systems including polygeneration energy conversion technologies. Ph.D. thesis n° 4018, Energy Systems Laboratory, Ecole Polytechnique Fédérale de Lausanne, **2008**.

3
Hydrogen-Based Energy Systems: The Storage Challenge
Eustathios S. Kikkinides

Keywords
hydrogen storage systems, compressed hydrogen storage (CGH2), liquid hydrogen (LH2), hydrogen storage by adsorption, metal hydrides, modeling and optimization of "solid" hydrogen storage units, heat management in metal hydride units

3.1
Introduction

Environmental problems related to the emission of greenhouse gases and to the depletion of fossil-fuel natural resources have led to significant research effort for alternative, environment friendly fuels. The simultaneous growth of the world population and of air-pollutant emissions produced by carbonaceous fuels imposes the replacement of gasoline by cleaner fuels such as hydrogen, which is well suited for electric-vehicle use [1–4].

Hydrogen is an attractive alternative fuel. However, unlike oil, coal, or natural gas, hydrogen is not a primary energy source but has a role that quite resembles electricity as a secondary "energy carrier," which must be first produced using energy from another source and then stored or transported for future use for producing energy. Hydrogen can be obtained from a variety of resources, both renewable and nonrenewable [3–8]. It can be stored as a fuel, and used in transportation and distributed heat and power generation systems using fuel cells, internal combustion engines, or turbines, with the only by-product being water. The ability of hydrogen to replace fossil fuels in the transportation sector could address one of the world's major environmental problems [9, 10]. Automotive exhaust emissions are among the largest single sources of air pollution in the world today, especially in urban areas, and they contribute significantly to the world's carbon dioxide emission, which is the primary source of greenhouse effect.

The importance of hydrogen as a potential energy carrier has increased significantly over the last decade, owing to rapid advances in fuel-cell technology, which produces electricity and water as a by-product. Thus, fuel cells are expected to be major factors in the transition to a future sustainable energy system with zero carbon dioxide emissions, particularly in the transportation sector where most of the

Process Systems Engineering: Vol. 5 Energy Systems Engineering
Edited by Michael C. Georgiadis, Eustathios S. Kikkinides and Efstratios N. Pistikopoulos
Copyright © 2008 WILEY-VCH Verlag GmbH & Co. KGaA, Weinheim
ISBN: 978-3-527-31694-6

Table 3.1 Energy densities for different fuels calculated on weight and volume basis.

Fuel	Energy density on weight basis (kWh/kg)	Energy density on volume basis (kWh/L)
Gasoline	12.7	8.7
Diesel	12.6	10.6
Coal	8.2	7.6
Natural gas	13.9	2.6
Liquefied natural gas (LNG)	13.9	5.6
Hydrogen gas (200 bar)	33.3	0.53
Hydrogen liquid	33.3	2.37
Methanol	5.6	4.4

world's major vehicle manufacturers make considerable investments in fuel-cell vehicle research and development programs [10, 11].

Worldwide conversion from fossil fuels to hydrogen requires the elimination of several barriers imposed along the different steps involved in hydrogen technology. Unlike gasoline or natural gas, hydrogen has no existing large-scale supporting infrastructure. Although hydrogen production, storage, and delivery technologies are currently used commercially by the chemical and petrochemical industry, these technologies are prohibitively expensive for a widespread use in energy applications. Commercially viable hydrogen storage is considered as one of the most crucial and technically challenging barriers to the widespread use of hydrogen as an effective energy carrier [4, 12, 13]. Hydrogen contains more energy on a weight-for-weight basis than any other substance. Unfortunately, since it is the lightest chemical element in the periodic table, it also has a very low energy density per unit volume as can be seen from Table 3.1 (adapted from [4]).

It is expected that the hydrogen economy will require two types of hydrogen storage systems, one for stationary and another for mobile (transportation) applications. Each system has its own constraints and requirements; however, it is evident that mobile applications are far more demanding facing the following requirements [3, 4]:

- High volumetric and gravimetric hydrogen densities due to space and weight limitations, especially in the automobile industry,
- Low operating pressures for safety reasons,
- Operating temperature in the range from -50 to $150\,°C$,
- Fast kinetics for hydrogen charging and discharging,
- Reversibility for many cycles during hydrogen charging–discharging,
- Reasonable cost of a storage system.

This set of requirements imposes several important scientific and technological challenges for the development of feasible hydrogen storage systems for mobile applications. Unfortunately, till date there are no hydrogen storage systems that can meet all these criteria. Stationary applications, on the other hand, do not have weight and space limitations, can operate at high pressures and temperatures, and have the extra capacity to compensate for slow kinetics.

Presently, commercially viable hydrogen storage technologies have been focused around high-pressure gas vessels (250–700 bar) or liquefied hydrogen at cryogenic temperatures (20–30 K). Underground hydrogen storage in depleted oil or natural gas reservoirs and/or mined salt caverns is a possible cost-effective alternative for stationary applications only [14, 15]. The use of advanced materials for hydrogen storage including adsorbents, metal and chemical hydrides, and clathrates may provide an interesting alternative; however, the need for excessively large quantities of these materials in conjunction with cost and reversibility issues limit this method to small-scale demonstration applications.

3.2
Underground Hydrogen Storage

The concept of hydrogen storage in underground structures is not new. Natural gas has been stored underground in depleted oil wells since 1916 and much of the experience is directly applicable to hydrogen. The storage of hydrogen in underground structures allows an extensive volume of hydrogen gas to be stored without the environmental impact of surface-built structures. In general, there are four basic categories of underground establishments that can be used to store gases under pressure [14]:

- Depleted oil or gas reservoirs,
- Aquifers,
- Excavated rock caverns,
- Mined salt caverns.

The first two categories are naturally formed structures, while the last two are man-made. The physical characteristics of each type of underground structure have a bearing on how it may be used for hydrogen storage. In addition, the location of potential salt caverns, in respect to current transport infrastructure, must also be considered for the distribution of stored H_2.

Initially, underground storage was confined to depleted oil and gas fields (Fig. 3.1(a)). These are porous rocks or sandstones with complex pore structures similar to those of aquifers (Fig. 3.1(b)).

Such storage facilities tend to be extremely large; volumes of gas stored exceed 10^9 m^3 at NTP and pressures can be up to 40 bar. The porosity of the porous rocks must be sufficiently high to provide a reasonable void space for an economically acceptable storage volume. In addition, the permeability must be high enough to provide an adequate rate of inlet and outlet flow of hydrogen. On the other hand, the caprock structure located on top of the reservoir (Fig. 3.1(a)) must be reasonably impermeable if it is to contain the gas.

An important point to note regarding the performance of the caprock structure is the mechanism involved in "sealing" the top of the underground reservoir, either a

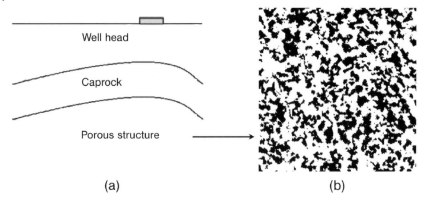

(a)   (b)

Fig. 3.1 (a) Depleted oil reservoir or aquifer.
(b) Two-dimensional, back-scattered SEM binary image section,
obtained from an oil reservoir. Pore space is indicated with
black while solid phase is shown in white.

depleted field or an aquifer. This sealing occurs because of water capillary action, in which water fills all the voids of the caprock structure and must be expelled by sufficiently high pressure to overcome the capillary resistances [14]. Below this pressure, the caprock will act as an effective barrier to the passage of any gas, provided it has a low solubility in water. The value of this pressure and the effectiveness of sealing are independent of the nature of the gas because it is a water-rock capillary effect, also related to the structural characteristics of the porous rock through percolation theory [16]. This observation indicates that hydrogen during underground storage will behave much like natural gas (the solubility of hydrogen is less than that of methane) as far as leakage issues are concerned [14].

It is important from an economic point of view for a storage facility to be able to release hydrogen to the withdrawal wells at relatively high rates. If the reservoir operates at isobaric conditions, similarly high rates of inflow of groundwater would be necessary in order to displace the gas; however, the permeability of the porous structure is generally too low to allow this. Consequently, the reservoir behaves more like a vessel whose gas density is constrained to change comparatively slowly, resulting in a progressively diminishing flow and reservoir pressure with time. Thus, at the beginning of the next charging period, there will still remain a residual quantity of gas in the reservoir known as the cushion gas, which is one of the main capital expenses of an underground storage facility [17].

In exploring the performance of stationary bulk storage systems, one needs to develop a model that is structured in two components: (a) the estimation of the specific costs associated with the construction and operation of such storage systems and (b) a set of generic hydrogen storage applications [15]. Venter and Pucher [15] have presented an economic model that explored the cost of bulk hydrogen storage alternatives. The model has been developed and employed to explore a specific case study in Sarnia, Ontario in Canada, by comparing the cost of salt cavern and depleted reservoir storage with that of liquid hydrogen (LH2) options within an emerging hydrogen infrastructure. Stone et al. [17] investigated the potential for large-scale hydrogen storage in the UK by considering the technical and geophysical problems of storage, the locations of salt deposits, and legal and socioeconomic issues. The results of their study showed that the UK has a number of potential locations where underground storage would provide a strategic reserve of hydrogen.

Results from these studies have demonstrated that for large quantities, it is unlikely that LH2 could economically compete with underground hydrogen gas storage. However, underground hydrogen storage is limited to stationary applications where there are no stringent constraints associated with space, weight, or transport limitations. Moreover, underground hydrogen storage requires further development of small- or large-scale delivery systems for a more widespread distribution.

3.2.1
Process Operation and Energy Issues

Initially the reservoir must be filled with hydrogen. In case of depleted oil or natural gas reservoirs, there may be residual gas/oil and/or water in the well and this has implications on the diffusion of the hydrogen gas storage with this residue.

If a porous structure is used, such as depleted fossil fuel well or aquifer, the limiting flow rate of gas through the structure is modeled following the well-known Darcy's law [19]:

$$\langle \mathbf{u} \rangle = - \left(\frac{\mathbf{k}}{\mu} \right) \cdot \nabla P \tag{3.1}$$

Here, μ is the viscosity of the fluid, $\langle \mathbf{u} \rangle$ is the average superficial velocity of H_2, ∇P is the pressure gradient (including the gravitational force), and $\mathbf{k}$ is the permeability tensor. The above definition of permeability applies in case of a single fluid. When two fluids are present in the porous medium (e.g., hydrogen–water), then the permeability of one fluid affects the permeability of the other. In such cases, a relative permeability of one fluid with respect to the other is defined, which is less than that of the single fluid. It has been shown that for a gas–liquid mixture, the permeability changes with the liquid content and drops to zero at certain liquid saturations, which are related to the pore structure characteristics of the reservoir [16].

A significant quantity of hydrogen may be lost by diffusion either into the surrounding rock or the naturally moving ground water, since hydrogen dissolves in water. Nevertheless, the diffusion rate will decrease with time because the surrounding water will gradually approach saturation, resulting in reduction of the dissolved hydrogen concentration gradient, which is the driving force for diffusion [18].

The capacity of a gas reservoir well-connected to the surface depends on the allowable flow rate into or out of the well. One of the limitations of flow rate is the pressure drop due to friction. The allowable frictional pressure drop depends on the ratio of the power loss through friction to the power being transmitted in the

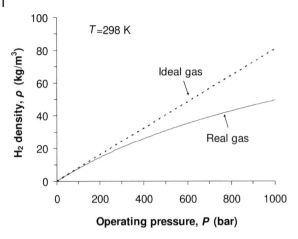

Fig. 3.2 Hydrogen gas density as a function of pressure at $T = 298$ K.

flowing gas. It has been shown in [18] that the frictional pressure drop must be limited to 13% of the nominal system pressure. In order to gain an insight into the magnitude of the components of frictional pressure drop, it is commonly assumed that this pressure drop is predominantly due to flow along the well liner. It can be shown, using the general equation for pipe flow [18]:

$$\Delta P = \frac{8}{\pi^2} f \frac{l}{d^5} \rho q^2 \tag{3.2}$$

where f is the friction factor, l and d are the length and diameter of the pipe, respectively, ρ is the H_2 gas density, and q is the volume flow rate. This equation implies that the H_2 reservoir should be as close to the surface as possible to limit the losses due to pipe length l, while an increase in the pipe diameter d results in a drastic decrease of the frictional pressure drop. An additional source of frictional pressure drop is due to flow through the permeable sedimentary layer, which depends on the permeability of the medium, the depth of drilling into the medium, and the diameter of the drilling [18].

3.3
Hydrogen Storage by Compression

Compressed hydrogen (CGH2) storage is a commercially available hydrogen storage technology. Since hydrogen has a low energy density, it must be compressed to very high pressures to store a sufficient amount of hydrogen, particularly for mobile applications. Moreover, hydrogen cannot be considered as an ideal gas for pressures above 150 bar [20], as can be seen from Fig. 3.2.

From Fig. 3.2, it is seen that at room temperature, the density of CGH2 at 350 bar is approximately 23 g/L, and at 700 bar it is around 40 g/L. Industry standards for CGH2 storage are currently set at 350 bar, with a future target of 700 bar. High-

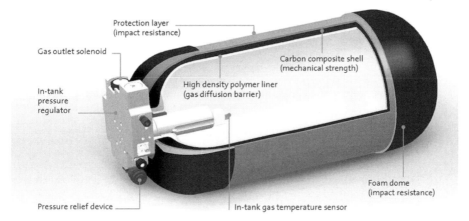

Fig. 3.3 A prototype vessel containing CGH2 gas. Reprinted from [10] with permission from Elsevier.

strength, carbon-fiber composite pressure vessels rated to 700 bar can achieve a gravimetric storage density of 6 wt% and a volumetric storage density of 30 g/L. However, they require the use of expensive materials or composites to achieve a set of different targets including minimal hydrogen leakage using gas diffusion barriers (polymer liner), maximum mechanical strength (carbon composite), and high impact resistance (foams). The combination of an assortment of different materials increases the cost of the storage tank significantly. A schematic of a prototype tank is shown in Fig. 3.3.

One of the main advantages of CGH2 storage is that it is a relatively simple and fast process since the filling of a vehicle tank can be completed in 3 min. Because of these advantages, CGH2 has been employed in many prototype fuel-cell vehicles [10]. The main disadvantages relate to the rather low volumetric and gravimetric densities compared to the other storage methods. Another important shortcoming of the widespread use of CGH2 is the public's awareness on safety issues associated with extremely high-pressure hydrogen tanks that a common passenger car must carry during its operation. Nevertheless, recent test results have indicated that 700 bar composite vessels may actually be safer than their low-pressure counterparts widely used in industry, primarily because the tank walls are much thicker [21].

3.3.1
Process Operation and Energy Issues

Compression of hydrogen is carried out the same way as in natural gas, and thus the procedure is well tested and readily available. It is sometimes even possible to use the same compressors, as long as appropriate Teflon-made gaskets are used and provided that the compressed gas is guaranteed to be oil free. Almost all common natural gas compressors can be easily modified to be suitable for hydrogen. New developments are mainly associated with the optimization of the individual units, with the primary application being the high-pressure compression at service stations in this case.

The process of compressing hydrogen from atmospheric pressure to a final pressure of 350 bar, or even 700 bar, requires the consumption of large amounts of energy. From the point of view of thermodynamics, isothermal reversible compression (cooled-chamber compression) consumes the least work, which can be calculated by the following equation (assuming ideal gas behavior):

$$W_0 = RT_1 \ln\left(\frac{P_2}{P_1}\right) \tag{3.3}$$

This logarithmic relationship between the work required and the compression level clearly shows that the initial pressure dominates the level of work required for compression. Thus, the compression from 1 to 300 bar needs only 10% more energy than the compression from 1 to 200 bar.

Usually, compression is carried out in multiple stages with the first stage providing a pre-pressurization going from one to several atmospheres. The choice of the highest pressure level depends primarily on the maximum permitted pressure that the storage tank can withstand. Assuming an ambient temperature of 300 K and a final pressure, P_2, of 700 bar, it follows from Eq. (3.3) that $W_0 = 8170$ kJ/kg.

In reality, the temperature rise during compression is substantial even with multistage, intercooling technology. Therefore, H_2 compression is more suitably described by a polytropic process [22, 23]. The energy consumption for such a process can be calculated from the following equation:

$$W = \frac{n}{n-1} RT_1 \left[\ln\left(\frac{P_2}{P_1}\right)^{(n-1)/n} - 1 \right] \tag{3.4}$$

where exponent $n = 1.36$ for H_2 gas. Because the pressure ratio for each stage is limited to approximately 7 for diaphragm compressors, and much less for other types, at least four compression stages would be necessary to obtain a pressure of 700 bar [21]. Assuming perfect intercooling between stages, the minimum compression work is $W = 10{,}161$ kJ/kg. This amount of work is 24% more than that of the isothermal process. Consequently, there is significant interest in achieving energy savings by means of enhanced heat transfer during the compression process.

Regarding the types of compressors employed in the above process, there are currently two types of compressors used in industry to raise hydrogen pressure to 150 bar: reciprocating piston compressors, and diaphragm compressors. Reciprocating piston compressors can be used for both large- and small-scale applications, while diaphragm compressors are mainly used for small-scale applications due to their limited flow rates imposed by the size of the diaphragm [21, 23].

Diaphragm compressors remove heat more effectively than piston compressors because of the relatively large surface area of the compression chamber, the presence of the cooled compression oil near the diaphragm, and the ability to include a cooling system in the cover of the compressor [21]. However, the compression ratio for diaphragm compressors is limited by the temperature increase during compression. Existing diaphragm compressors mostly employ a simple one-loop cooling system in the cover of the compressor, which can be improved significantly by increasing the number of cooling loops, implementing micro- or mini-channels, and/or coupling to a refrigeration loop [21].

Piston compressors, on the other hand, are cooled by lubricant oil, but this approach may not be feasible for hydrogen due to the high-purity hydrogen requirements by the fuel-cell operation. Yong [24] developed a cooling approach by which water spray is employed inside a compressed-gas cylinder to reduce the gas temperature by evaporation. Because PEM fuel cells require a humidified hydrogen stream, this method may be used to raise the pressure of hydrogen from pipeline level to the final charging pressure. A new approach in the design of piston compressors involves the use of a long, slow stroke in conjunction with a water-cooled gas cylinder to improve cooling performance. Unfortunately, the flow rates in this case are too small due to the slow motion of the piston [21].

An important heat-transfer issue for CGH2 storage is the temperature increase during fast tank-filling processes. It is known that hydrogen exhibits a reverse Joule–Thomson effect during expansion (throttling) at temperatures above 204 K (inversion temperature for hydrogen), resulting in the heating of the gas instead of cooling. During rapid filling (> 1 kg H_2/min), the temperature rise inside the tank can be as high as 50 °C and overheating can have harmful effects on the integrity of the composite tank. Furthermore, as the temperature inside the tank increases, the hydrogen capacity decreases at a given pressure. If this temperature rise is compensated by transient overpressurization, there is a 10% increase in the energy consumption due to the additional work of compression [21]. Another possible solution to overheating is precooling of the filling gas, but such an approach would require additional energy for refrigeration [25], which is quite significant because of the low inversion temperature of hydrogen. An energy-efficient solution could be to enhance both internal and external heat transfer rates during the filling process and to optimize the pressure throttling process. Enhancing the heat transfer characteristics requires cylinder frames made of materials with high thermal conductivities or installing heat pipes to transfer heat from inside the vehicle tank to an external heat sink, such as the vehicle frame [21].

To minimize throttling losses and resulting overheating during H_2 filling processes, multibank cascade systems can be employed. The overall temperature rise and the total fueling time in such a process can be optimized employing thermodynamic and heat transfer principles [21, 26].

3.4
Hydrogen Storage by Liquefaction

Liquid hydrogen storage is another commercially available technology. LH2 is stored at atmospheric pressure in cryogenic tanks at $T = 20.3$ K with a density of 70.8 g/L, which is nearly twice that of CGH2 at 700 bar. A 68-L cryogenic tank can carry approximately 5 kg LH2, which is sufficient to drive a fuel-cell passenger vehicle for 500 km. LH2 tanks can be filled in relatively short times and are much safer at appropriate cryogenic temperatures than high-pressure hydrogen tanks. The main disadvantages of LH2 storage are the high energy consumption associated with the liquefaction processes and continuous "boil-off" during storage [21]. The energy required to liquefy hydrogen is about 30–40% of the energy content of the gas, and hence it reduces the overall efficiency of the system significantly. The low operating temperatures of 20–30 K, compared to surrounding ambient temperatures around 300 K, lead to unavoidable heat flow due to the three basic heat transfer mechanisms: thermal conduction, convection, and radiation, among which the first and the third are the dominant mechanisms of heat inflow from the environment [10, 20].

The need to minimize energy loses and refueling time of LH2 has resulted in intensive research collaboration between BMW and SWB, a joint venture between Bayernwerk AG, BMW INTEC GmbH, Linde AG, and Siemens AG. Optimization studies on these issues have been proceeding at SWB since 1991 and have resulted in the reduction of refueling time for LH2 from > 1 h to less than 3 min and an almost complete elimination of the hydrodynamic losses of liquefaction energy [27]. In May 1999, the first public filling station for LH2 has started its operation and services at Munich Airport. The filling station operates fully automatically employing refueling robot technologies achieving fully automatic refueling of a vehicle within 2–3 min without emissions or odor, in a controlled and reliable system behavior [28].

3.4.1
Process Operation and Energy Issues

Liquefaction of hydrogen requires the removal of heat and subsequent rejection to the surroundings employing a combination of refrigeration processes that are well established in the cryogenics industry [29, 30]. Before liquefaction, hydrogen must be cleaned for removal of CO_2, CO, CH_4, and H_2O. This is usually done using a pressure-swing adsorption process. The simplest liquefaction cycle is the Joule–Thompson cycle in which hydrogen gas is first compressed at high pressures ($P \geqslant 40$ bar) and then throttled to a lower pressure through a nozzle or a valve, at adiabatic conditions, producing some liquid. The cooled gas is separated from the liquid and returned to the compressor through a heat exchanger. Because the inversion temperature of hydrogen is only 204 K, expansion at room temperature will not induce a cooling effect, and liquid nitrogen is often used to precool hydrogen at 77 K, which is below the inversion temperature before expansion [29, 30].

An additional issue in hydrogen liquefaction that must be taken into consideration is the quantum effect that hydrogen exhibits at cryogenic temperatures. More specifically, there exist two different types of hydrogen characterized by the difference in the orientation of the nuclear spin of the two H atoms that comprise the H_2 molecule. If the spins are symmetric, the molecule is referred as ortho-H_2, while if they are asymmetric, the molecule is referred as para-H_2. At the normal boiling point of H_2 (20.3 K), there is 99.79% para-H_2, which drops as temperature

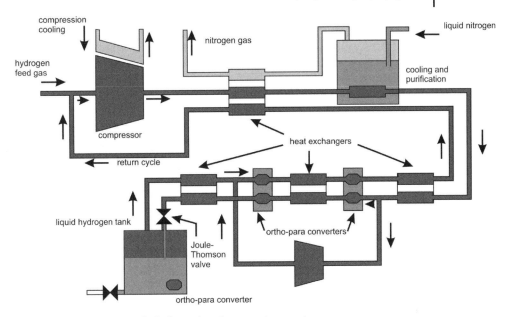

Fig. 3.4 Schematic flowchart for hydrogen liquefaction and conversion.

increases reaching an almost constant value of 25% para-H_2 at 300 K. The 75-25 ortho–para mixture is referred as normal H_2 [30]. Consequently, as normal H_2 is liquefied, there is an ortho–para conversion, which is very slow in the absence of a catalyst, and it is called self-conversion [31].

The conversion of ortho-hydrogen to para-hydrogen is an exothermic reaction. The heat of conversion is related to the change of momentum of the hydrogen nucleus when the direction of spin changes [31]. The amount of heat released during the conversion process is temperature dependent, and the heat of conversion is greater than the latent heat of vaporization of normal and para-hydrogen at the normal boiling point. If the unconverted normal hydrogen is placed in a storage vessel, the heat of conversion will be released within the container, which leads to the evaporation of the liquid. Because of these peculiarities in the physical properties of hydrogen, the boil-off of the stored liquid will be considerably larger than what one would be able to determine from calculations based on ordinary heat leak to the storage tank. In order to minimize the storage boil-off losses, the conversion rate of ortho- to para-hydrogen should be accelerated with a catalyst that converts the hydrogen during the liquefaction process [29–31].

Thermodynamically, liquefying hydrogen involves three heat transfer stages [21]: the first stage extracts sensible heat (4000 kJ/kg) to reduce the temperature from 300 K to 20.3 K. In the second stage, latent heat (450 kJ/kg) is extracted to condense hydrogen at 20.3 K. The third stage is the ortho–para H_2 conversion, which requires heat extraction of 703 kJ/kg. A schematic flowchart of the combined H_2 liquefaction–conversion process is presented in Fig. 3.4.

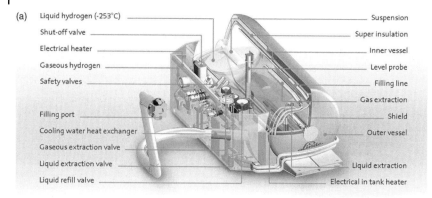

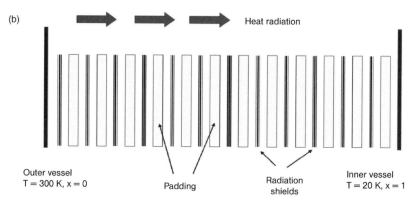

Fig. 3.5 (a) Liquid hydrogen tank and (b) multilayer vacuum superinsulation. Reprinted from [10] with permission from Elsevier.

The entire liquefaction process requires a heat removal of 5153 kJ/kg that must be provided by a liquefaction cycle. The minimum theoretical work to liquefy normal hydrogen to 99.79% para form is 14,280 kJ/kg [30]. In comparison, the minimum theoretical work to liquefy nitrogen is only 6% that of LH2. The actual energy consumption for hydrogen liquefaction is usually at least 2.5 times the theoretical minimum, depending on the capacity of the liquefaction plant [21, 32].

The LH2 storage requires highly insulated cryogenic tanks. Such a storage tank is shown in Fig. 3.5(a) [10]. To minimize the heat inflow, the three basic mechanisms of heat transfer should be considered. The rate of radiation heat transfer is proportional to the emissivity of the shields between the warm and cold surfaces and inversely proportional to the number of these shields. To minimize radiation heat transfer, multilayer insulation (MLI), also called superinsulation, is often used [10, 21], as can be seen in Fig. 3.5(b). The apparent thermal conductivity of MLI is of the order of 10^{-5} W/m K, which is approximately three orders of magnitude lower than the thermal conductivity of fiberglass-building insulation [21]. MLI consists of 30–80 layers of low-emissivity radiation shields made of metal (aluminum) foil, separated by low-conductivity spacers (e.g., fiberglass paper) [21]. For MLI, the apparent thermal conductivity parallel to the radiation shields is typically three to six orders of magnitude higher than that normal to the layers [33]. Such a large disparity in directional thermal conductivity presents considerable challenges in thermal insulation design for systems where structural supports and feed-lines penetrate the insulation [21]. An alternative to MLI is to use microsphere insulation, which exhibits isotropic thermal conductivity and can tolerate significantly higher compression; nevertheless, the apparent thermal conductivity of microsphere insulation is one order of magnitude higher than that of MLI [34].

A critical design variable for MLI-insulated vessels is the optimal number of insulation layers per unit thickness. If the layer density is too high, the increase in conductive heat loss through the spacers can outweigh the decrease in radiative heat flux provided by more insulation layers. Hence, the apparent thermal conductivity shows a minimum when plotted with the layer density for MLI as demonstrated in [21] and [35].

To minimize the convective losses, the volume between the inner vessel and the outer vessel is typically evacuated to high vacuum, for example, 1.3×10^{-8} bar [35]. Under such high vacuum conditions, the molecular mean free path, λ, is large relative to the distance, d, between the boundaries and the respective Knudsen number $Kn = \lambda/d$ is high. At these conditions, free molecular flow or diffusion dominates over viscous flow and the resulting heat flux varies linearly with pressure [36] leading to minimal convection effects at high vacuum.

Heat flow from the environment to the tank's cryogenic interior will inevitably cause hydrogen release to the environment by evaporation, or "boil-off," leading to an increase in the vessel pressure [21]. When the pressure increases to a prescribed level, hydrogen vapor must be vented through a pressure-relief valve. Because of heat inflow, LH2 can stratify inside the vessel, and this phenomenon increases the boil-off rate because the segregated warmer fluid vaporizes more readily, leading to a rise of the tank pressure to the saturation pressure of the warm layer. This pressure rise shortens the storage period between successive ventings and is especially severe with large vessels. For small-sized vehicle tanks, metal conducting rods can be used to redistribute the heat from the top to the bottom portions of the inner vessel [33].

Another source of boil-off is sloshing – the motion of liquid inside a vessel caused by acceleration and deceleration – which transforms some of the liquid's kinetic energy and impact energy to thermal energy and can be significant for vehicles moving in urban areas [31]. A potential solution is to insert antislosh baffles to restrain the motion of the liquid [31], but this approach will reduce the tank's gravimetric storage density [21].

The time period between putting a vehicle carrying an LH2 storage vessel into an idle mode and the venting process is normally called "dormancy" [10]. Both the length of the dormancy period and the H_2 "boil-off" losses are considered crucial for the evaluation of the LH2 storage systems [10].

A related issue is the cooling-down losses during hydrogen refilling. The complete transfer line must be cooled down to 21 K causing evaporative losses due to the heat flow from ambient air. Thus, the design of LH2 transfer lines involves several important heat transfer issues [21]. Two-phase flow is common in the transfer of LH2 as a result of heat transfer to the fluid. The presence of vapor in the transfer line can critically reduce its carrying capacity. It is, therefore, imperative to minimize evaporative losses using high-quality insulation. The insulation strategies for the transfer line are similar to those for the storage vessel [21].

Heat exchangers are used extensively in LH2 production. Because of the low operating temperature, cryogenic liquefiers cannot produce liquid if the heat exchanger effectiveness is less than approximately 85% [35]. Several measures can be utilized to improve heat exchanger effectiveness, including small temperature differences between inlet and outlet streams, large surface area-to-volume ratio, and high heat transfer coefficients [33, 35].

Evidently, the complexity of LH2 storage systems together with the challenge to minimize H_2 "boil-off" losses leads to systems' cost that is currently not favorable compared to CGH2 systems (especially for large-scale applications), despite design flexibility and much higher volumetric H_2 storage densities achieved with LH2 systems [10, 20].

3.5
"Solid" Hydrogen Storage

An alternative to the traditional albeit limited storage methods is proposed through the use of advanced solid materials as hosting agents for the storage of hydrogen in atomic or molecular form. This type of hydrogen storage is often called "solid" hydrogen storage since hydrogen becomes part of the solid material through some physicochemical bonding.

There are at present two fundamental mechanisms known for storing hydrogen in materials in a reversible manner: *absorption* and *adsorption*. In *absorptive* hydrogen storage, hydrogen is absorbed directly into the bulk of the material. In simple crystalline metal hydrides, absorption occurs by the incorporation of atomic hydrogen into interstitial sites in the crystallographic lattice structure [37]. *Adsorption* may be subdivided into physisorption and chemisorption, based on the energetics of the adsorption mechanism. Physisorbed hydrogen is more weakly bound to the internal surface of the adsorbent material than is chemisorbed hydrogen. A third mechanism for hydrogen storage is through chemical reactions used for both hydrogen generation and hydrogen storage. For reactions that may be reversible, hydrogen generation and hydrogen storage take place by a simple reversal of the chemical reaction as a result of modest changes in the temperature and pressure. Sodium alanate based complex metal hydrides are an example. In many cases, the hydrogen generation reaction is not reversible under modest temperature/pressure changes.

Storage by absorption as chemical compounds or by adsorption using porous adsorbents offers definite advantages from the safety perspective since it requires milder pressure and temperature conditions of operation compared to the traditional methods. Several comprehensive reviews summarize recent advances in hydrogen storage, describing progress made with carbon-based nanostructures, metal-organic frameworks (MOF) based on the physisorption process, and metal or chemical hydrides based on the chemisorption process [20, 37–42]. These materials should meet the US Department of Energy (DOE) target values for minimum hydrogen storage capacity of 6.5 wt% and 65 g/L at temperatures between 60 and 120 °C and pressures below 150 bar, for commercial viability [43]. Furthermore, the cost of a storage medium and its toxicity properties need to be carefully considered for the realization of the set goals [40].

3.5.1
Adsorbents

Hydrogen storage by physisorption at a relatively low pressure (150 bar or lower) and room temperature offers the advantage of considerably reduced risks compared to the traditional approaches that require very high pressure and additional energy for cooling purposes. In order to make this method viable, the adsorption capacity of the adsorbent must allow storage of a sufficient amount of hydrogen at a relatively short filling time. Temperature limitations may have to be considered during the charging process to take account of potential safety and performance considerations.

The fact that the normal boiling point and the critical temperature of H_2 are both very low (20.4 K and 33.25 K, respectively) precludes the condensation of H_2 inside the pores of an adsorbent at ambient conditions. Physisorption due to the interaction between the hydrogen molecules and the surface of the material by van der Waals forces results in the formation of a hydrogen monolayer on the adsorbent surface. For monolayer coverage at 77 K, the amount of hydrogen adsorbed scales linearly with the BET surface area of the adsorbents [37, 40]. Thus, several adsorbent candidates have been developed and explored along this direction [40, 42].

Carbon-based adsorbents such as activated carbons, carbon nanotubes, and carbon nanofibers have been the subject of intensive research over the past 15 years. The research on hydrogen storage in carbon materials was dominated by announcements of extraordinarily high storage capacities in carbon nanostructures. However, the experimental results on hydrogen storage in carbon nanomaterials scatter over several orders of magnitude, with reported storage capacities between 0.2 and 10 wt% [44, 45]. The experiments to date claiming very high values could not independently be reproduced in different laboratories. On the other hand, detailed simulation studies of hydrogen adsorption in carbon nanotubes or graphitic nanofibres, using molecular simulation methodologies [46–48], have also not been able to confirm these extraordinary findings. Hence, hydrogen storage in carbon nanostructures close to ambient conditions seems to be limited to values far below those set as a requirement of the DOE and the automotive industry due to physical reasons. One possible strategy to increase the hydrogen uptake is to either decrease the adsorption temperature to about 80 K and/or to increase the hydrogen pressure above 150 bar, both of which are quite prohibitive for mobile applications. Another way is to tune the properties of the adsorbent, which is mainly limited to an increase of the surface area and/or an increase of the micropore volume of the adsorbent. Other ideas involving doping of the carbon surface to suitable metal-oxide to induce the hydrogen-surface interaction [49] are still under investigation.

Metal-organic frameworks represent a new class of adsorbents that receive increasing attention due to their unique structure to enclave hydrogen and other gas molecules [50–54]. The preparation of these materials is a simple, inexpensive, and high-yield procedure. The storage capacities are not yet sufficiently high for practical applications, but they can be tuned by modifying the structure of the material with suitable linkers [55], as it is also supported by theoretical studies [56]. To date, more than 500 MOFs have been synthesized and structurally characterized and thousands more will be identified soon, as the periodic table is yet to be explored in the pursuit of practical storage targets [57]. These structures basically consist of inorganic units connected by organic linkers such as carboxylates forming a 3D network. The MOF structure is accessible from all sides of gas molecules, having an extraordinarily high surface area of above 3000 $m^2\,g^{-1}$. The exact mechanism of hydrogen sorption is not yet clear, but in general these materials have shown high hydrogen adsorption capacities with good reversibility properties and fast diffusion kinetics. In a recent study [58], Yaghi and coworkers presented an MOF structure with an estimated Langmuir surface area of 4500 $m^2\,g^{-1}$ and a storage capacity of 7.5 wt% at 7 MPa and 77 K, which is the highest surface area and hydrogen storage capacity achieved so far for MOFs.

From the process point of view, most of the literature studies present macroscopic models describing the dynamic behavior of hydrogen storage in carbon-based adsorbents [59, 60], although the extension of other more promising materials is straightforward. Lamari et al. [59] studied theoretically and experimentally the thermal effects in dynamic hydrogen storage by adsorption at room temperature and high pressure. A detailed 2D model describing adsorption-based hydrogen storage was developed by Delahaye et al. [60]. Ways to minimize heat effects in order to maximize the storage capacity were theoretically studied.

In a recent study [61], Bhatia and Myers have examined thermodynamically the adsorption–desorption cycle of hydrogen and methane in carbon-based porous adsorbents, and have set optimum conditions for adsorptive storage at the material level, based on the total adsorption–desorption cycle. Very little attention has been paid on exploring the synergistic benefits between material design, process operation, and design from the perspective of creating a process, which can operate safely on one hand and economically on the other. These considerations motivate the need for investigating multiscale modeling and optimization strategies.

In this context, a key issue in the hydrogen storage problem using carbon-based nanoporous materials is the material evaluation and design, which is systematically explored by the simulation of hydrogen storage at a molecular level to predict the hydrogen sorption capacity of the selected material. Furthermore, economic and operability considerations motivate the need for a maximum storage capacity while ensuring satisfaction of certain operating constraints. For example, large temperature differences in the process have detrimental effects on the storage performance and may also lead to material deterioration and potential explosions.

On the other hand, realistic charging times are required (of the order of several minutes) to achieve an economically and practically viable storage process.

Recently, Kikkinides et al. [62] presented an integrated approach that formally exploits the benefits between material and process design. Systematic simulation and optimization studies have been performed at two different length scales: adsorbent pore level and storage tank level. An interactive dynamic optimization approach determined the optimal pore size distribution of carbon-based adsorbents, maximizing H_2 storage under different temperature constrains. The outcome of this work was that carbon-based materials cannot be used (at least at normal temperature and pressure conditions) for the development of viable storage systems for mobile applications. Nevertheless, the above study sets up the general framework for the multiscale modeling and optimization methodology of hydrogen (or other gas) storage in nanoporous materials by adsorption and can serve to optimize material properties for maximum process performance under given operating constraints.

3.5.2
Metal and Chemical Hydrides

Metal hydrides have been seen as a potential means of hydrogen storage for some 35 years. The principle of hydrogen storage in metal hydrides is that atoms of hydrogen under low pressure bond chemically to the metal atoms. Adsorption of H_2 gas proceeds spontaneously in an exothermic process. Conversely, hydrogen can be desorbed upon gentle heating. Metal hydride compounds can be divided into three general categories: ionic, metallic, and covalent. This classification roughly reflects the position of the element M in the periodic table. Many different alloy types exist, displaying a wide range of properties and performance characteristics, such as hydrogen content, desorption rate, cycle life, and heat of reaction. Metal hydrides are composed of metal atoms that constitute a host lattice and hydrogen atoms. Metal and hydrogen usually form two different kinds of hydrides, α-phase at which only some hydrogen is absorbed and β-phase at which hydride is fully formed. Hydrogen storage in metal hydrides depends on different parameters and consists of several mechanistic steps. Metals differ in the ability to dissociate hydrogen, this ability being dependent on surface structure, morphology, and purity [40].

Binary metal hydrides (MH_n, $n = 1, 2, 3$) have limited use as storage materials because of their stability. They are also called interstitial hydrides as the H_2 occupies interstitial sites in the metal host lattice. The lattice expands during hydrogen sorption, often changing symmetry. Binary hydrides can be formed by rare earth and transition metals, none of which are in the pressure and temperature range attractive for mobile applications. For example, the light alkali metal hydrides LiH and NaH are even more stable, with desorption temperatures exceeding 500 °C. Considerable research efforts are currently invested on light/complex hydrides (alanates, borohydrides, amides, and mixtures) as well as regenerative chemical systems for hydrogen storage [37]. The main problems in these types of materials relate to reversibility, operating far away from the desired conditions for mobile applications, slow kinetics, and regeneration efficiency.

The mathematical modeling of hydrogen storage in metal hydride beds has received considerable attention over the past 10 years. Jemni and Nasrallah [63] presented a theoretical study of the mass and heat transfer dynamics in a metal-hydride reactor. The effect of several parameters on the overall hydrogen storage efficiency was illustrated. Some of the assumptions used in this study were validated in a subsequent contribution (Nasrallah and Jemni, [64]). Jemni et al. [65] presented an experimental approach to determine the reaction kinetics, equilibrium conditions, and transport properties in $LaNi_5–H_2$ system. Heat and mass transfer effects were also determined experimentally. Experimental data were compared with model predictions from previous work. Nakagawa et al. [66] presented a 2D model for the transient heat and mass transfer within a metal hydride bed. Several conclusions were made regarding the effect of the underlying operating condition on the storing process. Mat and Kaplan [67] developed a mathematical model to describe hydrogen absorption in a porous lanthanum metal bed. The model takes into account the complex mass and heat transfer and reaction kinetics. Model predictions were found to be in good agreement with experimental data. Aldas et al. [68] and Mat et al. [69] presented an integrated model of heat and mass transfer, reaction kinetics, and fluid dynamics in a hydride bed. An important conclusion of this study is that the flow conditions do not significantly affect the amount of hydrogen absorbed. Askri et al. [70] extended the previous studies [63–65] by investigating the effect of radiated heat transfer in $LaNi_5–H_2$ and $Mg–H_2$ hydride beds. Simulation results showed that radiation effects are negligible on the sorption process in the $LaNi_5–H_2$ bed, but they play an important role in the case of $Mg–H_2$ hydride bed. Recently, Kaplan and Veziroglu [71] extended the study of Mat et al. [69] and investigated numerically the hydrogen storage process in a 2D metal hydride bed including the full momentum balance equation, which is considered to be important when large pressure gradients exist in the system. The effect of the underlying operating conditions on the hydride formation rate was investigated and several conclusions were drawn up. Kikkinides et al. [72] developed effective heat management strategies and novel design options for hydrogen storage systems in $LaNi_5$ metal hydride reactors by performing systematic simulation and optimization studies. Optimization results indicated that significant improvements in the total storage time can be achieved under a safe and economically attractive operation [72].

3.5.3
Process Operation and Energy Issues

The process operation of a hydrogen storage tank containing adsorbents or metal hydrides is far more complex than that of compressed or liquefied hydrogen, because the storage mechanism of adsorption or absorption consists of complicated equilibrium and kinetics steps at the particle level, which are further coupled with the conservation laws for the interstitial hydrogen fluid at the tank level. The different length scales associated with the storage process require multiscale modeling considerations introducing additional complexity in the process design and optimization considerations. The conventional approach of addressing this complexity is to focus on each scale separately and to derive the best possible understanding of relevant phenomena by individual descriptions, extracting phenomenological equations that can then be incorporated in mathematical models. Couplings and interactions between different scales are handled via aggregate descriptions of physical behavior at each scale, subsequently used in modeling the behavior at higher scales. An important characteristic is the absence of computational integration of these descriptions, which are indirectly interfaced via ad hoc approaches in sequential, asynchronous patterns. This scale decomposition approach to process modeling has long been remarkably efficient, enabling design and simulation of numerous conventional unit operations. In case of modeling hydrogen storage in cylindrical tanks, packed with adsorbent or metal hydrides, 2D pseudohomogeneous macroscopic models are normally proposed based on mass, momentum, and energy balances for the interstitial fluid coupled with kinetic rate expressions and equilibrium isotherms relating pressure–temperature and concentration [59, 60, 62–72].

Standard mathematical descriptions of the modeling equations and respective boundary conditions are given next.

3.5.3.1 Material Balance

$$\varepsilon \frac{\partial \rho}{\partial t} + \nabla \cdot (\rho \mathbf{u}) + m_p = 0 \tag{3.5}$$

where ρ is the H_2 density in the gas phase, $\mathbf{u}$ is the superficial velocity vector, and m_p is the mass flux of hydrogen transported to the particle (either adsorbent or metal hydride). The porosity ε used in the model is the interparticle porosity of the bed, assuming that there is no compression in the micropores, for the case of adsorbents [60, 62].

3.5.3.2 Energy Balance

$$\frac{d}{dt}\left(\varepsilon \rho C_{pg} T + (1 - \varepsilon)\rho_s C_{ps} T - \varepsilon \frac{\rho R T}{M_{H_2}}\right)$$

$$+ \nabla(\rho C_{pg}\mathbf{u}T - \lambda_e \nabla T) + (-\Delta H)m_p = 0 \tag{3.6}$$

where T is the temperature of the bed, λ_e is the effective thermal conductivity, C_{ps} the solid heat capacity, ρ_s the bulk density of the adsorbent, and $(-\Delta H)$ is the heat of adsorption which is defined on a molar or mass basis to cope with the correct units. In the above equation, radiation effects have been neglected, although their inclusion is straightforward [70].

3.5.3.3 Momentum Balance

The steady-state momentum balance of laminar gas flow through a packed bed, which is a special type of a porous medium (unconsolidated), can be expressed again by Darcy's law:

$$\mathbf{u} = -\left(\frac{\mathbf{K}}{\mu}\right) \cdot \nabla P \tag{3.7}$$

where $\mathbf{u}$ is the superficial velocity of the gas flowing through the bed, $\mathbf{K}$ the permeability tensor of the bed, and P is the bed pressure. In many cases, the Blake–Cozeny equation, which is the equivalent expression for random sphere packs, is used [19, 72].

3.5.3.4 Adsorption Kinetics and Equilibrium

In case of adsorption, the H_2 molecules diffuse into the pores of the adsorbent before they finally adsorb at the surface of the pore walls. Thus, a rigorous approach should involve the solution of the diffusion equation at each scale required inside the porous adsorbent. Nevertheless, it has been shown that for sufficiently fast diffusion times, t_D, compared to the storage time, t_s ($t_s/t_D \geqslant 0.1$), the adsorption kinetics can be adequately approximated by the linear driving force (LDF) model [72]:

$$m_p = (1 - \varepsilon)\rho_s \frac{dq}{dt} = k(q^* - q) \tag{3.8}$$

where q^* is the concentration of the adsorbed gas in equilibrium with the gas phase and k is a given parameter potentially expressing a mass transfer coefficient in an aggregated level. In general, an equilibrium isotherm is employed to describe the adsorbate–adsorbent interactions at the macroscopic level. It was shown before that at ambient temperatures, the adsorption of H_2 in carbon-based materials is well described by the Langmuir isotherm [72]:

$$q^* = \frac{q_s b P}{1 + b P} = \frac{q_s b_0 \exp[-\Delta H/RT]P}{1 + b_0 \exp[-\Delta H/RT]P} \tag{3.9}$$

where q_s and b_0 are parameters depending on the selected material and P the total pressure.

It is customary to nondimensionalize q^* with q_s defining the surface coverage, θ, which reaches unity at infinitely large pressures. Typical plots of Langmuir isotherms are shown in Fig. 3.6.

Parameter $(-\Delta H)$ is the heat of adsorption. In general, ΔH is expected to vary with coverage due to energetic and structural heterogeneity effects, nevertheless, previous studies have shown that ΔH does not vary significantly with coverage at all scales [59, 60, 62].

3.5.3.5 Absorption Kinetics and Equilibrium

The kinetics of the hydride process can be described in terms of a shrinking core model, where the core is taken to be saturated α-phase metal, while hydride is the β-phase. The physicochemical mechanism in this case is far more complicated as it involves several distinct steps including (1) transport of H_2 molecules in the gas phase, (2) physisorption of H_2 molecules on the hydride surface, (3) dissociation of H_2 molecules to atomic H at the surface, (4) diffusion of H atoms through the

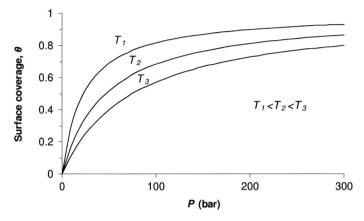

Fig. 3.6 Graphical representation of the Langmuir isotherm.

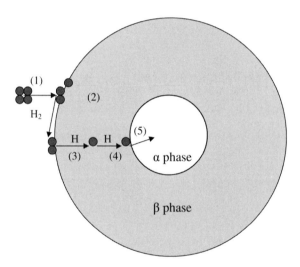

Fig. 3.7 Schematic model of the different mechanisms that take place during the formation of a metal hydride.

β-phase of the hydride, and (5) chemical reaction of the hydrogen atom with the α-phase of the metal. A schematic diagram of this processes is shown in Fig. 3.7.

If one of these steps is much lower than the others, then it becomes the rate-determining step and the kinetics of the whole absorption process can be described by relatively simple rate expressions.

In case of LaNi$_5$, the proposed kinetics model in the literature is described by a first-order reaction. In this case, the density of the material, ρ_s, is no longer constant and varies with time and space. The resulting form of the kinetic expression for the case of LaNi$_5$–H$_2$ system becomes [63–65]:

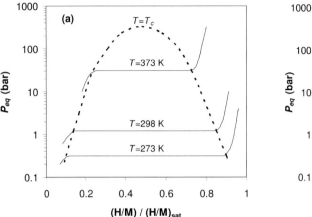

Fig. 3.8 (a) Phase diagram of a metal hydride (P–C–T curves). For a given compound, the plateau pressure corresponding to the phase transition between α and β phase is determined experimentally at different temperatures T. (b) Van't Hoff's plot used to calculate the heat of absorption.

$$m_p = (1 - \varepsilon) \cdot \frac{\partial \rho_s}{\partial \tau} = C_a \cdot \exp\left(-\frac{E_a}{R_g T}\right) \cdot \ln\left(\frac{P}{P_{eq}}\right) \cdot (\rho_{sat} - \rho_s) \qquad (3.10)$$

where P_{eq} is the equilibrium pressure, ρ_{sat} is the "saturated" bed density after complete reaction of the alloy with H_2, and E_a is the activation energy of the reaction kinetics process. The equilibrium pressure is determined from Van't Hoff equation that relates the plateau pressure with temperature as can be seen from Fig. 3.8.

$$\ln P_{eq} = -\frac{(-\Delta H)}{RT} \qquad (3.11)$$

In order to describe P_{eq} outside the plateau region, we fit the respective experimental data from pressure as function of the hydrogen to metal atomic ratio $f(H/M)$ using polynomial representations. In case of LaNi$_5$, available polynomial fits of 5th and 9th degree are given in [63] and [65], respectively.

The resulting equation for P_{eq} is a combination of the above:

$$P_{eq} = f(H/M) \times \exp\left[\frac{\Delta H}{R_g} \cdot \left(\frac{1}{T} - \frac{1}{T_{ref}}\right)\right] \qquad (3.12)$$

3.5.3.6 Boundary and Initial Conditions
The necessary boundary and initial conditions to complement the model equations are given next.

- At the tank inlet ($z = 0$)
  Dirichlet BCs:

$$\rho = \rho_f \tag{3.13a}$$

$$T = T_f \tag{3.13b}$$

$$P = P_f(t) \tag{3.13c}$$

where ρ_f, T_f, P_f are the inlet values of ρ, T, P, respectively. For the case of inlet pressure, P_f, a pressure history is normally employed to better control the inlet flow rate and hence the rate of charging of the tank with H_2. It has been shown [72] that pressure history can have a significant effect in the temperature rise inside the storage tank, as well as on the total storage time required at the process level.
• At the tank outlet ($z = L$)

$$\frac{\partial \rho}{\partial z} = 0 \tag{3.14a}$$

$$-\lambda_e \frac{\partial T}{\partial z} = h(T - T_c) \tag{3.14b}$$

$$u_z = u_r = 0 \tag{3.14c}$$

• At the tank center ($r = 0$)

$$\frac{\partial \rho}{\partial r} = 0 \tag{3.15a}$$

$$\frac{\partial T}{\partial r} = 0 \tag{3.15b}$$

$$\frac{\partial u_z}{\partial r} = \frac{\partial u_r}{\partial r} = 0 \tag{3.15c}$$

• At the tank wall ($r = R_t$)

$$\frac{\partial \rho}{\partial r} = 0 \tag{3.16a}$$

$$-\lambda_e \frac{\partial T}{\partial r} = h(T - T_c) \tag{3.16b}$$

$$u_z = u_r = 0 \tag{3.16c}$$

In the above model description, an overall heat transfer coefficient, h, has been used to represent the heat transfer through the walls. In general, the value of h should account for several resistances including the thermal conductivity and thickness of the wall (which can consist of several layers with different conductivities and thicknesses), and the heat transfer coefficient from the wall to the surrounding medium which can be atmospheric air, or a cooling fluid, or similar. Detailed models that take into account many of these features have been presented in the literature [59, 60].

The bed is initially assumed to contain no hydrogen and to be at a constant temperature equal to the inlet temperature (although this is not necessary). In practice, a small value for the initial pressure in the bed, P_0, is set for stability reasons ($P_0/P_f \sim 10^{-4}-10^{-6}$). This is totally justifiable from a practical point of view too, since in reality no pump can produce an absolute vacuum inside the storage tank [72]. In case of a metal hydride reactor, an additional initial condition is needed for the material density ρ_s which is $\rho_s = \rho_{s0}$ at $t = 0$, where ρ_{s0} is the density of the fresh material.

3.5.4
Solution of the Model

The model of hydrogen storage by adsorption comprises a set of integral, partial differential, and algebraic equations (IPDAEs). The solution method is normally based on a two-phase method-of-lines approach. In the first phase, the spatial dimensions (axial and radial) are discretized in terms of finite dimensional representations, and this reduces the IPDAEs into sets of differential algebraic equations (DAEs). In the second phase, the DAEs are integrated over the time horizon of interest using appropriate integration techniques. Several in-house or commercial software packages have been successfully employed to solve such problems for the case of H_2 storage [59, 60, 62–72].

The most important parameter used to assess the performance of the storage process is the amount of hydrogen stored in the adsorbent, m_t, defined as follows:

$$m_t = \frac{\int_{r=0}^{r=R} \int_{z=0}^{z=L} q(z,r) \cdot r \, dr \, dz}{\int_{r=0}^{r=R} \int_{z=0}^{z=L} r \, dr \, dz}$$

$$= \frac{\int_{r=0}^{r=R} \int_{z=0}^{z=L} (\rho_s(z,r) - \rho_{s0}) \cdot r \, dr \, dz}{\int_{r=0}^{r=R} \int_{z=0}^{z=L} r \, dr \, dz} \tag{3.17}$$

It is evident that m_t has the same units as the concentration of stored H_2 (kg-H_2/m^3 material).

3.5.5
Determination of Optimal Adsorption Properties in a Carbon-Based Adsorbent

The process model (Eqs. (3.8)–(3.17)) along with the boundary and initial conditions involve certain parameters characterizing the adsorbent material that must be optimally selected in order to achieve an economic and safe process performance. Such parameters are the macroscopic Langmuir constants q_s and b_0 whose values affect the maximum amount of hydrogen that can be stored in the bed for a specific charging time. Two main issues must be taken into consideration when establishing optimal control strategies for this system. The first is to ensure that the maximum process storage efficiency is achieved. The second is to ensure satisfaction of all operating and design constraints. This can be expressed by imposing an upper bound on the average bed temperature in order to account for potential safety concerns. The problem is posed as a dynamic optimization problem, which requires the maximization of stored hydrogen by controlling parameters q_s, b_0, ΔH, through the solution of the model equations (set of IPDAEs) with their respective boundary and initial conditions under the following constrains:

$$\Delta T_{avg} \leqslant \Delta T_b, \quad q_s^{min} \leqslant q_s \leqslant q_s^{max}, \quad b_0^{min} \leqslant b_0 \leqslant b_0^{max}$$

where ΔT_b, q_s^{min}, q_s^{max}, b_0^{min}, b_0^{max} are lower or upper bounds.

Since adsorption equilibrium properties can be related to the pore size distribution of the adsorption, this procedure can be extended to determine the optimum pore size distribution. The problem of determining the pore size distribution of microporous carbons by combining molecular simulation and experimental isotherms of different probe molecules at room temperature has received considerable attention over the last decade [73–79]. The micropore range (from 0.6 to 2.0 nm) is subdivided in N equidistant intervals (classes of pores) with 0.1 nm spacing between them. The fraction of the total pore volume associated with each interval is calculated on the basis of an assumed particle size distribution (PSD) and keeping the total pore volume equal to the measured one. Thus, the amount of gas adsorbed in every class at a certain pressure is evaluated by the simulation and, consequently, a computed isotherm is constructed. This, after comparison to its experimental counterpart, results in the optimum micropore size distribution provided by the best fit.

In a recent study [62], this procedure has been adopted to derive the optimal pore size distribution, using, instead of an "experimental" isotherm, the one obtained by the Langmuir equation where the parameters b_0, q_s, have been optimally determined from the macroscopic simulations using gPROMS dynamic optimization capabilities [62, 80]. The whole procedure was iterative due to the interdependence of the heat of adsorption and the bulk density of the adsorbent, employed in the macroscopic model, with the optimal pore size distribution, obtained from the microscopic model. Further details on this procedure can be found in [62].

Figure 3.9 presents the resulting values of the bulk density and heat of adsorption of the optimized material for different temperature constraints. It is seen that as the temperature constraints are relaxed (i.e., increase ΔT_b), we obtain optimized materials with higher bulk density and heat of adsorption. This is expected since both the values contribute to the heat release and hence in the temperature rise in the storage tank.

Figure 3.10 illustrates the resulting densities of H_2 stored in the carbon-based optimized materials on a per-volume and per-weight basis. It is interesting to observe that the two densities show a completely opposite trend as the temperature constraint changes. In particular, as ΔT_b decreases, so does the volumetric density of H_2 while its gravimetric density increases. The apparent contradiction is easily

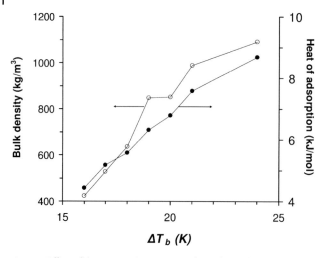

Fig. 3.9 Effect of the imposed temperature bounds on the converged value of the heat of adsorption and the bulk density of the material.

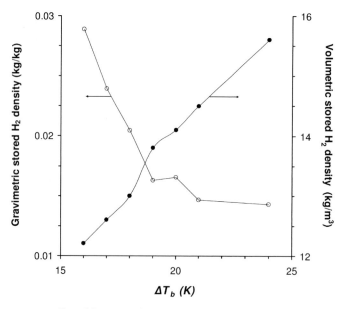

Fig. 3.10 Effect of the imposed temperature bounds on the optimum value of the gravimetric and volumetric hydrogen density stored in the adsorbent.

explained since as ΔT_b decreases, so does the density of the optimized material as already seen in Fig. 3.9. Thus, when evaluating the performance of the material, it is important to consider its density together with the maximum amount of H_2 ad-

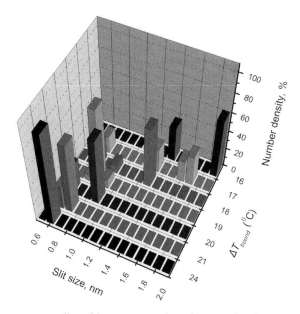

Fig. 3.11 Effect of the temperature bounds imposed at the process level on the optimal pore distributions determined at the adsorbent level.

sorbed if the comparison is on a per-weight basis, or alternatively, the comparison should be done on a per-volume basis.

The optimal pore size distributions that result from the optimization procedure are depicted in Fig. 3.11. It is evident that when loose bounds are imposed on the temperature rise in the bed (high ΔT_b), the optimum pore size distribution is very narrow and it is limited to sizes of 1 nm or lower (where the amount of adsorbed H_2 is maximum). On the other hand, as the temperature bounds are tightened, we observe a shift of a fraction of pores toward larger sizes, where both the volumetric hydrogen density and the heat of adsorption of the material are significantly lower as can be seen from Fig. 3.9. This shifting results in a decrease in the bulk density of the material and thus in a reduction of the volumetric density of H_2 stored in the adsorbent.

3.5.6
Effect of the Heat Conduction Coefficient

It was found that increasing the effective thermal conductivity of the material, λ_e, and more importantly the conductivity in the radial direction, results in a significant decrease of the generated temperature rise in the storage tank [59, 60]. Thus, the above parameter was varied in [62] up to about 10 times in order to study the effect of λ_e on the process performance. Note that such an increase in λ_e is feasible through a process that involves mechanical consolidation of the grains of activated carbon and subsequent mixing with expanded natural graphite [81, 82]. The re-

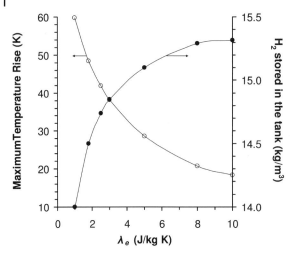

Fig. 3.12 Effect of the thermal conductivity of the storage tank on the process performance for an optimized adsorbent at a temperature bound (ΔT_b) of 21 K.

sults are presented in Fig. 3.12 for $\Delta T_b = 21$ K. As expected, one can decrease the maximum temperature rise in the bed by as much as 40 K and thus increase the volumetric density of hydrogen by 11%.

3.5.7
Optimal Design and Control of Hydrogen Storage in a Metal-Hydride Bed

The operation of hydrogen storage tank using metal hydrides presents distinct challenges such as the possible appearance of a maximum in the temperature profile (hot spot) and the possibility of temperature runaway. The occurrence of excessive temperatures (often due to parametric sensitivity) can obviously have detrimental consequences on the storage management of the tank such as potential explosions, limited storage capacity, etc. These considerations motivate the need for effective heat management strategies for such systems and novel design options (e.g., design of cooling systems, tank diameter, etc.). In this vein, control strategies that regulate the magnitude of the hot spot temperature while ensuring a maximum storage capacity at minimum total costs are of paramount important. Research indicates that attempts to address this problem directly and systematically are rather rare. Working along this direction, Kikkinides et al. [72] presented two novel cooling design options investigated by introducing additional heat exchangers in the metal hydride tank at a concentric inner tube and annular ring inside the tank (Fig. 3.13).

Design decisions included, among others, the radius of the inner tube and the radial position of the concentric annular ring in the tank. Optimization results indicated that significant improvements in the total storage time can be achieved under a safe and economically attractive operation. For example by placing an annular

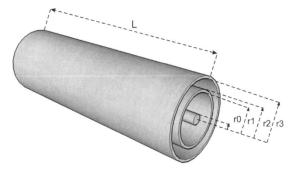

Fig. 3.13 A metal hydride reactor with heat exchangers inside and outside of the bed.

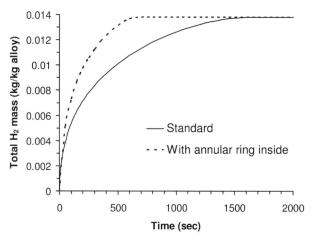

Fig. 3.14 Time evolution of the total H_2 mass stored in the metal hydride reactor using a standard outside heat exchanger and a proposed one with an additional annular ring inside the tank.

ring in the middle of the metal hydride tank, the time required for 99% storage drops from 1450 sec to about 620 sec which is close to 60% improvement, as can be seen from Fig. 3.14.

Similar results have recently been obtained experimentally as shown in [83], using a metal-hydride reactor with a spiral heat exchanger placed inside the tank. It is evident that the spiral heat exchanger can be represented to some extent by an annular ring, although more detailed models should be employed to obtain more accurate predictions of the process performance and heat management in this case.

3.5.8
Modeling the Heat Exchange Process

In the analysis so far, it has been assumed that the cooling medium is kept at constant temperature, T_c. Although this assumption can be valid when the heat exchange is between the reactor bed and the surrounding air, this is not necessarily true for cases where heat exchangers are placed inside the reactor. In this case, an energy balance for the cooling fluid must be employed in order to investigate in a systematic way, how its temperature varies with time and space. The addition of such an energy balance provides the necessary degrees of freedom manipulated variables (e.g., cooling medium flow-rate, inlet temperature, etc.) in order to derive close-loop control strategies that reduce the required time for full H_2 storage and suppress potential hot spots or thermal runaways. To account for the variation in the temperature of the cooling fluid T_f, one must introduce an energy balance for each heat exchanger inside the reactor [72].

3.6
Dynamic Optimization Framework

From the above analysis, it has been evident that there are certain parameters that must be optimally decided in order to achieve an economic and safe process performance. Such parameters can be related to the hydrogen storage tank characteristics and/or the heat-exchanging system. As a first step, the tank design (e.g., diameter and radius) is assumed to be fixed and attention is placed on the optimal design and control of the cooling system. Key parameters that control the heat exchange process can be, for instance, the volumetric flow rate of the cooling fluid, its inlet temperature, or the type of the cooling fluid. Because some of these parameters may affect the operation performance in many ways, we often end up with an optimization problem since the effect is not always monotonic [72].

Two main issues must be taken into consideration when establishing optimal control strategies for this system. The first is to ensure that the maximum process storage efficiency is achieved. The second is to seek the best economic performance which can be expressed by the total required storage time, since this is proportional to the high compression costs. However, because of the complexity of the underlying process, it is often difficult to define simple heuristic strategies to address these two issues and take into account all the operating constraints. The complexity arises primarily from the distributed and highly nonlinear nature of the system model. Kikkinides et al. [72] have performed a systematic optimization study on the effect of heat management on hydrogen storage performance in metal-hydride reactors, the main features of which are presented next.

3.6.1
Process Constraints

Some constraints must be satisfied at all times during the storage operation and they are known as path constraints. Presently, there is one type of path constraints related to the pressure drop in the heat exchange tube that must be less than an upper bound usually defined by design specifications. The pressure drop in the tube of diameter $D_t = 2R$ is calculated from the following expressions [84]:

$$\Delta p_f = f \frac{\rho_f u_f^2}{2 D_t} L \tag{3.18a}$$

$$f = 0.184 \, \text{Re}_{D_t}^{-0.2} \tag{3.18b}$$

The pressure drop constraints can be mathematically stated as follows:

$$\Delta p_f < \Delta p_{f,b} \quad \forall t \in [0, t_s] \tag{3.19}$$

where $\Delta p_{f,b}$ is the maximum allowed pressure difference in the heat exchange tube and it is related to certain mechanical and economic issues.

An additional constraint, which must be satisfied at the end of the storage operation, takes account of potential availability limitations on the cooling fluid. The total volume of the cooling fluid is calculated from the expression next:

$$V_f = \left(\pi R^2 \cdot u_f \right) \times t_s \tag{3.20}$$

where t_s is the total storage time. The constraint takes the following form:

$$V_f < V_{f,b} \tag{3.21}$$

where $V_{f,b}$ is the maximum allowed total volume of cooling fluid in the heat exchange tube. The exact value of this property can be further related to economic or environmental factors, including cost and type of cooling fluid, etc.

Finally, an additional constraint is imposed to ensure a minimum achievable storage capacity of the system:

$$m_t \geqslant m_{t,b} \tag{3.22}$$

Where $m_{t,b}$ is the minimum acceptable storage capacity and is usually 99% or more.

Constraints expressing a maximum average bed temperature can similarly be taken into account as path constraints [72]. The optimal control problem can then be mathematically formulated as the minimization of the storage time, t_s, by controlling parameters u_f, t_f, P, d, subject to constraints Eqs. (3.19), (3.21), and (3.22) imposing lower and upper bounds on the design vector d (radial positions of the cooling systems, bed length and radius, etc.) and path constraints on the average bed temperature.

3.6.2
Solution Approach

A serious complication in the optimization problem defined earlier is the fact that the model equations are variables to both time t, axial and radial position. As in the case of dynamic simulation, a spatial discretization method was applied to eliminate the last two independent variables, leaving a purely time-depended optimization problem. Nonetheless, it remains a complex infinite-dimensional, nonlinear optimization problem. The infinite dimensionality arises because the state and control variables (e.g., cooling medium flow rate) are functions of time rather than simple scalar quantities. There are several numerical methods to solve such dynamic optimal control problems. A class of methods widely used especially in process engineering applications is control vector parameterization [85]. This approach converts that infinite-dimensional optimization problem into a finite-dimensional one, allowing the use of standard nonlinear programming (NLP) solution techniques for the solution of the latter. The reduction in dimensionality is performed by using piecewise Lagrange polynomials to approximate the control vector. This allows the control variables such as cooling medium flow rate to be approximated by polynomials of various degrees and orders of continuity such as piecewise constant, and piecewise linear or quadratic functions.

Assume that $u(t)$ is a control variable in our problem (e.g., cooling medium flow rate and/or temperature, hydrogen charging rate, or inlet pressure, etc.). To make the dynamic optimization problem tractable first, the entire storage time horizon t_s is divided into N intervals with lengths $l_1, \ldots, l_k$:

$$t_s = \sum_{k=1}^{N} l_k \tag{3.23}$$

The values of l_k are not necessarily fixed or equal to each other. The control actions over each time interval are represented by

$$u(t) = \Phi_k(t, v_k) \quad t \in [l_{k-1}, l_k], \ k = 1, \ldots, N \tag{3.24}$$

where Φ_k are simple polynomial functions of time t (e.g., constant, linear, or quadratic). In some cases, additional constraints are imposed on the continuity of the values of the control variable at the boundaries of the intervals, or even their derivatives with respect to time. For example if we consider piecewise linear control profiles for the control variable u_f, then the following expressions are added in the model:

$$\frac{\partial u_f(t)}{\partial t} = \alpha$$

$$u_f(t) = 0 \quad t = 0 \tag{3.25}$$

where α is constant that is optimally selected and u_f is the cooling medium flow rate.

The precise form of the functions, Φ_k, is determined by the finite set of parameters v and l. It is important to note that if an optimization procedure determines the values for v and l, it will also have determined the optimal control u with the space of the allowable functions. It is also important to emphasize that in many cases, the choice of the form of the control variable is an engineering rather than a mathematical issue. It depends very much on the actual control strategy that can be implemented in practice. The number of intervals N depends on the practical problem to be solved. Thus, the parameterized control variable u in interval k at time t is given by

$$u_k(t) = u_k \quad k = 1, \ldots, N \tag{3.26}$$

In summary, the duration of each time interval and the parameterized control variable u_k are optimization decision variables where lower and upper bounds are specified:

$$u_k^{\min} \leqslant u_k \leqslant u_k^{\max} \quad k = 1, \ldots, N$$

$$l_k^{\min} \leqslant l_k \leqslant l_k^{\max} \quad k = 1, \ldots, N \tag{3.27}$$

The bounds $l_k^{\min}$ are set to a small value to avoid numerical problems that may occur if an interval k collapses to zero duration. The upper bounds $u_k^{\max}$ reflect the availability of hot utilities in the facility (e.g., maximum available cooling medium in the facility).

The aim of optimization is to determine the quantities u_k, l_k, t_s so as to minimize the objective function while satisfying the constraints. Note here that the same approach can be used to simultaneously address both the design (e.g., bed diameter and radius, design of cooling system, etc.) and control problem. More details of the formulation and solution strategy of the above optimization problem can be found elsewhere [72].

3.6.3
Optimization Results for Controlling the Cooling Medium Flow Rate

A dynamic optimization problem involving N discrete time intervals was studied in [72]. Optimization results are depicted in Figs. 3.15 and 3.16 for the parameters involved. The results indicate that the optimum time required for 99% H_2 storage in the bed, under the constraints imposed on Δp_f and V_f, is around 1281 sec, employing either $N = 4$ or $N = 10$ time intervals. Note that in order to satisfy all the operating constraints, an optimal cooling medium flow rate policy should be used as shown in Fig. 3.15.

It is seen that by increasing the refinement in the number of time intervals N, one can have a more detailed description of the flow rate schedule, without a significant effect in the basic features of cooling schedule, pressure drop history, and storage dynamics of the process.

The above results demonstrate that the dynamic optimization approach adopted, based on control vector parameterization techniques, identifies the optimal cooling medium control policy to achieve maximum storage efficiency at a minimum storage time horizon while satisfying complex operating constraints.

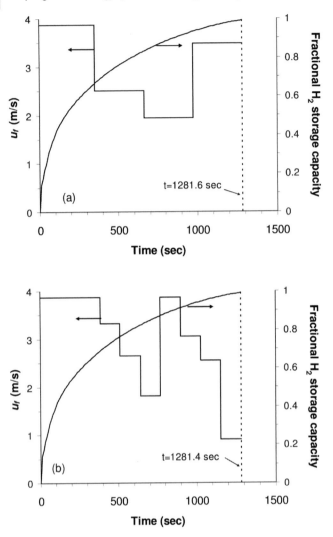

Fig. 3.15 Optimal cooling velocity profile and resulting H_2 mass uptake in the metal hydride reactor as a result of the optimization procedure. (a) $N = 4$ and (b) $N = 10$.

3.7
Conclusions

In this chapter, the important issues and challenges associated with hydrogen storage for the case of stationary and mobile applications have been presented. Different methods of storage have been analyzed in terms of process performance and heat management. The results of this analysis demonstrate that each method has its own advantages and disadvantages and the final selection depends on a variety of characteristics such as type of storage application, hydrogen supply re-

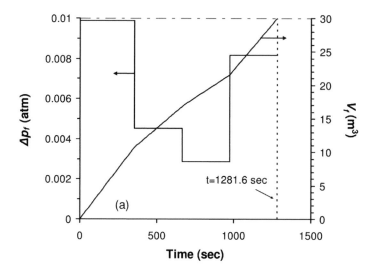

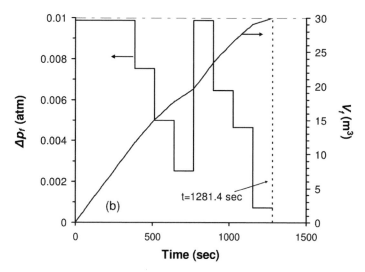

Fig. 3.16 Time evolution of the pressure drop in the optimized heat exchanger and the total volume of cooling fluid required as a result of the optimization procedure. (a) $N = 4$ and (b) $N = 10$.

quirements, etc. In the near term, CGH2 will compete with LH2 as the dominant storage method for fuel-cell vehicles at the demonstration level. Metal or chemical hydrides are expected to offer significant advantages when the current research and development efforts succeed in commercializing the required technology.

In all storage methods described and analyzed in this chapter, energy management issues impose important challenges that are in certain cases the determining factors in the feasibility of a storage process. Frictional pressure drop is important in underground hydrogen storage and must be limited to 13% of the nominal system pressure to limit energy losses in the wells. Improved thermal insulation is very important in LH2 systems due to the cryogenic temperatures involved. Enhanced cooling during hydrogen filling of the storage tank is found to be essential for compressed, liquefied, adsorbed, and absorbed hydrogen, regardless the different physicochemical mechanisms involved. In case of metal hydrides, enhanced heat transfer allows for faster tank filling while preventing the formation of large temperature gradients in the storage tank – a problem that is also met in adsorption units. For this reason, innovative design strategies for both heat transfer configurations and/or material microstructure should be developed to improve process performance.

References

1 Mohanty, K. K., The near-term energy challenge. *AIChE J.* 49 (**2003**), pp. 2454–2460.

2 Momirlan, M., Veziroglu, T. N., Current status of hydrogen energy. *Renew. Sustain. Energy Rev.* 6 (**2002**), pp. 141–179.

3 Agrawal, R., Offutt, M., Ramage, M. P., Hydrogen economy – an opportunity for chemical engineers. *AIChE J.* 51(6) (**2005**), pp. 1582–1589.

4 Edwards, P. P., Kuznetsov, V. L., David, W. I. F., Hydrogen energy. *Phil. Trans. R. Soc.* 365 (**2007**), pp. 1043—1056.

5 Sigfusson, T., Pathways to hydrogen as an energy carrier. *Phil. Trans. R. Soc.* 365 (**2007**), pp. 1025–1042.

6 Muradov, N. Z., Veziroglu, T. N., From hydrocarbon to hydrogen–carbon to hydrogen economy. *Int. J. Hydrog. Energy* 30 (**2005**), pp. 225–237.

7 Damen, K., van Troost, M., Faaij, A., Turkenburg, W., A comparison of electricity and hydrogen production systems with CO$_2$ capture and storage. *Progress in Energy and Combustion Science* 32 (**2006**), pp. 215–246.

8 Sherif, S. A., Barbir, F., Veziroglu, T. N., Wind energy and the hydrogen economy-review of the technology. *Sol. Energy* 78 (**2005**), pp. 647–660.

9 Jacobson, M. Z., Colella, W. G., Golden, D. M., Cleaning the air and improving health with hydrogen fuel-cell vehicles. *Science* 308 (**2005**), pp. 1901–1905.

10 Von Helmolt, R., Eberle, U., Fuel cell vehicles: Status 2007. *J. Power Sources* 165 (**2007**), pp. 833–843.

11 Mehta, V., Cooper, J. S., Review and analysis of PEM fuel cell design and manufacturing. *J. Power Sources* 114 (**2003**), pp. 32–53.

12 Crabtree, G. W., Dresselhaus, M. S., Buchanan, M. V., The hydrogen economy. *Phys. Today* 57 (**2004**), pp. 39–44.

13 Harris, R., Book, D., Anderson, P. A., Edwards, P. P., Hydrogen storage: The grand challenge. *Fuel Cell Rev.* 1 (**2004**), 17–23.

14 Taylor, J. B., Alderson, J. E. A., Kalyanam, K. M., Lyle, A. B., Phillips, L. A., Technical and economic assessment of methods for the storage of large quantities of hydrogen. *Int. J. Hydrog. Energy* 11 (**1986**), pp. 5–22.

15 Venter, R. D., Pucher, G., Modelling of stationary bulk hydrogen storage systems. *Int. J. Hydrog. Energy* 22 (**1997**), pp. 791–798.

16 Sahimi, M., Flow phenomena in rocks: From continuum models to fractals, percolation, cellular automata, and simulated annealing. *Rev. Modern Phys.* 65(4) (**1993**), pp. 1393–1534.

17 Stone, H. B. J., Veldhuis, I., Richardson, R. N., An investigation into large-scale hydrogen storage in the UK. *Proceedings International Hydrogen Energy Congress and Exhibition IHEC*, Istanbul, Turkey, 13–15 July **2005**.

18 CARDEN, P. O., PATERSON, L., Physical, chemical and energy aspects of underground hydrogen storage. *Int. J. Hydrog. Energy* 4 **(1979)**, pp. 559–569.

19 DULLIEN, F. A. L., *Porous Media: Fluid Transport and Pore Structure*, Academic Press, New York, London, **1979**.

20 FELDERHOFF, M., WEIDENTHALER, C., VON HELMOLT, R., EBERLE, U., Hydrogen storage: The remaining scientific and technological challenges. *Phys. Chem. Chem. Phys.* 9 **(2007)**, pp. 2643–2653.

21 ZHANG, J., FISHER, T. S., RAMACHANDRAN, P. V., GORE, J. G., MUDAWAR, I., A review of heat transfer issues in hydrogen storage technologies. *J. Heat Transf.* 127 **(2005)**, pp. 1391–1399.

22 CENGEL, Y. A., BOLES, M. A., *Thermodynamics: An Engineering Approach*, McGraw-Hill Higher Education, New York, **2002**.

23 TZIMAS, E., FILIOU, C., PETEVES, S. D., VEYRET, J. B., Hydrogen Storage: State-of-the-art and Future Perspective. Institute for Energy, Directorate General Joint Research Centre, The Netherlands, **2003**. www.jrc.nl/publ/P2003-181=EUR20995EN.pdf

24 YONG, K., Calculation of the heat and mass transfer in reciprocating compressor with spraying water into cylinders. *Proceedings of the 1988 International Compressor Engineering Conference, Purdue, West Lafayette, IN*, 2, **1988**, pp. 472–476.

25 SCHNEIDER, J., SUCKOW, T., LYNCH, F., WARD, J., CALDWELL, M., TILLMAN, J., MATHISON, S., STEPHANIAN, G., RICHARDS, M., LISS, B., QUONG, S., DURÁN, A., FRIEDLMEIER, G., MAUS, S., KING, J., CANTEEN WALLA, Z., MOORHEAD, B., ADLER, R., CHERNICOFF, W., SLOANE, C., STEELE, M., CHERRY, J., Optimizing hydrogen vehicle fueling. *Proceedings of the National Hydrogen Association Annual Conference, Washington, DC*, **2005**.

26 NILSEN, S., ANDERSEN, S. H., HAUGOM, G. P., RIKHEIM, H., Risk assessments of hydrogen refueling station concepts based on onsite production. *The 1st European Hydrogen Energy Conference & Exhibition (EHEC), Grenoble, France*, **2003**.

27 WETZEL, F. J., Improved handling of liquid hydrogen at filling stations: Review of six years' experience. *Int. J. Hydrog. Energy* 23(5) **(1998)**, pp. 339–348.

28 PEHR, K., SAUERMANN, P., TRAEGER, O., BRACHA, M., Liquid hydrogen for motor vehicles-the world's first public LH2 filling station. *Int. J. Hydrog. Energy* 26 **(2001)**, pp. 777–782.

29 NEWTON, C. L., Hydrogen production, liquefaction and use. *Cryogen. Eng. News* 2(8) **(1967)**, pp. 50–60.

30 BAKER, C. R., SHANER, R. L., A study of the efficiency of hydrogen liquefaction. *Int. J. Hydrog. Energy* 3(3) **(1978)**, pp. 321–334.

31 SHERIF, S. A., ZEYTINOGLU, N., VEZIROGLU, T. N., Liquid hydrogen: Potential, problems, and a proposed research program. *Int. J. Hydrog. Energy* 22(7) **(1997)**, pp. 683–688.

32 BOSSEL, U., ELIASSON, B., TAYLOR, G., The future of the hydrogen economy: Bright or bleak? *2003 European Fuel Cell Forum, Lucerne, Switzerland*, **2003**. www.efcf.com/reports/E08.pdf

33 TIMMERHAUS, K. D., FLYNN, T. M., *Cryogenic Process Engineering*, Plenum Press, New York, **1989**.

34 FLYNN, T. M., *Cryogenic Engineering*, Marcel Dekker, New York, **1997**.

35 BARRON, R. F., *Cryogenic Heat Transfer*, Taylor & Francis, Philadelphia, PA, **1999**.

36 JEANS, J., *An Introduction to the Kinetic Theory of Gases*, Cambridge University Press, London, **1948**.

37 SCHLAPBACH, L., ZÜTTEL, A., Hydrogen-storage materials for mobile applications. *Nature* 414 **(2001)**, pp. 353–358.

38 DORNHEIM, M., EIGEN, N., BARKHORDARIAN, G., KLASSEN, T., BORMANN, R., Tailoring hydrogen storage materials towards application. *Adv. Eng. Mater.* 8(5) **(2006)**, pp. 377–385.

39 SEAYAD, A. M., ANTONELLI, M., Recent advances in hydrogen storage in metal containing inorganic nanostructures and related materials. *Adv. Mater.* 16 **(2004)**, pp. 765–777.

40 SAKINTUN, B., LAMARI-DARKRIM F., HIRSCHER, M., Metal hydride materials for solid hydrogen storage: A review. *Int. J. Hydrog. Energy* 32 **(2007)**, pp. 1121–1140.

41 BERUBE, V., RADTKE, G., DRESSELHAUS M., CHEN, G., Size effects on the hydrogen storage properties of nanostructured metal hydrides: A review. *Int. J. Energy Res.* 31 **(2007)**, pp. 637–663.

42 Thomas, K. M., Hydrogen adsorption and storage on porous materials. *Catal. Today* 120 (**2007**), pp. 389–398.

43 DOE: US Department of Energy. Website: www.doe.gov

44 Darkrim, F. L., Malbrunot, P., Tartaglia, G. P., Review of hydrogen storage by adsorption in carbon nanotubes. *Int. J. Hydrog. Energy* 27 (**2002**), pp. 193–202.

45 Hirscher, M., Becher, M., Hydrogen storage in carbon nanotubes. *J. Nanosci. Nanotech.* 3 (**2003**), pp. 3–17.

46 Darkrim, F., Levesque, D., Monte Carlo simulations of hydrogen adsorption on single-walled carbon nanotubes. *J. Chem. Phys.* 109 (**1998**), pp. 4981–4984.

47 Wang, Q., Johnson, J. K., Molecular simulation of hydrogen adsorption in single walled carbon nanotubes and idealized carbon slit pores. *J. Chem. Phys.* 110 (**1999**), pp. 577–586.

48 Wang, Q., Johnson, J. K., Hydrogen adsorption on graphite and in carbon slit pores from path integral simulations. *Mol. Phys.* 95 (**1998**), pp. 299–309.

49 Li, Y. W., Yang, R. T., Hydrogen storage on platinum nanoparticles doped on superactivated carbon. *J. Phys. Chem. C* 111 (**2007**), pp. 11086–11094.

50 Li, H., Eddaoudi, M., O'Keeffe, M., Yaghi, O. M., Design and synthesis of an exceptionally stable and highly porous metal-organic framework. *Nature* 402 (**1999**), pp. 276–279.

51 Eddaoudi, M., Kim, J., Rosi, N., Vodak, D., Wachter, J., O'Keeffe, M., Yaghi, O. M., Systematic design of pore size and functionality in isoreticular MOFs and their application in methane storage. *Science* 295 (**2002**), pp. 469–472.

52 Eddaoudi, M., Li, H., Yaghi, O. M., Highly porous and stable metal-organic frameworks: Structure design and sorption properties. *J. Am. Chem. Soc.* 122 (**2000**), pp. 1391–1397.

53 Chen, B., Ockwig, N. W., Millward, A. R., Contreras, D. S., Yaghi, O. M., High H_2 adsorption in a microporous metal-organic framework with open metal sites. *Angew. Chem. Int. Ed.* 44 (**2005**), pp. 4745–4749.

54 Panella, B., Hirscher, M., Püttner, H., Müller, U., Hydrogen adsorption in metal-organic frameworks: Cu-MOFs and Zn-MOFs compared. *Adv. Funct. Mater.* 16 (**2006**) pp. 520–524.

55 Rowsell, J., Millward, A. R., Park, K. S., Yaghi, O. M., Hydrogen sorption in functionalized metal-organic frameworks. *J. Am. Chem. Soc.* 126 (**2004**), pp. 5666–5667.

56 Hübner, O., Glöss, A., Fichtner, M., Klopper, W., On the interaction of dihydrogen with aromatic systems. *J. Phys. Chem. A* 108 (**2004**), pp. 3019–3023.

57 Rowsell, J. L. C., Yaghi, O. M., Strategies for hydrogen storage in metal-organic frameworks. *Angew. Chem. Int. Ed.* 44 (**2005**), pp. 4670–4679.

58 Wong-Foy, A. G., Matzger, A. J., Yaghi, O. M., Exceptional H_2 saturation uptake in microporous metal-organic frameworks. *J. Am. Chem. Soc.* 128 (**2006**), pp. 3494–3495.

59 Lamari, M., Aoufi, A., Malbrunot, P., Thermal effects in dynamic storage of hydrogen by adsorption. *AIChE J.* 46 (**2000**), pp. 632–646.

60 Delahaye, A., Aoufi, A., Gicquel, A., Pentchev, I., Improvement of hydrogen storage by adsorption using 2-D modelling of heat effects. *AIChE J.* 48 (**2002**), pp. 2061–2073.

61 Bhatia, S. K., Myers, A. L., Optimum conditions for adsorptive storage. *Langmuir* 22 (**2006**), pp. 1688–1700.

62 Kikkinides, E. S., Konstantakou, M., Georgiadis, M. C., Steriotis, T. A., Stubos, A. K., Multi-scale modeling and optimization of H_2 storage using nanoporous adsorbents. *AIChE J.* 31(6) (**2006**), pp. 737–751.

63 Jemni, A., Nasrallah, S. B., Study of two-dimensional heat and mass transfer during absorption in a metal-hydrogen reactor. *Int. J. Hydrog. Energy* 20(1) (**1995**), pp. 43–52.

64 Nasrallah, S. B., Jemni, A., Study of two-dimensional heat and mass transfer during desorption in a metal-hydrogen reactor. *Int. J. Hydrog. Energy* 22(1) (**1997**), pp. 67–76.

65 Jemni, A., Nasrallah, B., Lamloumi, J., Experimental and theoretical study of a metal-hydrogen reactor. *Int. J. Hydrog. Energy* 24 (**1999**), pp. 631–644.

66 Nakagawa, T., Inomata, A., Aoki, H., Miura, T., Numerical analysis of heat and mass transfer characteristics in the metal hydride bed. *Int. J. Hydrog. Energy* 25 (**2000**), pp. 339–350.

67 MAT, M. D., KAPLAN, Y., Numerical study of hydrogen absorption in an La-Ni5 hydride reactor, *Int. J. Hydrog. Energy* 26 (**2001**), pp. 957–963.

68 ALDAS, K., MAT, M. D., KAPLAN, Y., A three-dimensional model for absorption in a metal hydride bed. *Int. J. Hydrog. Energy* 27 (**2002**), pp. 1049–1056.

69 MAT, M. D., KAPLAN, Y., ALDAS, K., Investigation of three-dimensional heat and mass transfer in a metal hydride reactor. *Int. J. Energy Res.* 26 (**2002**), pp. 973–986.

70 ASKRI, F., JEMNI, A., NASRALLAH, S. B., Study of two-dimensional and dynamic heat and mass transfer in a metal-hydrogen reactor. *Int. J. Hydrog. Energy* 28 (**2003**), pp. 537–557.

71 KAPLAN, Y., VEZIROGLU, T. N., Mathematical modelling of hydrogen storage in LaNi5 hydride bed. *Int. J. Hydrog. Energy* 27 (**2003**), pp. 1027–1038.

72 KIKKINIDES, E. S., GEORGIADIS, M. C., STUBOS, A. K., On the optimisation of hydrogen storage in metal hydride beds. *Int. J. Hydrog. Energy* 31(6) (**2006**), pp. 737–751.

73 LASTOSKIE, C., GUBBINS, K. E., QUIRKE, N., Pore-size distribution analysis of microporous carbons – A density-functional theory approach. *J. Phys. Chem.* 97 (**1993**), pp. 4786–4796.

74 GUSEV, V. I., O'BRIEN, J. A., SEATON, N. A., A self-consistent method for characterization of activated carbons using supercritical adsorption and grand canonical Monte Carlo simulations. *Langmuir* 13 (**1997**), pp. 2815–2821.

75 DAVIES, G. M., SEATON, N. A., VASSILIADIS, V. S., Calculation of pore size distributions of activated carbons from adsorption isotherms. *Langmuir* 15 (**1999**), pp. 8235–8245.

76 RAVIKOVITCH, P. I., VISHNYAKOV, A., RUSSO, R., NEIMARK, A. V., Unified approach to pore size characterization of microporous

carbonaceous materials from N_2, Ar and CO_2 adsorption isotherms. *Langmuir* 16 (**2000**), pp. 2311–2320.

77 SAMIOS, S., STUBOS, A., PAPADOPOULOS, G. K., KANELLOPOULOS, N. K., RIGAS, F., The structure of adsorbed CO_2 in slitlike micropores at low and high temperature and the resulting micropore size distribution based on GCMC simulations. *J. Colloid Interface Sci.* 224 (**2000**), pp. 272–290.

78 SAMIOS, S., PAPADOPOULOS, G., STERIOTIS, T. A., STUBOS, A. K., Simulation study of sorption of CO_2 and N_2 with application to the characterization of carbon adsorbents. *Mol. Simul.* 27 (**2001**), pp. 441–456.

79 SWEATMAN, M. B., QUIRKE, N., Characterization of porous materials by gas adsorption at ambient temperatures and high pressure. *J. Phys. Chem. B* 105 (**2001**), pp. 1403–1411.

80 Process Systems Enterprise Ltd., gPROMS *Advanced Users Guide*, London, UK, **2004**.

81 OLIVES, R., MAURAN, S., A highly conductive porous medium for solid-gas reactions: Effect of dispersed phase on the thermal tortuosity. *Transp. Porous Media* 43 (**2001**), pp. 377–394.

82 BILOE, S., GOETZ, V., MAURAN, S., Dynamic discharge and performance of a new adsorbent for natural gas storage. *AIChE J.* 47 (**2001**), pp. 2819–2830.

83 MELLOULI, S., ASKRI, F., DHAOU, H., JEMNI, A., NASRALLAH, S. B., A novel design of a heat exchanger for a metal-hydrogen reactor. *Int. J. Hydrog. Energy* 32 (**2007**), pp. 3501–3507.

84 FLEMING, W. H., KHAN, J. A., RHODES, C. A., Effective heat transfer in a metal-hydride-based hydrogen separation process. *Int. J. Hydrog. Energy* 26 (**2001**), pp. 711–724.

85 VASSILIADIS, V. S., SARGENT, R. W. H., PANTELIDES, C. C., Solution of a class of multistage dynamic optimization problems: 2. Problems with path constraints. *Ind. Eng. Chem. Res.* 33 (**1994**) pp. 2123–2134.

4
Hydrogen Energy Systems

Zheng Li, Le Chang, Dan Gao, Pei Liu and Efstratios N. Pistikopoulos

Abstract

Driven by concerns over urban air quality, global warming caused by greenhouse gas (GHG) emissions, and energy security, a transition from the current global energy system is receiving serious attention. Increasingly alternative economies are being suggested, whereby the growing energy demand of the future is met with greater efficiency and with more renewable energy sources such as wind, so-lar, and biomass. This implies a gradual shift from the reliance on conventional hydrocarbon-driven technologies toward more innovative, carbon-neutral, sustain-able ones.

Using hydrogen in fuel-cell applications offers a number of advantages over ex-isting fuels and other emerging competitors. It is a high-quality, carbon-free energy carrier that can achieve improved efficiencies with reduced or zero GHG emissions over the entire "well-to-wheel" (WTW) life cycle. These benefits are even further underpinned by the fact that hydrogen can be manufactured from a number of primary energy sources, such as natural gas, coal, biomass, and solar energy, con-tributing toward greater energy security and flexibility. Based on these attributes, a number of long-term strategic initiatives have been undertaken to promote the development of national and regional hydrogen economies [1].

This chapter focuses on hydrogen-related issues, starting from the introduction of hydrogen production technologies and their integration into a polygeneration energy system, and followed by a section of model-based strategic planning of hy-drogen infrastructure.

Keywords

greenhouse gas (GHG) emissions, hydrogen production technologies, polygener-ation energy systems, integrated gasification combined cycle (IGCC), solid oxide fuel cell (SOFC) combined cycle, water-gas shift reaction (WGS)

Process Systems Engineering: Vol. 5 Energy Systems Engineering
Edited by Michael C. Georgiadis, Eustathios S. Kikkinides and Efstratios N. Pistikopoulos
Copyright © 2008 WILEY-VCH Verlag GmbH & Co. KGaA, Weinheim
ISBN: 978-3-527-31694-6

4.1

Introduction of Hydrogen Production Technologies and their Integration into Polygeneration Energy Systems

4.1.1

Introduction of Hydrogen Production Technologies

Although hydrogen is the most abundant element in the universe, it does not naturally exist in its elemental form on Earth. It must be produced from other compounds such as water, biomass, or fossil fuels. Each method of production from these constituents requires energy in some form, such as heat, light, or electricity, to initiate the process.

Steam reforming is a thermal process, typically carried out over a nickel-based catalyst, which involves reaction between natural gas or other light hydrocarbons and steam. This is a three-step process that results in a mixture of hydrogen and carbon dioxide, which is then separated by pressure-swing adsorption to produce pure hydrogen. Steam reforming is the most energy-efficient, commercialized technology currently available and most cost-effective when applied to large, constant loads. Research is being conducted on improving catalyst life and heat integration, which would lower the temperature needed for the reformer and make the process even more efficient and economical.

Partial oxidation (autothermal production) of fossil fuels is another method of thermal production. It involves the reaction of fuel with a limited supply of oxygen to produce a hydrogen mixture, which is then purified. Partial oxidation can be applied to a wide range of hydrocarbon feedstocks, including light hydrocarbons as well as heavy oils and hydrocarbon solids. However, it has a higher capital cost because it requires pure oxygen to minimize the amount of gas that must later be treated. In order to make partial oxidation cost-effective for the specialty chemical market, lower cost fossil fuels must be used. Current research is aimed at improving membranes for better separation and conversion processes in order to increase efficiency, and thus decrease the consumption of fossil fuels.

Hydrogen can also be produced from renewable and nuclear resources by extracting hydrogen from water, but these methods are currently not as efficient or cost-effective as using fossil fuels. Biomass can be thermally processed through gasification or pyrolysis to produce hydrogen. Research on nuclear-based hydrogen production is mostly conducted on thermochemical processes, which make use of high reactor exit temperatures. Both are continuing to be developed. Creation of more efficient, less expensive electrolyzers using renewables and nuclear power is also ongoing [2].

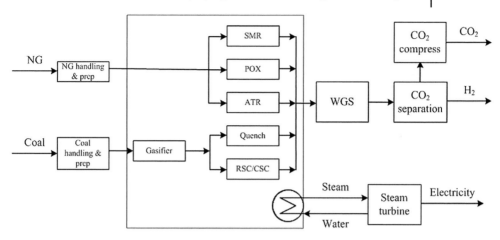

Fig. 4.1 Hydrogen production systems by fossil energy with CO_2 mitigation.

4.1.2
Integration of Hydrogen Production into Polygeneration Energy Systems

4.1.2.1   Integration into a Hydrogen-Electricity Polygeneration System
Most of the polygeneration systems are derived from current technologies, and therefore according to the origin of polygeneration systems, hydrogen-electricity polygeneration systems can be divided into three types: based on hydrogen production system, based on integrated gasification combined cycle (IGCC), and based on solid oxide fuel cell (SOFC) combined cycle.

Hydrogen-Electricity Polygeneration System Based on Hydrogen Production System

To realize the life cycle zero emissions of CO_2 in the hydrogen system, CO_2 must be captured and sequestrated in the dominant sector of fossil fuel based hydrogen production.

Figure 4.1 shows the typical flow sheet of polygeneration system with CO_2 mitigation, including coal handling and preparation, gasification, syngas purification, water-gas shift reaction (WGS), CO_2 separation, pressure-swing adsorption hydrogen production, heat recovery, and power generation. As the key component of the system, with the effect of catalyst, water-gas shift reaction in WGS transforms CO into CO_2, which helps to enhance the hydrogen output. It is a weak exothermic reaction almost independent of pressure, and the determinant factor for the reaction is temperature. The main components of output are H_2 and CO_2. Thus, CO_2 is to be separated, compressed, and condensed to get high-pressure liquid CO_2. Carbon dioxide can be used for the enhancement of oil recovery, coal bed methane production, chemical production, future geological sequestration, etc.

According to the different feedstock and syngas preparation processes, the systems can be divided into the following categories: natural gas steam methane reforming (SMR), natural gas autothermal reforming (ATR), natural gas (NG) par-

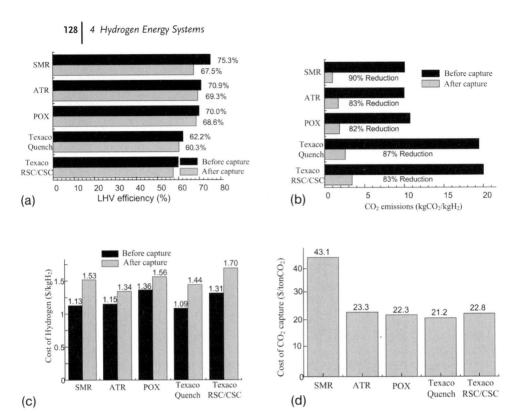

Fig. 4.2 Comparison of cost and efficiency of polygeneration systems producing hydrogen.

tial oxidation (POX), and coal gasification. The selected Texaco coal gasification in the article has two basic process schemes: Quench and radition syngas coller (RSC)/convention syngas coller (CSC). Thus, we will explore these five hydrogen-production processes.

In this section, we use Aspen Plus to simulate the five systems. Each scheme includes two scenarios: with CO_2 mitigation and without CO_2 mitigation.

Based on the principle of achieving CO_2 mitigation as much as possible to realize zero emissions of CO_2 in hydrogen production process, this section compares the 3E performances (energy, environment, economy) of fossil fuel based hydrogen production system with CO_2 mitigation on the same benchmark and explores the effect of CO_2 mitigation; and the result is presented in Fig. 4.2.

From Fig. 4.2 it can be concluded:

• First, for fossil fuel-based with hydrogen production systems mitigation, the reduction of the energy efficiency is only 2%, but the CO_2 emissions are largely decreased by 80–90%. It shows that large-scale CO_2 mitigation has limited influence on the energy efficiency.

• Second, the cost of producing hydrogen with mitigation is increased. The mitigation cost of SMR systems is the largest, while the cost of the other four miti-

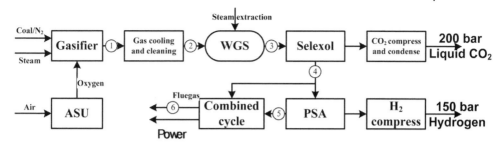

Fig. 4.3 IGCC-based hydrogen and electricity coproduction with CO_2 mitigation.

gation systems, amongst which, the Texaco quench system has the lowest cost, about half that of SMR systems.

- Last, by considering the 3E performances, the ATR system is the best among the three NG-based systems, because its hydrogen production cost is less than the other two, while their energy efficiency and CO_2 mitigation performances have little difference; the Texaco quench system is the better of the two coal-based systems in each aspect of the 3E performances.

Hydrogen-Electricity Polygeneration System Based on IGCC

IGCC is an innovative power generation process that combines coal gasification with combined cycle power generation technologies. The main equipment includes air separation units, coal gasifiers, gas cleaners, gas turbines, heat-recovery boilers, steam turbines, etc.

Compared with current advanced power generation technologies, such as supercritical power generation, IGCC shows more advantages in energy and environment aspects. But the high cost of electricity prevents the development of IGCC. Compared with the pure IGCC, IGCC-based hydrogen-electricity coproduction system has more advantages, such as reaching maximum environment and economic benefits due to the advantages of large-scale gasification. And it also indicates the development direction of future gasification technology.

The CO_2 mitigation costs of IGCC-based cogeneration system and IGCC power plants are almost the same, the only difference being that in cogeneration system, part of H_2-rich gas is fed into PSA hydrogen-producing device and the rest is used for power generation, as shown in Fig. 4.3.

Hydrogen-Electricity Polygeneration System Based on SOFC Combined Cycle

Carnot cycle has the highest energy conversion efficiency of thermodynamic cycles. In the current thermodynamic conditions, IGCC has reached a quite high level of efficiency for coal power generation, and it is difficult to further improve the efficiency through technology progress. Therefore, it is necessary to adopt new technology to break the Carnot cycle restrictions in efficiency. The emergence of fuel cells will realize this idea and eventually bring the coal power industry to hydrogen era.

Current worldwide power production is based on a centralized, grid-dependent network structure. This system has several disadvantages such as high emissions, transmission losses, long lead times for plant construction, and large and long-term financing requirements. Distributed generation is an alternative.

Fuel cells have the potential to provide distributed power generation, and it is considered to be an advanced power generation technology for the 21st century. It is highly efficient, and environmentally friendly. Its greatest feature is that it directly converts chemical energy into electrical energy through electrochemical reaction rather than combustion and therefore breaks Carnot cycle constraints in energy conversion efficiency. It reaches a high efficiency of 60–80%, which is two to three times of the internal combustion engine. Thus, fuel cells have many advantages over conventional power-generating equipment: high efficiency, low emissions, high flexibility, high reliability, low maintenance, excellent part-load performance, etc.

Several different types of fuel cells are suitable for applications in the commercial market and will compete with one another. However, it is difficult to predict which type of fuel cell will succeed given their immature development and commercialization, high cost, and uncertainties in cell performances. High-temperature fuel cell can be fed on syngas that is mainly composed of CO and H_2. After dust removal and desulphurization, the cleaned syngas is sent to the fuel cell for power generation. CO will not cause catalyst poisoning and can react in electrochemical oxidation reaction like hydrogen. Currently, high-temperature fuel cell family mainly includes molten carbonate fuel cell (MCFC) and SOFC, both of which can be fueled by synthetic gas, natural gas, and hydrogen.

SOFC technology can potentially span all of the traditional power-generating markets (residential, commercial, industrial/on-site generation, and utilities) but is likely to penetrate niche markets first, such as small portable generators and remote or premium power applications. Fuel cells for the distributed power market segment will supply power in the range of 3 to 100 MW. High-temperature fuel cells such as SOFC will serve this market, which includes traditional utilities, unregulated subsidiaries, municipal utilities, and energy service providers. Fuel cells for this market may be integrated with coal gasification after the year 2015.

Because SOFC has the similar function as water-gas shift to increase CO_2 concentration for easy separation, according to the position of SOFC and shift section, and the coproduction mode of hydrogen and electricity, a polygeneration system has several configurations: SOFC preposed serial mode, SOFC preposed parallel mode, and SOFC postposed parallel mode, as shown in Figs. 4.4–4.6.

4.1.2.2 Economic Analysis of Hydrogen-Electricity Polygeneration Systems

This section will give an economic analysis of IGCC-based hydrogen-electricity polygeneration system and SOFC combined cycle based hydrogen-electricity polygeneration system.

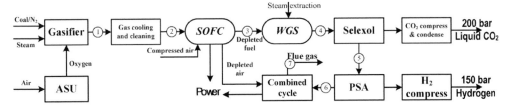

Fig. 4.4 SOFC preposed serial mode.

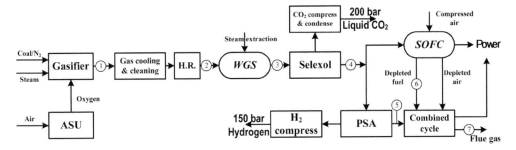

Fig. 4.5 SOFC preposed parallel mode.

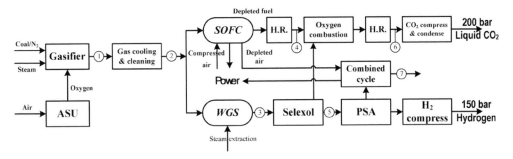

Fig. 4.6 SOFC postposed parallel mode.

Capital Investment

Compared with power generation system, pressure-swing adsorption and hydrogen compressor are added in the hydrogen-electricity polygeneration system. The equivalent specific capital is defined as the ratio of overnight capital costs[1] to equivalent power output, which is the sum of net electricity output and low heat value of hydrogen. The equivalent specific capital is abbreviated below as specific capital. Figure 4.7 summarizes the specific cost for different polygeneration systems.

1) Overnight capital costs refer to the cost of erecting the plant, including construction contingencies, but not considering construction financing, or long-term financing costs. The overnight capital cost is sometimes referred to as the total plant cost or engineering, procurement, and construction cost (EPC).

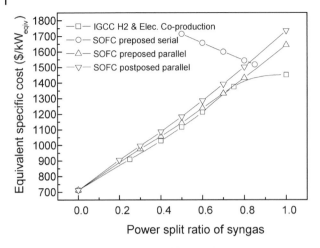

Fig. 4.7 Equivalent specific cost of different polygeneration systems.

Polygeneration systems have a lower specific capital than the sole production systems. The specific capital increases with the power split ratio, while the efficiency decreases with the power split ratio. The specific capital level in the case of IGCC is the lowest among the four cogeneration cases, while the SOFC preposed mode is the lowest among the three SOFC hybrid cases, slightly higher than IGCC case.

Cost of Electricity (Taxes Included)

In polygeneration systems, hydrogen and electricity are two main products (if sold, the recovered CO_2 will be the third product). Therefore when calculating the cost of electricity, hydrogen selling price must be specified. The current market price of hydrogen varies in the range of 0.8–1.5 RMB/Nm^3. Other economic parameters are identical to the basic economic assumptions.

Figure 4.8 compares the production cost of electricity (COE) of IGCC polygeneration system and SOFC preposed parallel mode case. Figure 4.8(a) shows that when hydrogen price is 0.8 RMB/Nm^3, the power split ratio must be less than 33% so that the COE could be lower than IGCC with CO_2 mitigation. This ratio could be increased to 65% or even 100%, when hydrogen price is 1.0 RMB/Nm^3 and 1.2 RMB/Nm^3, respectively. However when compared with the COE of conventional PC without CO_2 mitigation, which is 0.22 RMB/kWh, power split ratio must be lower than 38%, 50%, and 59% with hydrogen price of 1.0, 1.2, and 1.4 RMB/Nm^3, respectively.

Figure 4.8(b) shows that SOFC preposed parallel mode could improve the economy of coal gasification SOFC hybrid power plants even when hydrogen price is 0.8 RMB/Nm^3. Compared with IGCC with CO_2 mitigation, this case is more competitive when power split ratio of syngas is less than 35% with hydrogen price of 0.8 RMB/Nm^3. Higher hydrogen price of up to 1.0 and 1.2 RMB/Nm^3 would raise the upper limit of power split ratio to 70% and 80%, respectively. However compared

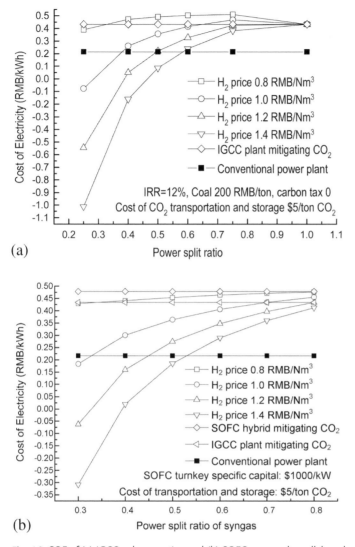

Fig. 4.8 COE of (a) IGCC polygeneration and (b) SOFC preposed parallel mode.

with conventional PC, the power split ratio must be less than 33% (1.0 RMB/Nm³), 45% (1.2 RMB/Nm³), and 54% (1.4 RMB/Nm³).

Cost of Hydrogen (Tax Included)

The goal of hydrogen cost in FutureGen programme is about $4/MMBtu (whole-sale price), equivalent to 0.4 RMB/Nm³. According to Fig. 4.9(a), when electricity is sold at 0.7 RMB/kWh, the power split ratio should be higher than 57% to achieve

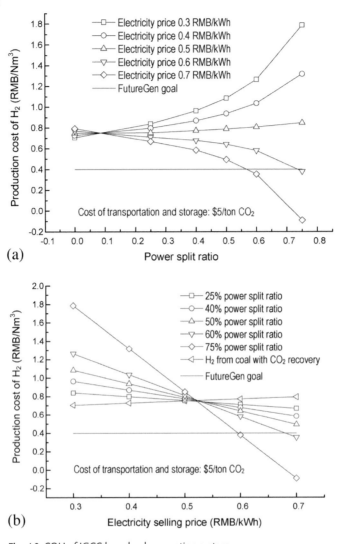

Fig. 4.9 COH of IGCC-based polygeneration system.

the FutureGen goal, and a price of 0.6 RMB/kWh requires 74% or more power split ratio. *The lower the selling price of electricity, the higher the power split ratio.*

Figure 4.9(b) shows the minimum required electricity selling price. In IGCC-based polygeneration system, when the power split ratio of syngas is 75%, the required electricity price is 0.59 RMB/kWh and power split ratio of 60% will increase the price to 0.68 RMB/kWh. *The lower the power split ratio, the higher the required selling price of electricity.* The IGCC case with zero power split ratio is the hydrogen production system with CO_2 mitigation. And the corresponding hydrogen production cost is 0.7–0.8 RMB/Nm^3. IGCC-based coproduction scenarios could be more

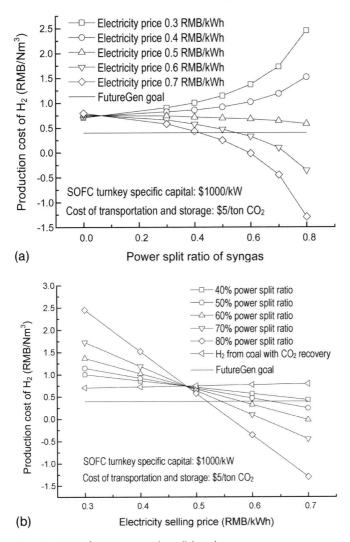

(a)

(b)

Fig. 4.10 COH of SOFC preposed parallel mode.

competitive, only if electricity selling price is over 0.52 RMB/kWh, without the limit of power split ratio.

Figure 4.10(a) shows the minimum power split ratio of SOFC preposed parallel mode for achieving the FutureGen goal decreases to 42% (57% in IGCC) with electricity price of 0.7 RMB/kWh, while 0.6 RMB/kWh requires a minimum ratio of 55% (74% in IGCC). That means that SOFC preposed parallel mode could provide much cheaper hydrogen than IGCC. From another point of view, SOFC preposed parallel mode has greater flexibility in choosing power split ratio of syngas.

Similarly, 80% power split ratio to achieve the FutureGen goal requires minimum electricity selling price of 0.52 RMB/kWh, while 60% corresponds to 0.58

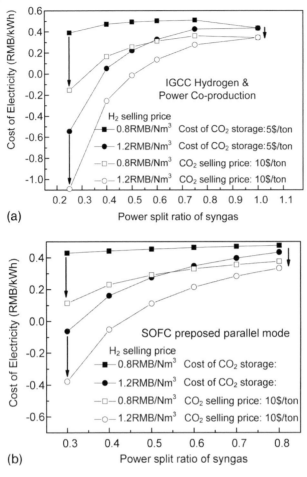

Fig. 4.11 Influence of CO_2 sales on COE of polygeneration systems.

RMB/kWh. Only if the electricity price is over 0.48 RMB/kWh, the hydrogen production cost could be lower than that of hydrogen from coal gasification with CO_2 mitigation, as shown in Fig. 4.10. These limit levels are considerably lower than their counterparts in IGCC case.

CO_2 as Product

Although it may take a relatively long time before carbon taxes become true, the commercial sales of recovered CO_2 is developing rapidly for chemical material, enhancement of oil recovery, and so on. Therefore, this section analyses the impact of CO_2 sale on the economics of coproduction systems before CO_2 taxes come true.

Figure 4.11 shows the impact of CO_2 sale on COE in cogeneration cases (all the other parameters are identical to basic economic assumptions). When the price is $10/ton, the COEs of the two polygeneration systems are lower than

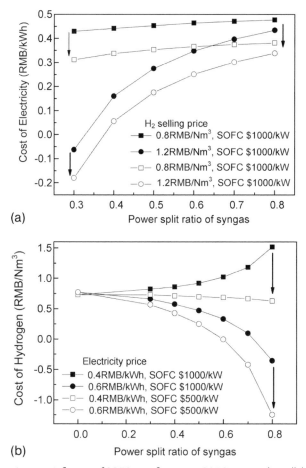

(a)

(b)

Fig. 4.12 Influence of SOFC specific cost on SOFC preposed parallel mode case.

0.4 RMB/kWh. In addition, where hydrogen is sold at 0.8 RMB/Nm^3, the COE of SOFC preposed parallel mode is competitive with power split ratio lower than 70%, and drops to 0.3445 RMB/kWh. The COEs of IGCC polygeneration systems are all less than 0.3445 RMB/kWh.

Impact of Technology Progress – Specific Cost of SOFC

Considering that it will take about 10 years or more for China to reduce CO_2 emission in large scale, it is when the "turnkey" cost of SOFC system is expected to be reduced dramatically due to the technology progress. The Siemens's target is to reduce the specific capital cost to \$400/kW around 2012. This section analyzes the impact of the reduced specific capital cost on the production cost of hydrogen (Fig. 4.12). It is assumed that CO_2 is transported and stored at a cost of \$5/ton, and the other parameters are identical to the basic assumptions.

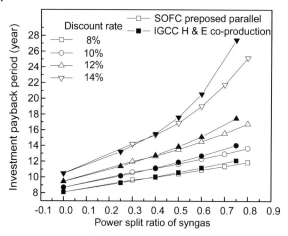

Fig. 4.13 Investment payback period of hydrogen-electricity polygeneration systems.

As shown in Fig. 4.12, with the specific capital cost of SOFC decreasing from $1000/kW to $500/kW, the COE drops to about 0.1 RMB/kWh in spite of power split ratio of syngas, while the COH increases by 0.10 RMB/kWh (30% power split ratio) to 0.90 RMB/kWh (80% power split ratio).

Payback Period

Here, hydrogen price is set to be 1.2 RMB/Nm^3, and electricity price 0.55 RMB/kWh, and the other parameters are identical to the basic assumptions. Figure 4.13 compares the payback periods of IGCC polygeneration system and SOFC preposed parallel mode.

The higher the power split ratio, the longer the payback period. When power split ratio is 1%, the two systems become the IGCC with CO_2 mitigation case and the SOFC hybrid case (mitigation rate is 93%), respectively. Figure 4.13 shows that the hydrogen and electricity cogeneration cases have a distinctly shorter payback period than the other two cases.

When the power split ratio is less than 50%, the payback periods of IGCC case and SOFC preposed parallel mode are almost the same. Otherwise in more than 50% cases, the payback period of IGCC case is longer than the SOFC preposed parallel mode. And the difference is enlarged as the power split ratio and discount rate increase.

4.2
Methodology of Hydrogen Infrastructure Strategic Planning Model

4.2.1
Hydrogen Infrastructure Pathway Options

A hydrogen infrastructure is defined as the supply chain required to manufacture, store, and deliver hydrogen to the consumer. Like any supply chain, it con-

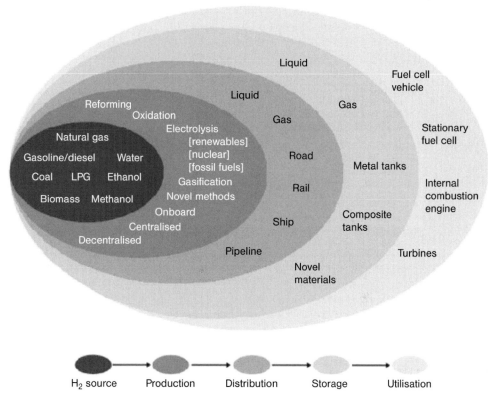

Fig. 4.14 Potential technology components within hydrogen infrastructure pathways (taken from BP hydrogen website: http://www.bp.com/hydrogen).

sists of several distinct components. Production processes are required to convert primary energy resources into hydrogen. Storage units and terminals are needed to compensate for fluctuations in demand. Distribution systems are essential for transporting hydrogen from the production facilities to the point of sale. Finally, dispensing/refueling technologies allow transfer of hydrogen to users at forecourt retail stations.

At each of these stages along the infrastructure, a wide variety of potential technological options exist, as is represented in Fig. 4.14. Not only the hydrogen can be manufactured from a variety of primary energy feedstocks, but it can also be distributed in a variety of forms using different technologies. Gaseous hydrogen, for example, can be distributed in dedicated pipelines over long distance (as is in place in the Rhein-Ruhr area in Germany over a distance of 200 km), while liquefied hydrogen can be transported by rail, ship, or road in tankers. An additional dimension exists when defining the location of production within the supply chain. Unlike most other fuel infrastructures, hydrogen can be produced either centrally or distributed. A centralized production option would be analogous to current gasoline supply chains, where the economies of scale are capitalized upon within an industrial context and large quantities are produced at a central site and then distributed. Alternatively, through the use of small-scale reformers and electrolyzers, hydrogen can be produced closer to the point of use, i.e., *on-site*, in smaller quantities. Such a scenario would exploit the existing natural gas and electricity grid to produce hydrogen at the forecourt refueling stations, thereby alleviating the significant cost of distribution.

Each of the pathway options has its own unique advantages and disadvantages. Cost, operability, reliability, environmental impacts, safety, and social implications are all performance measures that should be considered when assessing and comparing the different pathways. It can also be expected that tradeoffs between these metrics will exist and each option will have its own attributes. Selection of the "best" delivery pathway, therefore, involves comparison of the various technological options in terms of multiple performance criteria, with the ultimate goal being to define a strategy whereby the infrastructure investment can be planned with confidence.

Also of importance is the local market conditions and how the regional primary energy feedstock availabilities can be utilized. For example, Iceland is using electricity from their geothermal resources to generate hydrogen by electrolysis to initiate their transition to a hydrogen economy. In China, however, the use of polygeneration using coal as a feedstock may create an economic source of hydrogen. Most advocates agree to the fact that there is no single supply chain solution template for investing in a hydrogen infrastructure. Instead, it is necessary to have a generic framework that can analyze and compare the performance of the various integrated pathway options on a consistent basis.

The dynamic changes to infrastructure over time and how transitions from one pathway to another should take place as market conditions change are the central issues for hydrogen infrastructure planning. Given the expected changes in hydrogen demand levels, fuel-cell vehicle (FCV) geographical distribution patterns, energy prices, GHG mitigation legislation, and technology performances, it is crucial to accommodate the timing of the investment when analyzing the various pathway options.

The aim of this chapter is to present a generic optimization based model to facilitate the design and planning of hydrogen infrastructures. This model utilizes formal optimization techniques to allow advanced decisions such as the timing of the investment to be captured and to provide comprehensive integrated solutions for investment recommendations. The model is also able to assess the performance of different infrastructure scenarios involving different technologies and raw material feedstocks. The model is generic and can be applied on a case-specific basis across different regions, e.g., Southern California, Greater London Area, Germany, or Japan, to take account of their unique characteristics. Since multiple performance indices are of interest, the model assesses options both in terms of investment and environmental impact criteria to identify optimal infrastructure pathways and investment strategies.

Fundamental to the model is the use of mixed-integer linear programming (MILP) techniques to capture the interactions between the various components of the hydrogen infrastructure. Most readers will be familiar with the linear programming (LP) model, which has a long established history of providing operational, management, and investment decision support in the processing and energy industries. The standard LP problem can obtain an additional degree of functionality when some of the decision variables are limited to a discrete/integer domain, giving rise to the MILP problem. While computationally more intensive, MILP allows various propositional, logical operations associated with strategic decision making to be modeled. For example, an integer variable can be defined such that it determines whether a processing unit should be invested in or not. Because of its capability to naturally capture logical conditions, applications of MILP have been widespread in areas of investment planning, supply chain and logistics management, energy industry planning, engineering design, and production scheduling [3, 4].

4.2.2
Model Overview

To apply a tool such as MILP to model the strategic investment decisions associated with developing a future hydrogen infrastructure, it is necessary to explicitly consider some of the unique features of hydrogen supply chains. More specifically, the model must be able to accommodate the following:

1. A long-term future planning horizon
2. State of the existing infrastructure, especially the natural gas distribution network, electricity grid, and existing mercantile hydrogen production facilities (e.g., any excess reforming capacity at refineries)
3. Multiple and diverse primary energy feedstocks and production technologies
4. Both large-scale centralized production and small-scale distributed/on-site/ forecourt production
5. Both gaseous/liquid hydrogen and hydrogen-carrier agent distribution
6. Economies of scale of large-scale production and distribution technologies
7. Transitions from one supply chain structure to another over time, involving the decommissioning of certain technologies and reinvestment in others
8. Geographical site allocation of technologies
9. Multiple performance indicators – both financial and environmental – that can drive the decision making

In Fig. 4.15, the superstructure representation that forms the basis of the hydrogen supply model is shown [5]. The superstructure acts as the overriding model, capturing all the possible alternatives and interactions between the various supply chain components. Then from this superstructure, the optimization algorithm searches for the best combinations by eliminating the existence of units and the links between them. The superstructure starts with a set of primary energy resources:

$$r \in R := \{\text{Natural Gas, Coal, Biomass, Renewable Electricity,} \ldots\}$$

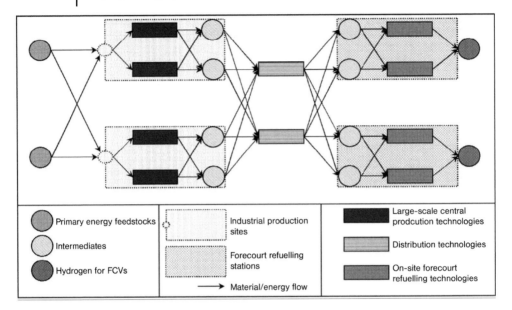

Fig. 4.15 Model superstructure.

which can be used as feedstocks for producing hydrogen at a set of $s \in S$ geographical industrial sites – such as refineries – using any of the large-scale centralized manufacturing technologies:

$$j \in J := \{\text{Steam Methane Reforming, Gasification, Electrolysis}, \ldots\}$$

Each of these production technologies are defined such that they can perform conversion of primary energy feedstocks into an intermediate that is suitable for distribution:

$$l \in L := \{\text{Compressed Natural Gas, Liquid H}_2,$$

$$\text{Compressed Gaseous H}_2, \ldots\}$$

These intermediates are then delivered from the production sites to the set of forecourt refueling stations (markets), $m \in M$, using a corresponding distribution technology:

$$p \in P := \{\text{Natural Gas Pipeline, Liquid H}_2 \text{ Truck},$$

$$\text{Compressed Gaseous H}_2 \text{ Tube-Trailer}, \ldots\}$$

At the refueling stations, the intermediates are dispensed as the final product, namely hydrogen for fuel-cell vehicles, using the appropriate forecourt technology, $q \in Q$. This mathematical representation also allows distributed on-site production to be explicitly considered as a pathway option. This is achieved by defining the set of forecourt technology options to include both technologies for dispensing hydrogen received from the central production facilities as well as technologies for small-scale production:

$$q \in Q := \{\text{Liquid H}_2 \text{ Dispensing, Small-Scale Reforming,}$$

$$\text{Small-Scale Electrolysis,} \ldots\}$$

The primary objective of the model is to support the optimal strategic investment planning and asset management of hydrogen supply chain networks over a long-term future horizon, $t \in T$. The model achieves this by making optimal decisions in terms of four levels:

- Level 1 – Strategic supply chain design
  - Selection of primary feedstocks
  - Allocation of conversion technologies to production sites – where to install which production technologies
  - Assignment of distribution technologies to link production sites to forecourt markets – which markets to supply with the selected sites
- Level 2 – Capacity and shutdown master planning
  - Capacity expansion planning of production, distribution, and refueling technologies – when to expand which technologies
  - Shutdown planning of production, distribution, and refueling technologies – when to switch production technologies
- Level 3 – Production planning
  - Estimation of how much of each primary energy feedstock the selected technologies require and what are the rates of H_2 production, distribution, and refueling at each stage along the supply chain
- Level 4 – Performance index assessment and tradeoff analysis
  - Computation of financial and ecological objectives
  - Multiobjective optimization to establish the set of optimal compromise solutions

Using MILP modeling techniques, constraints can be formulated representing these various decisions. For brevity, the underlining mathematical model is not presented here. In addition to the constraints describing the physical phenomena, objective functions have to be formulated for the financial and ecological performance criteria. These objective functions drive the optimization in search of the best investment strategy and supply chain design. Since a long-term future investment horizon is of interest, the net present value (NPV) is chosen as the financial performance measure, thereby capturing both the capital expenditure and operating cost requirements of the supply chain as well as the time value of the investment over the future horizon. Assessment of the environmental performance of the competing hydrogen infrastructures can be executed using the cumulative WTW GHG emissions that result from delivering hydrogen to the FCV consumer.

To derive the GHG emissions objective function over the entire supply chain (life cycle), from well to wheel, at the beginning it is necessary to define the set of chemicals known to contribute toward the greenhouse effect:

$$e \in E := \{CO_2, CH_4, N_2O, \ldots\}$$

Next, using the Intergovernmental Panel for Climate Change (IPCC) guidelines, a vector of corresponding global warming potential (GWP) factors v_e, expressed as CO_2 equivalents, needs to be constructed [6]. These characterization factors are expressed relative to the GWP of CO_2 and depend on the time horizon over which the global warming effect is assessed. Short time periods (20–50 years) consider the more immediate effects of GHG on the climate, while longer periods (100–500 years) are used to predict cumulative effects of gases on the global climate. For example, when considering the effect over 100 years,

$$v_{CO_2} = 1; \qquad v_{CH_4} = 21; \qquad v_{N_2O} = 310$$

It is also necessary to determine the inventory of GHG emissions associated with the unit reference flow of each supply chain activity. For example, η_{er} is the amount of GHG emitted by e during the unit extraction, processing, and delivery of primary energy feedstock r, while μ_{ej} is the amount of GHG emitted by e during the unit hydrogen production using technology j.

Then, assuming that emissions are linearly proportional to the production, delivery, and dispensing rates, the WTW GHG emissions objective function is simply given as the cumulative sum over the entire planning horizon, over all the individual supply chain activities.

In its entirety, the model for the optimal planning and designing of hydrogen infrastructures is then formulated as a multi-objective MILP problem, summarized as follows:

$$\underset{x,y}{\text{minimize}} \quad U \begin{cases} f_1 = -Net\ Present\ Value\ [\$] \\ f_2 = WTW\ GHG\ Emissions\ [\text{kg } CO_2\ \text{eq}] \end{cases}$$

s.t.

$$\begin{aligned} h(x, y) &= 0 \\ g(x, y) &\leqslant 0 \end{aligned} \left\{ \begin{array}{l} PRIMARY\ ENERGY\ SELECTION \\ SITE\ ALLOCATION \\ DISTRIBUTION\ NETWORK\ DESIGN \\ CAPACITY\ EXPANSION\ \&\ SHUT\text{-}DOWN\ PLANNING \\ TECHNOLOGY\ SELECTION \\ COST\ CORRELATIONS \\ MATERIAL\ \&\ ENERGY\ BALANCES \\ DEMAND\ SATISFACTION \\ GHG\ EMISSIONS\ IMPACT\ ASSESSMENT \end{array} \right.$$

$$x \in \mathfrak{R}^n, \quad y \in Y = \{0, 1\}^m$$

where the goal is to find the values of the operational ($x \in \Re^n$) and strategic ($y \in Y = (0, 1)^m$) decision variables, subject to the set of equality ($h(x, y) = 0$) and inequality ($g(x, y) \leq 0$) constraints such that the utility function (U) is optimized in terms of the two objective functions (f_1, f_2). In our formulation, the continuous operational variables capture decisions related to, for example, production and distribution rates, while the discrete strategic variables model the capacity expansion/shutdown and investment decisions. The two objective functions chosen are the NPV of the investment evaluated over the long-term planning horizon and the cumulative WTW GHG emissions.

It can be expected that there is a conflict between these objectives, i.e., the most profitable infrastructure is not necessarily the least environmentally damaging at the same time. Because of this tradeoff, there is not a single solution to this class of problem. Instead, the solution is a set of multiple compromises known as the *Set of Efficient* or *Pareto Optimal Solutions* (also referred to as noninferior and nondominant solutions). Each solution within the set represents an alternative supply chain configuration and corresponding investment strategy, each achieving a unique combination of environmental and economic performance. A solution is said to be efficient (Pareto optimal) if it is not possible to find another feasible solution so as to improve one objective without worsening at least one of the others.

The value of formulating the decision-making process within a multicriteria optimization framework is that it does not require the a priori articulation of preferences by the decision maker. Instead, the aim is to generate the full set of tradeoff solutions and not to present only one single "best" alternative. From the set of alternatives, the decision maker can then further investigate interesting tradeoffs and ultimately select a particular strategy that satisfies his/her willingness to compromise.

4.3
Case Study Application

To illustrate the features of the model, the results of a China case study conducted are presented here.

4.3.1
Hydrogen Energy System Taking Methanol as an Agent

A key challenge for hydrogen energy is the high cost of transportation and storage of hydrogen. Although the R&D work, including the research on some novel methods, has been undergoing, there is still no commercially available means to solve this issue effectively.

This section presents an innovative idea aiming to solve the issue of high transportation cost: instead of dealing with hydrogen during the transportation, we introduce methanol as the hydrogen-carrier to transport. In this context, the newly introduced "methanol pathway" means using natural gas/coal gasification technology to produce methanol centrally, transporting it by truck to the hydrogen demand areas, and then applying on-site methanol-reforming technology to produce hydrogen at forecourt station.

The inspiration of this innovative thinking is derived from three facts:

1. The recent progress of small-scale methanol-reforming process made in the Dalian Institute of Chemical Physics (DICP) of China Academy of Science [7]. The main part of the process is steam reforming of methanol integrated with H_2 separation using Pd membrane. This new process offers low-cost, high methanol conversion ratio and high hydrogen purity through the process integration of hydrogen production and purification, application of innovative catalyst, and Pd purification membrane.
2. The low cost of methanol transportation compared with hydrogen transportation, especially in a long-distance transportation situation [8].
3. The potential continuity of the current and future fuel supply chain. Suffering from high dependency on import oil, some countries, like China, are developing and promoting alternative fuels seriously including coal-derived methanol fuel. The vision this section sets reveals the potential picture that current methanol fuel activity is valuable not only for relieving near- and medium-term energy security problem, but also for a smooth transition to a possible hydrogen era, with all the infrastructures that can be used continuously.

4.3.2
A Case Study of China Hydrogen Infrastructure Strategic Planning

The case study problem specification is depicted in Fig. 4.16. It consists of a China map where six production sites have been identified for the potential installation of central production technologies. Demand for hydrogen by FCV drivers is expected at 21 major cities with three different demand level (early demand, mid demand, and late demand), acting as the markets in the formulation. Of the six central production sites, C1, C2, C3, and C4 take coal as main primary energy; also C1, C2, and C3 have some reserve of natural gas; N1 and N2 take natural gas as main primary energy. Furthermore, model assumes that all the production sites have plenty of biomass resources and nonrenewable electricity, while C3, C4, N1, and N2 have power resources derived from renewables; C4 also has nuclear power sources. These primary energy availability conditions limit the technologies that are allowed to be installed there.

In Fig. 4.17, the hydrogen demand forecast for the geographical region over the planning horizon is represented. It shows both the expected number of hydrogen FCVs and corresponding hydrogen consumption requirement per year during each of the planning intervals. The long-range planning horizon is defined as the period from 2010 to 2034 divided into five intervals of 5 years each.

The hydrogen infrastructure technology superstructure is presented in Fig. 4.18. As stated before, we include the "methanol pathway" as an alternative. In order to express the impact of introducing this alternative, two different scenarios are set:

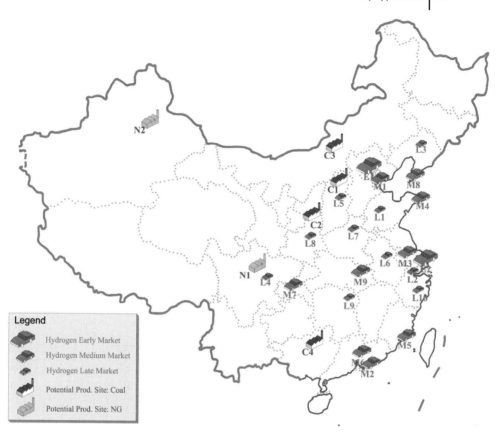

Fig. 4.16 Geographical problem specification for the case study.

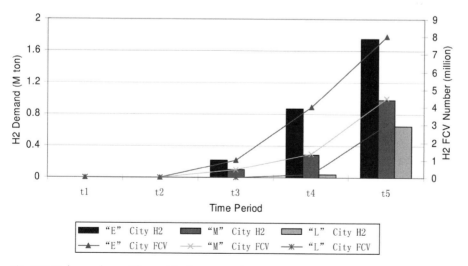

Fig. 4.17 Hydrogen demand forecast for the case study.

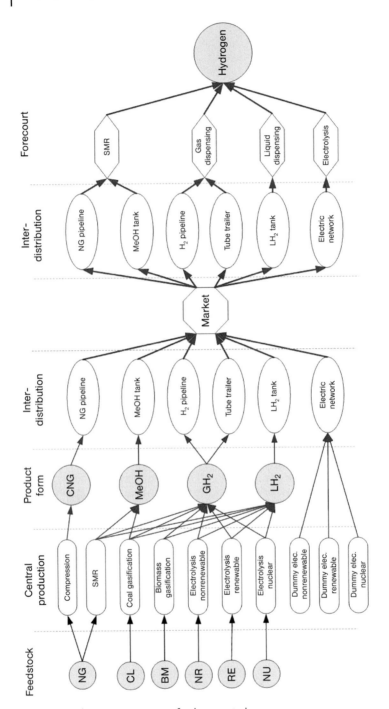

Fig. 4.18 Hydrogen superstructure for the case study.

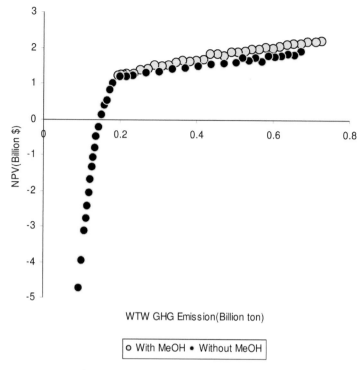

WTW GHG Emission(Billion ton)

| o With MeOH • Without MeOH |

Fig. 4.19 Optimal tradeoff results for the case study.

Scenario A to include "methanol pathway" and Scenario B to exclude "methanol pathway." Applying the multi-objective optimization approach to these two scenarios results in two sets of tradeoff solutions as presented in Fig. 4.19.

The two sets of tradeoff solutions show that the largest NPV Scenario A can achieve is larger than that of Scenario B. And the two sets of solutions merge into one when the GHG emission is less than around 0.2 billion tonnes, because when the environmental emission constraint becomes stricter, the "methanol pathway" will be eliminated from the supply chain components, as "methanol pathway" can no more meet the emission constraint. As a result, there is no difference between Scenario A and Scenario B. When the GHG emission is more than 0.2 billion tonnes, Scenario A will offer a larger NPV than Scenario B for the same GHG emission constraint. This means that the introduction of "methanol pathway" could improve the economic competitiveness of hydrogen infrastructure within a scope of relatively gentle emission constraint. However, this advantage will disappear when the emission constraint becomes stricter.

Each of the two tradeoff curves has two extremes, one stands for the maximum NPV solution and the other stands for the least GHG emission solution. For Scenario A, the corresponding infrastructure of maximum NPV solution is based upon coal, utilizing centralized gasification at optimally selected central production sites to manufacture methanol. Accordingly, the optimum distribution network delivers

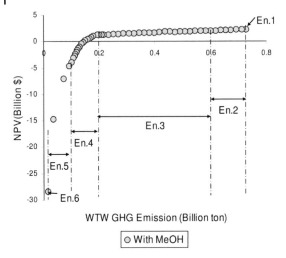

WTW GHG Emission (Billion ton)

☐ With MeOH

Fig. 4.20 Critical enterprise breakdown of the tradeoff front of Scenario A.

the methanol in trucks to the forecourt markets where on-site methanol-reforming process and hydrogen dispensing take place. Additionally, the small hydrogen demand during the market-introduction period is met by on-site nonrenewable water electrolysis installed in the forecourt markets. At the other extreme, the minimum GHG emission solution corresponds to an infrastructure based entirely upon on-site hydrogen production from renewable electricity through electrolysis of water. For Scenario B, due to the absence of "methanol pathway," the maximum NPV solution corresponds to an infrastructure based upon coal, utilizing centralized gasification at optimally selected central production sites to produce liquid hydrogen. Accordingly, the optimum distribution network delivers the liquid hydrogen in trucks to the forecourt markets where liquid hydrogen dispensing takes place. Additionally, the small hydrogen demand during the market-introduction period is met by on-site nonrenewable water electrolysis installed in the forecourt markets. At the other extreme, the minimum GHG emission solution corresponds to an infrastructure totally based on renewable-derived water electrolysis to produce hydrogen on-site.

Noting that each solution within the set represents an alternative infrastructure design and investment strategy, the extent of the compromise between the solutions achieving maximum return on the investment and minimum GHG emissions can be explicitly quantified. Moving along the tradeoff front from one extreme to the other involves a series of distinct infrastructures. The optimal tradeoff front can be broken into critical enterprises based upon the different feedstock, production, distribution, and refueling components of the supply chain that are consistent over a specific region of the curve (as Scenario A shown in Fig. 4.20). Table 4.1 contains the detailed supply chain component descriptions corresponding to these critical enterprises of Scenario A.

Starting with the maximum NPV strategy (Enterprise 1) involving only coal to methanol investment, the optimal transition toward reducing emissions requires

Table 4.1 Components of the critical enterprises in the tradeoff front of Scenario A.

		Supply Chain Components		
	Feedstock	Production	Distribution	Refueling
En. 1	Coal	Coal to MeOH	Truck	Onsite MeOH reforming Onsite nonrenewable water electrolysis
En. 2	Coal Biomass	Coal to MeOH Coal to liquid H_2 Biomass to liquid H_2	Truck	Onsite MeOH reforming Liquid H_2 refueling Onsite nonrenewable water electrolysis
En. 3	Coal Biomass	Coal to MeOH Biomass to liquid H_2	Truck	Onsite MeOH reforming Liquid H_2 refueling Onsite nonrenewable water electrolysis
En. 4	Biomass	Biomass to liquid H_2 Biomass to gaseous H_2	Truck Pipeline	Liquid H_2 refueling Gaseous H_2 refueling Onsite renewable water electrolysis
En. 5	Biomass	Biomass to gaseous H_2	Pipeline	Gaseous H_2 refueling Onsite renewable water electrolysis
En. 6				Onsite renewable water electrolysis

the introduction of biomass gasification as a complimentary production technology and coal to liquid hydrogen production (Enterprise 2). Further reductions in emissions can be achieved while remain cost competitive (Enterprise 3) by eliminating coal to liquid hydrogen production while increasing the biomass to liquid hydrogen proportion. Progressively, coal-based production is totally taken placed by biomass-based technology in order to achieve the desired level of emissions mitigation (Enterprise 4), the distribution including both liquid and gaseous hydrogen truck delivery. Any further emission reduction requires the distribution component of supply chain transforming to gaseous pipeline delivery from liquid delivery (Enterprise 5). The least emission strategy involves only on-site water electrolysis by renewable power (Enterprise 6).

Figure 4.21 and Table 4.2 show the critical enterprise division result of Scenario B. The evolution pattern of enterprises with emission constraint is similar with Scenario A; therefore, we will not express it here.

When analyzing the features of the optimal enterprises in more detail, one realizes that certain technologies and primary energy feedstocks are not present in the optimal tradeoff front. The multi-objective optimization framework, therefore, not only facilitates the identification of the most promising candidates but also assists the elimination of inferior ones. More specifically, under the specifications of the case study, neither centralized water electrolysis technology nor nuclear electricity appears in the set of efficient solutions. The reason is that the absent candidates do not offer either competitive financial returns or the environmental benefits relative to the others. Of course, as technologies develop at different rates in the future, the structure of the optimal solutions may change radically. For example, introduction

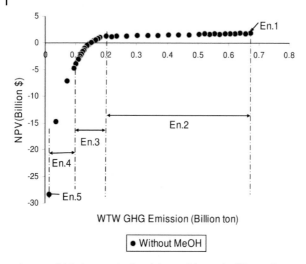

Fig. 4.21 Critical enterprise breakdown of the tradeoff front of Scenario B.

Table 4.2 Components of the critical enterprises in the tradeoff front of Scenario B.

			Supply Chain Components	
	Feedstock	Production	Distribution	Refueling
En. 1	Coal	Coal to liquid H_2	Truck	Liquid H_2 refueling Onsite nonrenewable water electrolysis
En. 2	Coal Biomass	Coal to liquid H_2 Biomass to liquid H_2	Truck	Liquid H_2 refueling Onsite nonrenewable water electrolysis
En. 3	Biomass	Biomass to liquid H_2 Biomass to gaseous H_2	Truck Pipeline	Liquid H_2 refueling Gaseous H_2 refueling Onsite renewable water electrolysis
En. 4	Biomass	Biomass to gaseous H_2	Pipeline	Gaseous H_2 refueling Onsite renewable water electrolysis
En. 5				Onsite renewable water electrolysis

of carbon capture and sequestration, improvement in nuclear technologies, and reduced renewable electricity costs can all drastically change the shape of the optimal tradeoff front. The solutions presented here for the case study, though, are based upon the best data presently available of technologies considered proven to date.

To highlight the characteristics of the solution obtained from the model, one of the optimal compromise solutions is isolated and presented in Fig. 4.22. It corresponds to the maximum NPV solution of Scenario A. The infrastructure evolution is as following: (1) when $t = t1$, while the hydrogen demand is low and only occurs at E1 and E2, the super metropolis Beijing and Shanghai, it is not

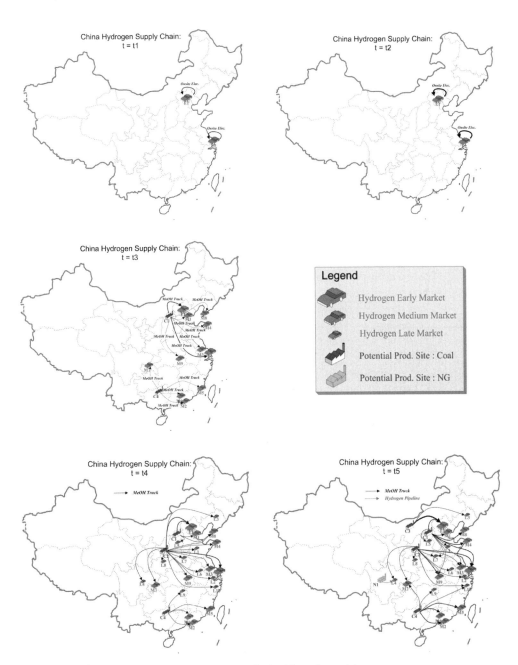

Fig. 4.22 Sample compromising investment strategy obtained from the model.

Table 4.3 Assumption of hydrogen prices.

Time	t1	t2	t3	t4	t5
Price ($/GJ)	19.09	20.12	21.09	22.15	23.46

economic to apply centralized production. Instead, the on-site water electrolysis is utilized in market places; (2) when $t = t2$, no other markets become active, E1 and E2 remain the same hydrogen production pattern to meet the demand, namely on-site water electrolysis, with an increased capacity; (3) when $t = t3$, besides early markets E1 and E2, all the mid-demand markets become active. The growing demand allows the economies of scale to be exploited by decommissioning the forecourt on-site water electrolysis and switching to centralized manufacturing of methanol through coal gasification. Both C1 and C4 are optimally selected as the central production sites, and the distribution network is optimized based on the flow rates and distances between the sites and markets. All the markets provide hydrogen to customers by on-site methanol-to-hydrogen reforming technology; (4) when $t = t4$, all the late-demand markets become active. Another site C2 is selected to produce methanol from coal centrally. The distribution network is reorganized through optimization because of the join of C2 site. For instance, M3 and M9 are not supplied by C1 as they were in $t3$, and now C2 supplies them instead; (5) when $t = t5$, with increasing demand, all the existing production sites increase their capacity. At the same time, two other sites C3 and N1 start manufacturing methanol and hydrogen, respectively, and C3 becomes the main supplier of E1, while C1 shifts its supply to the other markets. Because of the availability of natural gas resources and short distribution distance to L4, the newly built N1 supplies natural gas derived gaseous hydrogen to L4 by pipeline.

The above result of model running is based on the assumption of hydrogen price as presented in Table 4.3. While setting $\pm10\%$ and $\pm15\%$ fluctuation to the basic price, Figs. 4.23 and 4.24 present the sensitivity analysis results of Scenario A and Scenario B, respectively. A linear relationship between hydrogen price and NPV is shown. The situation when all the NPV of optimal solutions falls below zero will appear if hydrogen price is low to a certain extend, such as in the case of Scenario B with -15% fluctuation. The hydrogen price, therefore, has a big impact on the economic performance of the infrastructure, and a comprehensive estimation of hydrogen price should be undertaken before the practical construction.

Additionally, the figures show that the tradeoff curve presents a much sharper slope with the GHG emissions being lower than 0.2 billion tonnes, compared with the rest part. This means that the point with 0.2 billion tonnes of emissions can achieve great emission reduction while remaining competitive economically. And this point corresponds to an infrastructure entirely based upon biomass as the

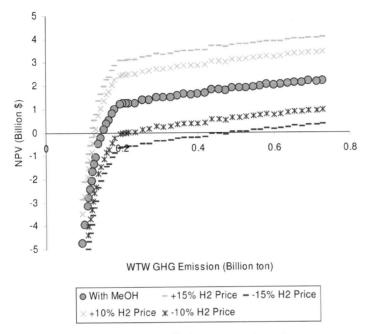

Fig. 4.23 Sensitivity analysis result of hydrogen price in Scenario A.

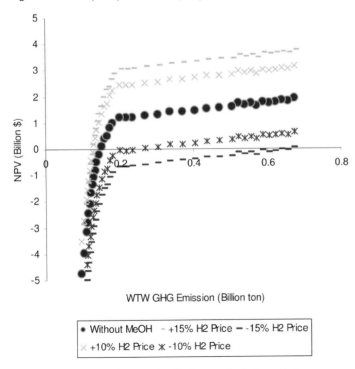

Fig. 4.24 Sensitivity analysis result of hydrogen price in Scenario B.

primary energy; therefore, biomass could play a special role during the hydrogen transition of China.

4.4
Conclusions

For hydrogen to succeed as the fuel of a sustainable future, a commitment is required to create an entirely new fuelling infrastructure, from production through storage and distribution, to dispensing. Any investment strategy for building up a hydrogen supply chain needs to be supported by rigorous quantitative analysis that takes into account all possible alternatives, interactions, and tradeoffs. To assist this strategic decision-making process, this chapter presents a generic model for the optimal long-range planning and designing of future hydrogen supply chains for FCVs. The model presented here utilizes mixed-integer optimization techniques to provide optimal integrated investment strategies across a variety of supply chain decision-making stages.

Key high-level decisions addressed by the model are the optimal selection of the primary energy feedstocks, allocation of conversion technologies to either central or distributed production sites, design of the distribution technology network, and selection of refueling technologies. At the strategic planning level, capacity expansions as well as technology shutdowns are captured to explicitly address the dynamics of the infrastructure and the timing of the investment. Low-level operational decisions addressed include the estimation of primary energy feedstock requirements and production, distribution, and refueling rates. Realizing that both financial and ecological concerns are driving the interest in hydrogen, formal multi-objective optimization techniques are used to establish the optimal tradeoff between the NPV of the investment and the WTW GHG emissions.

To illustrate the capabilities of the model, the results of the China case study have been presented here. Through the study, it was shown how the model can identify optimal supply chain designs, capacity expansion policies, and investment strategies for a given geographical region. In particular, the set of tradeoff solutions allows the most promising pathways to be isolated and the inferior ones to be eliminated from further consideration.

References

1 INTERNATIONAL ENERGY AGENCY, *Hydrogen & Fuel Cells: Review of National R&D Programs*, **2004**.

2 DOE, *A national vision of America's transition to a hydrogen economy – to 2030 and beyond*, **2002**.

3 WILLIAMS, H., *Model Building in Mathematical Programming*, 3rd ed., Wiley, New York, **1993**.

4 FLOUDAS, C. A., *Nonlinear and Mixed Integer Optimization*, Oxford University Press, New York, **1995**.

5 BIEGLER, L. T., GROSSMANN, I. E., WESTERBERG, A. W., *Systematic Methods of Chemical Process Design*, Prentice-Hall, New Jersey, **1997**.

6 IPCC. *Impacts, Adaptation and Mitigation of Climate Change: Scientific–Technical*

Analyses, Intergovernmental Panel on Climate Change, Cambridge University Press, Cambridge, **1996**.

7 PAN, L. W., WANG, S. D., Methanol steam reforming in a compact plate-fin reformer for fuel-cell systems. *Int. J. Hydrog. Energy.* 30 (**2005**) 973–979.

8 CHANG, L., LI, Z., GAO, D., HUANG, H., NI, W., Pathways for hydrogen infrastructure development in China: Integrated assessment for vehicle fuels and a case study of Beijing. *Energy* 32 (**2007**) 2023–2037.

5

Integrated Optimization of Oil and Gas Production
Michael C. Georgiadis and Efstratios N. Pistikopoulos

Keywords

sequential quadratic programming (SQP), well management routines, gas-oil ratio (GOR), pipeline network simulators, successive linear programming (SLP) method, linear programming (LP)

5.1
Introduction

A typical offshore oilfield is shown in Fig. 5.1 and consists of (i) the reservoir, which is defined as an accumulation of oil, gas, and water in porous permeable rock, (ii) the production wells, which transfer reservoir fluids to the manifold, (iii) the manifold, where the well streams are mixed, (iv) the surface flow lines, which connect the manifolds to the surface facilities, and (v) the surface facilities where the reservoir fluids are separated into oil, gas, and water and gas is compressed and injected back to the reservoir. Each well consists of two pipe segments, the well tubing and the well flow line. Between them there is a valve, the choke, which is used to control the well flow rate. The region of the well inside the reservoir is called the well bore. There are two types of production wells: (i) naturally flowing and (ii) gas lift. The first are able to provide flow naturally to the surface, while the second require an injection of high-pressure gas to reduce the pressure drop in the well tubing and therefore facilitate extraction. Finally, the wells, manifolds, flow lines, and surface facilities define the pipeline network, as shown in Fig. 5.1.

In petroleum fields, oil production is often constrained by the reservoir conditions, flow characteristics of the pipeline network, and the capacity of the surface facilities [1–3]. Consequently, proper determination of the daily optimal operating conditions requires simultaneous consideration of the interactions between the reservoir, the wells, and the surface facilities.

Various methods for daily oil production optimization have been presented in the literature. These can be divided into three categories: (i) sensitivity analysis using simulation tools, (ii) heuristic methods, and (iii) mathematical programming methods.

Process Systems Engineering: Vol. 5 Energy Systems Engineering
Edited by Michael C. Georgiadis, Eustathios S. Kikkinides and Efstratios N. Pistikopoulos
Copyright © 2008 WILEY-VCH Verlag GmbH & Co. KGaA, Weinheim
ISBN: 978-3-527-31694-6

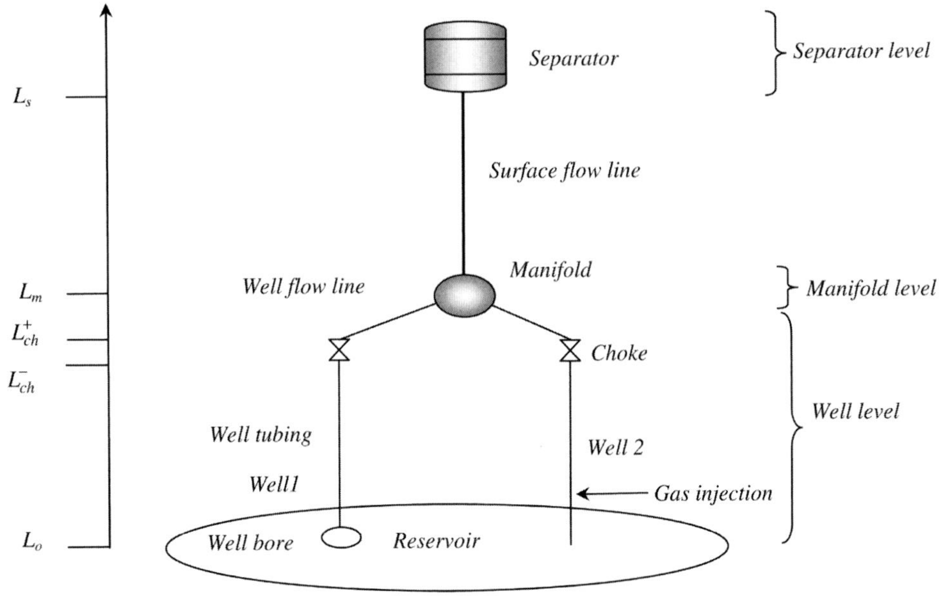

Fig. 5.1 A production operation system.

Traditionally, the petroleum industry has applied NODAL®[1] analysis [4], which is a simulation method to determine the daily optimal operating policy, by repetitively varying the optimization variables and simulating the underlying system. Therefore, NODAL analysis is by its trial and error nature limited to oilfields with a small number of wells.

The majority of publications [5–7] tackle the daily oil production optimization using heuristic rules, which are incorporated in software tools known as well management routines [8]. Well management routines decompose the pipeline network into levels, usually: (i) the well level, (ii) the manifold level, and (iii) the separator level as shown in Fig. 5.1 and heuristic rules are applied sequentially. For instance, at the well level rules such as close a well if it violates an upper bound in the gas-oil ratio (GOR), which is defined as the ratio of gas-to-oil flow rate, are applied. At the manifold level upper bound constraints on oil, gas, and water flow rates are imposed. If any of the upper bounds is violated the well production rate is scaled appropriately until the constraint is satisfied. At the separator level, oil production targets are imposed. If these are not satisfied, then decisions such as gas lift initiation are enabled. It is obvious that well management routines, while accounting for network constraints, are formulated in an *ad hoc* manner and do not systematically address the maximization of oil production. The current major commercial reservoir simulators [9, 10] are based on similar heuristic rules and gas lift allocation is considered separately from well rate optimization. One of the most widely applied heuristic rules for allocation of gas-to-gas lift wells is known as the incremental

---

1) Trademark of Schlumberger

gas-oil ratio (IGOR) rule. *IGOR* is defined as the amount of gas required by a gas-lift well to produce an additional barrel of oil. It was proposed originally by Redden et al. [11] but Weiss et al. [7] derived a necessary condition for optimal allocation of gas-to-gas lift wells. The condition states that all wells tied to a common manifold must operate at the same *IGOR*. The *IGOR* heuristic rule has also been applied by Barnes et al. [12]; Stoisits et al. [13] in the Prudhoe Bay and Kuparuk River fields, respectively. However, it must be noted that the necessary condition was derived by analyzing a pipeline network where all the wells are tied directly to a fixed pressure separator and therefore, interactions between the wells that share a common flow line are not considered.

Mathematical programming techniques have also been applied in production operations optimization. Carrol and Horne [14] applied gradient-based and polytope optimization techniques to a single well. Fujii and Horne [15] proposed a two-stage optimization strategy where a pipeline network simulator acted as an evaluation tool for a series of suggested values of control variables produced by an optimization algorithm. Martinez et al. [16], Palke and Horn [17]; Wang et al. [18] applied genetic algorithms (GA). Although genetic algorithms are robust, they are especially computationally intensive when the optimization problem subjects to nonlinear equality constraints. Fang and Lo [2] proposed a linear programming model to optimize lift-gas is subject to multiple nonlinear flow rate constraints. Dutta-Roy and Kattapuram [19] analyzed a gas lift optimization problem with two wells sharing a common flow line. They pointed out that when flow interactions among wells are significant, nonlinear optimization tools are needed. They coupled a pipeline network simulator with an optimization tool that is based on the sequential quadratic programming (SQP) method. However, due to commercial confidentiality there is no information regarding the optimization model. Handley-Schahler et al. [20] developed a commercial optimization tool for optimization of production operations. The resulting optimization problem was solved using sequential linear programming (SLP) method. Their tool was applied to a gas production network and to gas-lift wells that are tied directly to a separator. However, due to commercial confidentiality there is no information available about the optimization model. Wang et al. [18] proposed a linear programming (LP), and a MILP for optimization of production operations. The LP formulation was similar to that proposed by Lo et al. [21], which is based on the premise that for a pipeline network of naturally flowing wells by setting the well chokes fully open and simulating the pipeline network the resulting well oil rates are maximum. These well oil rates were used as upper bounds to the LP formulation that incorporated only mass balance and surface capacity constraints. The MILP formulation extended the work of Fang and Lo [2]. Both the LP and MILP formulations do not handle flow interactions among wells that share a common flow line.

Van den Heever and Grossmann [22] presented a mixed-integer nonlinear model for oilfield infrastructure that involves design and planning decisions. The nonlinear reservoir behavior is directly incorporated into the formulation. For the solution of this model an iterative aggregation/disaggregation algorithm is proposed according to which time periods are aggregated for the design problem, and subsequently disaggregated for the planning subproblem. Van den Heever et al. [23] addressed the design and planning of offshore oilfield infrastructure focusing on business rules and complex economic objectives. A specialized heuristic algorithm that relies on the concept of Lagrangean decomposition was proposed by Van den Heever et al. [24] for the efficient solution of this problem. Ierapetritou et al. [25] studied the optimal location of vertical wells for a given reservoir property map. The problem is formulated as a large scale MILP and solved by a decomposition technique that relies on quality cut constraints.

Alarcon et al. [26] proposed a nonlinear programming model for the gas allocation problem to a group of wells. A salient feature of this approach is a new mathematical fit to the gas lift performance curve. Comparison of results with other approaches from the literature illustrates the efficiency of the proposed approach. Camponogara and Nakashima [27] presented a mixed-integer linear programming model for the problem of maximizing oil filed profit under multiple facility constraints such as limited gas lift, fluid handling, and storage capacities. Optimization results reveal that the proposed approach yields to an increase in the oil production rates as compared to other approaches from the literature and commercial software tools. Bangerth et al. [28] presented a comparison of several optimization algorithms for the reservoir oil well placement problem. It was shown that none of the tested algorithms can find the optimal solutions. Camponogara and Nakashima [29] extended the gas-lift optimization problem to account for uncertainties in the oil outflow and activation of wells. A dynamic programming algorithm was proposed and its efficiency was illustrated by using several example problems. Camponogara and Nakashima [30] investigated a piecewise linear formulation of the gas lift production of oil wells problem, by exploring some properties of the polyhedron associated with the space of feasible solutions and delivering families of valid inequalities that are shown to improve the performance of the optimization problem. Hernandez-Barragan et al. [31] presented a new approach for the simulation and optimization of gas and oil production systems which integrates the whole process from reservoirs to surface facilities where all units are rigorously modeled. Several example problems illustrated the efficiency and robustness of the proposed approach.

In summary, the open literature does not provide systematic optimization techniques that takes into account, in details, the interactions in a production system and simultaneously optimizes the well and gas-lift rates. Either only a part of the problem is addressed such as wells tied directly to the separator or *ad hoc* rules are used that lead to suboptimal operating policies. The objective of this chapter is to develop a robust and efficient optimization algorithm that simultaneously optimizes all the relevant decisions and is applicable to any pipeline structure.

In Sections 5.2 and 5.3, we present the problem statement and a model for the production operations optimization. Section 5.4 contains a motivating example showing the need for a more efficient solution method. In Section 5.5, a novel approximate optimization model is presented. The resulting optimization model is based on a degree of freedom analysis, well upper and lower bounds and separability programming techniques and is solved as a sequence of MILP problems. In

Section 5.6, two motivating examples are used to demonstrate the accuracy and robustness of the proposed formulation. The performance of the algorithm is demonstrated in Section 5.7 with three example problems. Conclusions are discussed in Section 5.8.

5.2
General Problem Statement

The production optimization problem is based on a time period of one day and can be stated as follows: given is a reservoir, a set of naturally flowing wells, a set of gas-lift wells, a surface pipeline network, a set of separation facilities, and a compressor plant (see Fig. 5.1). The problem involves the maximization of the profit from the sales of oil minus the cost of gas compression subject to a set of constraints. The constraints include mass and momentum balances in the wells, minimum and maximum pressure and flow rate constraints at the inlet and outlet of the pipelines, maximum oil, gas and water handling capacity constraints, and gas compression constraints. The following key decisions are considered:

- how to control the well oil rates by manipulating appropriately the well chokes, and
- how to distribute gas among gas-lift wells.

5.3
Mathematical Model

The mathematical model of the system has been developed at Imperial College [32, 33] and depends on the type of reservoir such as dry gas and oil and gas reservoirs and the type of well such as naturally flowing and gas lift. In this section, we present the mathematical model of the two-well network shown in Fig. 5.1, which consists of a naturally flowing and a gas-lift well. The two-well network model contains all the components that allow modeling of more complex pipeline networks. The model is based on the following assumptions [32, 33]:

- The system is under steady-state conditions since the dynamics of the reservoir is in order of weeks, the pipeline network is in order of minutes, and while the time horizon of the production operation problem is one day.
- The thermodynamic description of the reservoir fluids is based on the black-oil model [34], an empirical approach widely applied in the petroleum industry. However, the proposed method can also be extended to the compositional model.
- A continuous and differentiable multiphase pressure drop model was applied [35].
- The energy balance in the pipeline network is considered assuming linear temperature profiles along the pipes [4].

For simplicity in the presentation of the paper, we describe in the following the optimization model of a two-well network problem as shown in Fig. 5.1.

Table 5.1 Consistent initialization of the first stage of naturally flowing well.

Equations	Unknowns
1. $q_{o,i} = PI_i(P_{R,i} - P_i(L_0))$	$q_{o,i}, P_i(L_0)$
2. $q_{g,i} = GOR_i q_{o,i}$	$q_{g,i}$
3. $q_{w,i} = WOR_i q_{o,i}$	$q_{w,i}$
4. $q_{L,i} = q_{o,i} + q_{w,i}$	$q_{L,i}$
5. $\frac{dP_i(L_0)}{dL} = f_P\left(P_i(L_0), q_{o,i}, q_{w,i}, q_{g,i}\right)$	$\frac{dP_i(L_0)}{dL}$
Number of equations 5	Number of unknowns 6
DoF 1	

1. A well-bore model is the one that describes the multiphase fluid flow from the reservoir to the well bores of a production well. In this work, Peaceman's well model was applied [36]. This model assumes steady-state radial flow and can be expressed as

$$q_{o,i} = PI_i\left(P_{R,i} - P_i(L_0)\right), \quad \forall i \in I \tag{5.1}$$

where I is the set of wells i, $q_{o,i}$ is the well oil rate under stock tank conditions, PI_i is the productivity index of the well, $P_{R,i}$ is the reservoir block pressure that contains the well and can be assumed constant for a time period of one day, and $P_i(L_0)$ is the bottom hole pressure of the well i. For naturally flowing wells, the gas flow rate $q_{g,i}$ under stock tank conditions is given by the following relation:

$$q_{g,i} = GOR_i q_{o,i}, \quad \forall i \in N \tag{5.2}$$

where N is the set of naturally flowing wells. For gas-lift wells, the gas flow rate is given by

$$q_{g,i} = q_{g,i}^{inj} + GOR_i q_{o,i}, \quad \forall i \in G \tag{5.3}$$

where G is the set of gas-lift wells and $q_{g,i}^{inj}$ is the gas injection rate. The water $q_{w,i}$ and liquid $q_{L,i}$ flow rates are given by

$$q_{w,i} = WOR_i q_{o,i}, \quad \forall i \in I \tag{5.4}$$

$$q_{L,i} = q_{o,i} + q_{w,i}, \quad \forall i \in I \tag{5.5}$$

where WOR_i is the water-oil ratio of well i defined as the ratio of water-to-oil flow rate.

The productivity index PI_i, GOR_i, and WOR_i are generally nonlinear functions of the well oil flow rate $q_{o,i}$. However, they may be assumed constant for a period of one day with their values given from a reservoir simulator.

2. The well tubing model simulates the multiphase fluid flow from the well bore to the wellhead of a production well and is described by the following ordinary differential equation (ODE) which is valid at $L \in [L_0, L_{ch})$

$$\frac{dP_i}{dL} = f_P(P_i, q_{o,i}, q_{g,i}, q_{w,i}), \quad \forall i \in I, \; L \in [L_0, L_{ch}) \tag{5.6}$$

3. The well choke model is used to control the well flow rate. The characteristic of the choke valve is the transition between the critical and subcritical flow regimes. A generic choke model is as follows:

$$q_{L,i} = f_c\left(d_i, P_i(L_{ch}^-), P_i(L_{ch}^+), y_i, GOR_i, WOR_i\right), \quad \forall i \in I \tag{5.7}$$

$$y_i = \max\left(y_c, \frac{P_i(L_{ch}^-)}{P_i(L_{ch}^+)}\right), \quad \forall i \in I \tag{5.8}$$

where d_i is the choke valve diameter, $P_i(L_{ch}^-)$, $P_i(L_{ch}^+)$ is the pressure upstream and downstream of the choke, y_i is the well choke pressure ratio, and y_c is a constant known as critical pressure ratio. In this paper, the choke model proposed by Sachdeva et al. [37] is used since it satisfies continuity at the transition point. The choke model is not differentiable due to the existence of the max operator, which is approximated using the smoothing function proposed by Samsatli et al. [38].

4. The well flow line model simulates the multiphase fluid flow from the wellhead to the manifold and is described by an ODE similar to Eq. (5.6) but valid at $L \in [L_{ch}, L_m)$. The combination of the well bore, well tubing, choke, and well flow line models will be referred in the rest of the paper as the well model.

5. The manifold model is the one where the well streams are mixed. The mass balance for each phase p is given by the following linear equation:

$$\sum_{i \in I_m} q_{p,i} = q_{p,m}, \quad \forall m \in M, \; p \in \{o, g, w, L\} \tag{5.9}$$

where I_m is the set of the wells tied to the manifold m and $q_{p,m}$ is the flow rate of phase p in manifold m. Moreover, any network point must have unique pressure. Therefore,

$$P_i(L_m) = P_m(L_m), \quad \forall m \in M, \; i \in I_m \tag{5.10}$$

where $P_i(L_m)$ is the pressure of the well i at the manifold level and $P_m(L_m)$ is the manifold pressure.

6. The surface flow line model simulates the multiphase flow from the manifold to the surface facilities and is described by the following ODE valid at $L \in [L_m, L_s]$:

$$\frac{dP_m}{dL} = f_P(P_m, q_{o,m}, q_{g,m}, q_{w,m}), \quad \forall m \in M, \; L \in [L_m, L_s] \tag{5.11}$$

7. The separator s, which operates at pressure P_s and has capacity $C_{p,s}$ for each phase p

$$P_m(L_s) = P_s, \quad \forall s \in S, m \in M_s \tag{5.12}$$

$$\sum_{m \in M_s} q_{p,m} \leqslant C_{p,s}, \quad \forall p, s \in S \tag{5.13}$$

where M_s is the set of manifold flow lines m that are connected to separator s, S is the set of separators and $P_m(L_s)$ is the pressure of the surface flow line at the separator level.

8. Operational constraints that involve (a) well oil rate lower bounds to satisfy stable flow rate [4], (b) well oil rate upper bounds to prevent sand production, (c) lower bound pressure constraints at the inlet of the flow lines to prevent development of erosional velocities [39], (d) choke diameter design constraints, and (e) upper bound constraints on gas lift availability

$$q_{o,i}^L \leqslant q_{o,i} \leqslant q_{o,i}^U, \quad \forall i \in I \tag{5.14}$$

$$P_m^L \leqslant P_m(L_m), \quad \forall m \in M \tag{5.15}$$

$$0 \leqslant d_i \leqslant 1, \quad \forall i \in I \tag{5.16}$$

$$\sum_{i \in G} q_{g,i}^{inj} \leqslant C \tag{5.17}$$

9. An objective function which is defined as maximization of profit from the sales of oil minus the cost of gas compression

$$\max w_o \sum_{i \in I} q_{o,i} - w_g \sum_{i \in G} q_{g,i}^{inj} \tag{5.18}$$

where the coefficients w_o, w_g are the price of oil and gas compression, respectively.

The mathematical model of the system involves (i) ordinary differential equations (ODEs) that are applied over particular length interval or stage [40] and (ii) boundary conditions between the stages. Considering, for instance, the two-well network of Fig. 5.1 then the first stage of the system is defined by the well tubing model (Eq. (5.6)) with boundary conditions described by the well bore model (Eqs. (5.1)–(5.6)). The second stage of the system is defined by the well flow line model (Eq. (5.6)) with boundary conditions imposed by the choke model (Eqs. (5.7) and (5.8)). Finally, the third stage of the system is described by the surface flow line model (Eq. (5.11)). Consistent initialization of the system of ODEs that describe the first stage of a naturally flowing well involves the solution of the nonlinear equations shown in Table 5.1. By counting the number of equations and unknowns in

Table 5.2 Initialization of the second stage of a well.[a,b]

Equations	Unknowns
$q_{L,i} = f_c(d_i, P_i(L_{ch}^-), P_i(L_{ch}^+), y_i, GOR_i, WOR_i)$	$d_i, P_i(L_{ch}^+), y_i$
$y_i = \max\left(y_c, \dfrac{P_i(L_{ch}^-)}{P_i(L_{ch}^+)}\right)$	
$\dfrac{dP_i(L_{ch}^+)}{dL} = f_P\left(P_i(L_{ch}^+), q_{o,i}, q_{w,i}, q_{g,i}\right)$	$\dfrac{dP_i(L_{ch}^+)}{dL}$
Number of equations 3	Number of unknowns 4
DoF 1	

a) $q_{o,i}, q_{w,i}, q_{g,i}$ were calculated from initialization of the first stage.
b) $P_i(L_{ch}^-)$ was calculated from integration of the first stage.

Table 5.3 Initialization of the third stage.

Equations	Unknowns
$\dfrac{dP_m(L_m)}{dL} = f_P\left(P_m(L_m), q_{o,m}, q_{w,m}, q_{g,m}\right)$	$\dfrac{dP_m(L_m)}{dL}, P_m(L_m), q_{o,m}, q_{w,m}, q_{g,m}$
Number of equations 1	Number of unknowns 5
DoF 4	

Table 5.1, it is observed that the first stage of a naturally flowing well has only one degree of freedom. By selecting the well oil rate $q_{o,i}$ as degree of freedom and specifying its value, the nonlinear system of equations in Table 5.1 becomes well posed and its solution provides initial conditions for the integration of the ODE for the first stage. Therefore, the first stage of a naturally well model can be replaced by the following algebraic equation:

$$P_i\left(L_{ch}^-\right) = n(q_{o,i}), \quad \forall i \in N \tag{5.19}$$

where $n(.)$ is a functional relation, whose closed form is unknown, but whose value can be determined after initialization and integration of the corresponding ODE.

Initialization of the second stage involves the system of nonlinear equations shown in Table 5.2. By counting the number of equations and the number of unknowns, it is observed that the second stage adds one extra degree of freedom to the system. The choke diameter d_i is selected as degree of freedom since in practice this is the control variable of a well. In summary, by selecting $q_{o,i}$ and d_i as degrees of freedom, the ODEs in the first and second stage can be integrated from the well bore up to the manifold. Then, the pressure of the well at the manifold level $P_i(L_m)$ is calculated from the following algebraic equation:

$$P_i(L_m) = n(q_{o,i}, d_i) \tag{5.20}$$

The manifold pressure of a gas-lift well is given by the following relation as shown in Appendix A.

$$q_{o,i} = g\left(P_i(L_m), q_{g,i}^{inj}\right) \tag{5.21}$$

Initialization of the third stage involves the solution of the system of nonlinear equations shown in Table 5.3. By counting the number of equations and the number of unknowns, it is observed that the third stage has four degrees of freedom. Selection of fluid flow rates $q_{p,m}$ and manifold pressure $P_m(L_m)$ allows the integration of the ODEs for the third stage from the manifold to the separator. Then, the pressure of the surface flow line at the separator level is calculated from the following relation:

$$P_m(L_s) = f\left(P_m(L_m), q_{o,m}, q_{g,m}, q_{w,m}\right) \tag{5.22}$$

Taking into account the above developments, the optimization formulation for the two-well network can be written as follows:

$$\max w_o \sum_{i \in N} q_{o,i} - w_g \sum_{i \in G} q_{g,i}^{inj}$$

s.t.

$$P_m(L_s) = f\left(P_m(L_m), q_{o,m}, q_{g,m}, q_{w,m}\right), \quad \forall m \in M$$

$$q_{p,m} = \sum_{i \in I_m} q_{p,i}, \quad \forall p, m \in M$$

$$P_m(L_m) = P_i(L_m), \quad \forall m \in M, i \in I_m$$

$$P_m(L_s) = P_s, \quad \forall s \in S, m \in M_s$$

$$P_m^L \leqslant P_m(L_m), \quad \forall m \in M$$

$$\sum_{i \in G} q_{g,i}^{inj} \leqslant C$$

$$P_i(L_m) = n(q_{o,i}, d_i), \quad \forall i \in N \tag{P1}$$

$$q_{o,i} = g\left(P_i(L_m), q_{g,i}^{inj}\right), \quad \forall i \in G$$

$$q_{w,i} = WOR_i q_{o,i}, \quad \forall i \in I$$

$$q_{L,i} = q_{o,i} + q_{w,i}, \quad \forall i \in I$$

$$q_{g,i} = GOR_i q_{o,i}, \quad \forall i \in N$$

$$q_{g,i} = GOR_i q_{o,i} + q_{g,i}^{inj}, \quad \forall i \in G$$

$$q_{o,i}^L \leqslant q_{o,i} \leqslant q_{o,i}^U, \quad \forall i \in I$$

$$0 \leqslant d_i \leqslant 1, \quad \forall i \in I$$

However, formulation (P1) can cause severe computational difficulties when the choke operates in the critical flow regime, since in this case the choke upstream pressure $P_i(L_{ch}^+)$ becomes independent of the well flow rate and the nonlinear system of equations in Table 5.2 becomes infeasible. Mathematically, this implies that the Jacobian matrix of the nonlinear system of equations in Table 5.2 becomes rank deficient. A way to avoid this difficulty is to replace the choke model with a positive pressure drop. The rational behind this is the following. The choke valve is used to control the well flow rate by creating a suitable pressure drop in the well tubing. Therefore, the choke upstream pressure is the new degree of freedom that can replace the choke diameter, with this (P1) becomes

$$\max w_o \sum_{i \in N} q_{o,i} - w_g \sum_{i \in G} q_{g,i}^{inj}$$

s.t.

$$P_m(L_s) = f\left(P_m(L_m), q_{o,m}, q_{g,m}, q_{w,m}\right), \quad \forall m \in M$$

$$q_{p,m} = \sum_{i \in I_m} q_{p,i}, \quad \forall p, m \in M$$

$$P_m(L_m) = P_i(L_m), \quad \forall m \in M, \ i \in I_m$$

$$P_m(L_s) = P_s, \quad \forall s \in S, \ m \in M_s$$

$$P_m^L \leqslant P_m(L_m), \quad \forall m \in M$$

$$\sum_{i \in G} q_{g,i}^{inj} \leqslant C$$

$$P_i(L_m) = n\left(q_{o,i}, P_i(L_{ch}^+)\right), \quad \forall i \in N \qquad\qquad \text{(P2)}$$

$$P_i(L_m) = g\left(q_{o,i}, q_{g,i}^{inj}\right), \quad \forall i \in G$$

$$q_{w,i} = WOR_i q_{o,i}, \quad \forall i \in I$$

$$q_{g,i} = GOR_i q_{o,i}, \quad \forall i \in N$$

$$q_{g,i} = GOR_i q_{o,i} + q_{g,i}^{inj}, \quad \forall i \in G$$

$$q_{o,i}^L \leqslant q_{o,i} \leqslant q_{o,i}^U, \quad \forall i \in I$$

Formulation (P2) may again not be robust for the following reasons:

- when a well is weak and cannot flow given the manifold pressure selected by the optimizer, the well integration becomes infeasible and (P2) will fail to converge, since the objective function and the constraints cannot be evaluated,
- some gas-lift wells known as noninstantaneous cannot flow without the appropriate amount of gas lift. In this case (P2) will fail to converge since the well integration becomes infeasible, and
- multiple solutions exist in multiphase network problems and not all of them are stable.

The above limitations can be avoided by implementing a pre-processing step where weak, noninstantaneous gas-lift wells and unstable solutions are identified and treated appropriately prior to the solution of the optimization problem.

5.4
Separable Programming for Optimization of Production Operations

Fang and Lo [2] applied separability programming techniques to the optimization of production operations. To simplify the optimization problem, they assumed that (i) the manifold pressure is constant and known, (ii) the gas-lift wells respond instantaneously to gas lift, (iii) the GOR and WOR of each well is constant, and (iv) the multiphase flow network model has a unique solution.

Since a naturally flowing well has two degrees of freedom, by fixing the manifold pressure and setting the choke fully open the well model equations can be solved from the well bore up to the manifold. Moreover, the resulting oil flow rate is the maximum since the choke was set fully open and any oil flow rate less than the maximum can be achieved by reducing the choke diameter. In the case of gas-lift wells, by fixing the manifold pressure, the oil flow rate becomes a function of the gas injection rate as can be seen by Eq. (5.21). Figure 5.2 shows a graphical representation of the oil and gas flow rate relations for a naturally flowing and gas-lift well. Fang and Lo [2] approximated the gas oil relation for gas-lift wells with piecewise linear functions. First, the gas injection rate was discretized at j points for each well i, $q_{g,i,j}^{inj,d}$ and the oil flow rate $q_{o,i,j}^{d}$ was calculated after simulation of each well for each discrete value of the gas injection rate. Then the gas oil relation was approximated with piecewise linear functions using the following constraints [41]:

$$\left.\begin{array}{c} q_{o,i} = \sum_j \lambda_{i,j} q_{o,i,j}^{d} \\ q_{g,i}^{inj} = \sum_j \lambda_{i,j} q_{g,i,j}^{inj,d} \\ \sum_j \lambda_{i,j} = 1 \\ \lambda_{i,j} \geq 0 \\ \lambda_{i,j}, SOS \end{array}\right\} \quad \forall i \in G \tag{5.23}$$

where J is the set of discrete points j, $|J|$ is the cardinality of the set J, $\lambda_{i,j}$ are positive variables, known as a special order set (SOS), where at most two must be

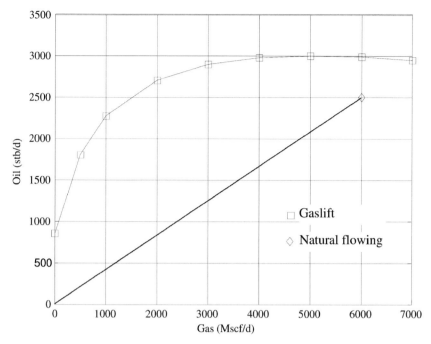

Fig. 5.2 Oil gas relation for a naturally flowing and gas lift well.

adjacent. The adjacency condition is imposed with the binary variables $y_{i,j}$ (see Appendix B).

Due to the concavity property of the gas oil relations as shown in Fig. 5.2 the adjacency condition can be automatically satisfied without using binary variables $y_{i,j}$ and the problem can be formulated as a linear programming problem [2]:

$$\max w_o \sum_{i \in I} q_{o,i} - w_g \sum_{i \in G} q_{g,i}^{inj}$$

s.t.

$$\sum_i q_{p,i} \leqslant C_p, \quad \forall p$$

$$\left.\begin{aligned}
q_{o,i} &= \sum_j \lambda_{i,j} q_{o,i,j}^d \\
q_{g,i}^{inj} &= \sum_j \lambda_{i,j} q_{g,i}^{inj,d} \\
\sum_j \lambda_{i,j} &= 1 \\
\lambda_{i,j} &\geqslant 0 \\
q_{g,i} &= GOR_i q_{o,i} + q_{g,i}^{inj}
\end{aligned}\right\} \quad \forall i \in G \qquad \left.\begin{aligned} q_{o,i} &\leqslant q_{o,i}^{max} \\ q_{g,i} &= GOR_i q_{o,i} \end{aligned}\right\} \quad \forall i \in N \quad \text{(P3)}$$

$$\left.\begin{aligned}
q_{w,i} &= WOR_i q_{o,i} \\
q_{L,i} &= q_{w,i} + q_{o,i} \\
q_{o,i}^L &\leqslant q_{o,i} \leqslant q_{o,i}^U
\end{aligned}\right\} \quad \forall i \in I$$

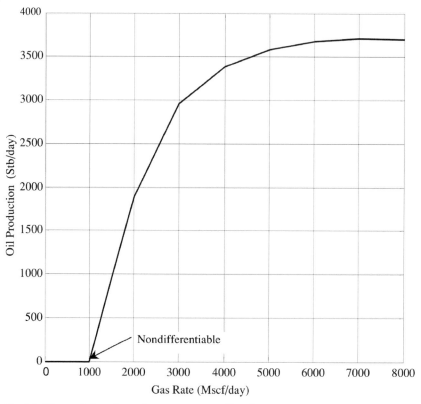

Fig. 5.3 Oil gas relation of a noninstantaneous gas lift well.

However, the oil and gas relation for a noninstantaneous gas-lift well is not concave as shown in Fig. 5.3, where the well starts flowing only after a certain amount of gas is injected. In this case, the adjacency condition must be imposed using binary variables and the formulation (P3) becomes a MILP problem which can be solved by a standard branch-and-bound method [42]. (P3) ignores the network constraints from the manifold to the surface facility, which could lead to suboptimal or even in-feasible solutions depending on the value selected for the manifold pressure. Next, we propose an optimization formulation, which extends the work of Fang and Lo [2] and overcomes the computational difficulties of formulation (P2). The algorithm is based on (i) well oil rate upper and lower bounds and (ii) approximation of each well with piecewise linear functions.

5.4.1
Well Oil Rate Upper and Lower Bounds and Piecewise Linear Approximation

Since a naturally flowing well has two degrees of freedom, by discretizing the man-ifold pressure at j points $P_{m,j}^d$ and setting the choke of each well i that is connected to the manifold m fully open, the maximum well oil rate $q_{o,i,j}^{\max,d}$ is determined for each discrete value of the manifold pressure after solution of the well model. If

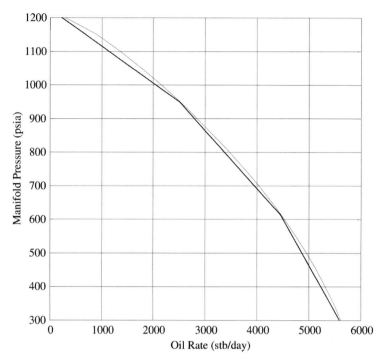

Fig. 5.4 Piecewise linear approximation of the maximum well oil rate in naturally flowing wells.

for a given value of manifold pressure $P_{m,j}^d$ the well model is infeasible, $q_{o,i,j}^{\max,d}$ is set equal to zero. The resulting flow rate is the maximum since the well was set fully open and any flow rate $q_{o,i}$ can be achieved by reducing the choke diameter. Figure 5.4 shows a typical schematic representation of the maximum well oil rate $q_{o,i}^{\max}$ as a function of the manifold pressure. The maximum well oil rate $q_{o,i}^{\max}$ can be approximated with piecewise linear functions applying the following constraints:

$$
\left.
\begin{aligned}
q_{o,i}^{\max} &= \sum_j \lambda_{m,j} q_{o,i,j}^{\max,d} \\
P_m &= \sum_j \lambda_{m,j} P_{m,j}^d \\
P_m^L &\leqslant P_m \leqslant P_m^U \\
q_{o,i} &\leqslant q_{o,i}^{\max} \\
\sum_j \lambda_{m,j} &= 1 \\
\lambda_{m,j}, &\ \text{SOS}
\end{aligned}
\right\} \quad \forall m, \ \forall i \in I_m
\tag{5.24}
$$

Constraint (5.24) sets an upper bound on the well flow rate for any manifold pressure. Next, well oil rate lower bounds for stable flow are determined. Lea and Tighe [43] proposed the following criterion for stability:

$$
\frac{\partial P_i(L_o)}{\partial q_{o,i}} \geqslant 0, \quad \forall i \in I
\tag{5.25}
$$

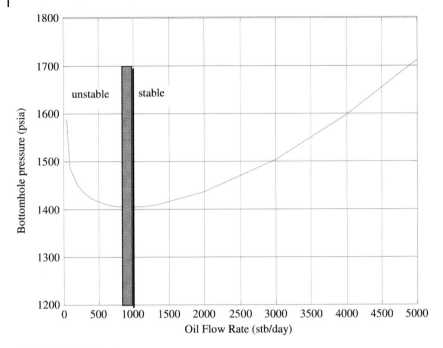

Fig. 5.5 Well stable region.

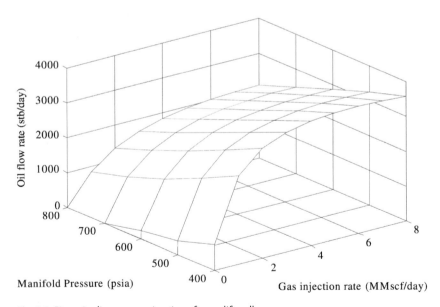

Fig. 5.6 Piecewise linear approximation of a gas lift well.

Condition (5.25) states that the partial derivative of the bottom hole pressure of the well tubing with respect to the well oil flow rate must be nonnegative. A typical form of the bottom hole pressure as a function of the well oil rate is shown in Fig. 5.5. The stability condition can be transformed to the following optimization problem whose solution determines the minimum well oil rate $q_{o,i}^L$ for stable production:

$$\min_{q_{o,i}} q_{o,i}$$

s.t.

$$\left.\begin{array}{c} \frac{dP_i}{dL} = f(P_i, q_{o,i}, q_{w,i}, q_{g,i}), \ \forall L \in [L_0, L_{ch}) \\ q_{g,i} = GOR_i q_{o,i} \\ q_{w,i} = WOR_i q_{o,i} \\ P_i^L \leqslant P_i(L_{ch}) \leqslant P_i^U \end{array}\right\} \ \forall i \in I \qquad \text{(PM)}$$

(PM) can be solved with standard nonlinear programming techniques after discretization of the ODE. Since a gas-lift well i has two degrees of freedom, by discretizing the manifold pressure at j points $P_{m,j}^d$ and the gas injection rate of each well at k points $q_{g,i,k}^{inj,d}$ the well oil flow rate $q_{o,i,j,k}^d$ can be calculated by solving the well model equations for each grid point (j, k). If the simulation of the gas-lift well is infeasible then the corresponding well oil rate $q_{o,i,j,k}^d$ is set equal to zero. Therefore, the well oil flow rate $q_{o,i}$ for each gas-lift well can be approximated by the following piecewise linear relations:

$$\left.\begin{array}{c} q_{o,i} = \sum_j \sum_k \mu_{i,j,k} q_{o,i,j,k}^d \\ q_{g,i}^{inj} = \sum_j \sum_k \mu_{i,j,k} q_{g,i,j,k}^{inj,d} \\ P_m = \sum_j \sum_k \mu_{i,j,k} P_{m,j}^d \\ \sum_j \sum_k \mu_{i,j,k} = 1 \\ \eta_{i,j} = \sum_k \mu_{i,j,k} \\ \xi_{i,k} = \sum_j \mu_{i,j,k} \\ \zeta_{i,t} = \sum_j \mu_{i,j,j+t} \\ \eta_{i,j}, \xi_{i,k}, \zeta_{i,t}, \ SOS \end{array}\right\} \ \forall m, \ \forall i \in I_m \qquad (5.26)$$

Figure 5.6 depicts the oil flow rate as a function of the manifold pressure and gas injection rate for a gas-lift well. Taking into account the above developments the mathematical programming formulation for optimization of production operations can be written as follows:

$$\max w_o \sum_{i \in N} q_{o,i} - w_g \sum_{i \in G} q_{g,i}^{inj}$$

s.t.

$$P_m(L_s) = f\left(P_m(L_m), q_{o,m}, q_{g,m}, q_{w,m}\right), \quad \forall m$$

$$q_{p,m} = \sum_{i \in I_m} q_{p,i}, \quad \forall p, m$$

$$q_{p,m} \leqslant C_{p,m}, \quad \forall p, m$$

$$P_m(L_s) = P_s, \quad \forall s, m$$

$$P_m^L \leqslant P_m(L_m), \quad \forall m$$

$$\sum_{i \in G} q_{g,i}^{inj} \leqslant C \tag{P4}$$

$$\left.\begin{array}{l} q_{o,i}^{max} = \sum_j \lambda_{m,j} q_{o,i,j}^{max,d} \\ P_m = \sum_j \lambda_{m,j} P_{m,j}^d \\ q_{o,i} \leqslant q_{o,i}^{max} \\ \sum_j \lambda_{m,j} = 1 \\ \lambda_{m,j}, \; SOS \\ q_{o,i} \leqslant q_{o,i}^{max} \\ q_{g,i} = GOR_i q_{o,i} \end{array}\right\} \quad \forall m, \; \forall i \in N$$

$$\left.\begin{array}{l} q_{o,i} = \sum_j \sum_k \mu_{i,j,k} q_{o,i,j,k}^d \\ q_{g,i}^{inj} = \sum_j \sum_k \mu_{i,j,k} q_{g,i,j,k}^{inj,d} \\ P_m = \sum_j \sum_k \mu_{i,j,k} P_{m,j}^d \\ \sum_j \sum_k \mu_{i,j,k} = 1 \\ \eta_{i,j} = \sum_k \mu_{i,j,k} \\ \xi_{i,k} = \sum_j \mu_{i,j,k} \\ \zeta_{i,t} = \sum_j \mu_{i,j,j+t} \\ \eta_{i,j}, \xi_{i,k}, \zeta_{i,t}, \; SOS \end{array}\right\} \quad \forall m, \; \forall i \in G$$

$$\left.\begin{array}{l} q_{w,i} = WOR_i q_{o,i} \\ q_{L,i} = q_{o,i} + q_{w,i} \\ q_{o,i}^L \leqslant q_{o,i} \leqslant q_{o,i}^U \end{array}\right\} \quad \forall i \in I$$

(P4) involves nonlinear equations and 0–1 binary variables; therefore, it belongs to the class of MINLP [44] problems. Moreover, it has two characteristics (i) the number of nonlinear equality constraints is equal to the number of surface flow lines and independent of the number of wells, and (ii) the binary variables are used to linearize nonlinear constraints. The most efficient way to solve (P4) is a sequence of MILP problems following a successive linear programming (SLP) method.

5.5
Solution Procedure

SLP methods solve nonlinear optimization programming problems via a sequence of linear programming problems. Since their introduction by Griffith and Stewart [45], many variants of SLP algorithms have appeared in the literature [46–48]. In this work, the penalty successive linear-programming algorithm of Zhang et al. [47] was applied. The formulation (P4) can be written in compact form as follows (P5):

$$\max f(x) \tag{5.27}$$

s.t.

$$h(x) = 0 \tag{5.28}$$

$$Ax + By \leqslant c \tag{5.29}$$

$$x^L \leqslant x \leqslant x^U \tag{5.30}$$

where x is the vector of continuous variables, y is the vector of binary variables, $f(x)$ is the linear objective function of (P4), $h(x)$ is the nonlinear equality constraints of (P4) which represent the surface flow line model and constraint (5.29) represents the mixed-integer linear constraints which approximate the well models. The algorithm of Zhang et al. [47] involves the following steps:

1. *Initialization.* Set iteration counter $l = 1$ and select a starting point x_l after deleting the nonlinear constraints ($h(x)$) in (P5) and solving the corresponding MILP problem. Select a large positive weight ϕ and scalars $0 < \rho_0 < \rho_1 < \rho_2 < 1$ (typically $\rho_0 = 10^{-6}$, $\rho_1 = 0.25$, $\rho_2 = 0.75$). Let Δ_l be a bound vector on continuous variables and $\Delta_{LB} > 0$ its lower bound.
2. *MILP subproblem.* Solve the following MILP to obtain a new point x

$$\max f(x) - \phi\left[\sum_r (z_r^+ - z_r^-)\right]$$

s.t.

$$z_r^+ - z_r^- = h_r(x_l) + \nabla h_r(x_l)^t(x - x_l)$$

$$Ax + By \leqslant c \tag{P6}$$
$$x^L \leqslant x \leqslant x^U$$

$$-\Delta_l \leqslant x - x_l \leqslant \Delta_l$$

$$z_r^+, z_r^- \geqslant 0$$

where r is the number of nonlinear equality constraints in (P4).

3. *Trust region.* Calculate Trust region. Calculate the quantities $\Delta F_{E,l}$ and $\Delta F_{EL,l}$

$$\Delta F_{E,l} = F_E(x_l) - F_E(x) \tag{5.31}$$

$$\Delta F_{EL,l} = F_E(x_l) - F_{EL}(x) \tag{5.32}$$

where

$$F_E(x_l) = f(x_l) + \phi \left[\sum_r |h_r(x_l)| \right] \tag{5.33}$$

$$F_{EL}(x) = f(x) + \phi \sum_r |h_r(x_l) + \nabla h_r(x_l)^t (x - x_l)| \tag{5.34}$$

If $\Delta F_{EL,l} = 0$ or

$$|f(x) - f(x_l)| < \varepsilon(1 + |f(x_l)|) \tag{5.35}$$

stop: the optimum is found. Otherwise compute the ratio $R_l = \frac{\Delta F_{E,l}}{\Delta F_{EL,l}}$

- If $R_l < \rho_0$ reject the current solution x and shrink Δ_{l+1} to $0.5\Delta_l$
- If $\rho_0 < R_l < \rho_1$ accept the current solution x, shrink Δ_{l+1} to $0.5\Delta_l$, set counter $l = l + 1$
- If $\rho_1 < R_l < \rho_2$ accept the current solution x, set counter $l = l + 1$
- If $R_l > \rho_2$ accept the current solution x, $\Delta_{l+1} = 2\Delta_l$ and set counter $l = l + 1$

  go to step 2

Remarks

1. The MILP sub problems were modeled in CPLEX.
2. To find a near global optimal solution the relative optimality criterion of MILPs was set 0.0001%, while the convergence criterion for the SLP, which is defined by Eq. (5.35), was set equal to $\varepsilon = 10^{-3}$.
3. The adjacency condition can also be enforced in a Simplex type LP solver[37] with restricted basis entry rule, which allows only adjacent SOS variables to enter the Simplex tableau.
4. A similar algorithm was proposed by Bullard and Biegler [48] but in the context of solving nonsmooth nonlinear system of equations, where the nonsmooth equations were handled using binary variables and the nonlinear system of equations was solved with an SLP method.

The solution procedure for the optimization of production operations involves the following steps:

- Preprocessing step, where the reservoir block pressure $P_{R,i}$, the productivity index PI_i, the GOR_i and the WOR_i of each well are extracted from a reservoir simulator and the manifold pressure and the gas injection rate are discretized. For each discrete point the well model is solved and the gas/oil relations are approximated with piecewise linear functions.

- Processing step, which involves the solution of (P4) as described previously.
- Postprocessing step, which involves the determination of each well choke setting. The well choke setting is determined by fixing the manifold pressure, the well oil rate and gas injection rate at the values calculated from step (ii) and solving the corresponding well model.

Remark

Very often the behavior of models such as momentum balances in well tubing and well flow lines, chokes, pumps in the petroleum industry are given in the form of discrete performance tables [39] rather than analytical functions. In this case, instead of simulating each well, piecewise linear approximations can be constructed by extracting suitable data from the discrete tables, which have been constructed by a simulation tool and are valid throughout the life of the field.

5.6
Accuracy and Robustness of the Formulation

The accuracy and robustness of the optimization formulation (P4) has been investigated by comparing it with the exact optimization formulation (P2) in two example problems. The first example involves the two-well network shown in Fig. 5.1, where the reservoir fluid is a dry gas. The corresponding exact optimization formulation is given in Appendix C. The reservoir block pressure, the productivity index, the well design parameters, separator pressure, and capacity are summarized in Table 5.3. The proposed algorithm was applied. First, the manifold pressure was discretized from 1800 to 2300 psia every 100 psia and the corresponding maximum gas well rates are summarized in Table 5.5. The objective function was maximization of gas production and the resulting MINLP problem involved one nonlinear equation since the system has one flow line, and five binary variables. The MILP subproblems were solved using CPLEX and the optimality criterion was set equal to 0.0001%. Finally, the well choke settings were calculated by simulating each well after fixing the manifold pressure and the well gas rate to the optimal values. The choke settings are presented in Table 5.6 where it is observed that both wells have their chokes fully open. Then the exact optimization formulation (P2) was solved using the SQP solver of MATLAB [50]. The values of the objective function, manifold pressure, and gas well rate calculated from application of both methods are summarized in Table 5.7. The results suggest that the proposed method is very accurate. In Table 5.8, the computational performance of both methods in terms of number of iterations and function evaluations are reported. The results suggest that the SQP method requires more iterations and function evaluations compared to the proposed method. This is due to the fact that linearization of highly nonlin-

Table 5.4 Reservoir, well, and separator data for the two-well dry gas network.

	Well 1	Well 2
Reservoir pressure (psia)	3660	3000
α, b coefficients	793, 0.004	700, 0.003
Well vertical length (ft)	5000	5000
Well horizontal length (ft)	2000	2000
Flow line length (vertical) (ft)	2000	
	Pressure (psia)	Capacity (lbm/s)
Separator	1500	10

Table 5.5 Interpolation data for the two-well dry gas network.

Manifold pressure (psia)	Well 1 (lbm/s)	Well 2 (lbm/s)
1800	4.97	2.80
1900	4.65	2.38
2000	4.30	1.92
2100	3.92	1.43
2200	3.52	0.91
2300	3.10	0.35

Table 5.6 Choke settings for the two-well dry gas network.

	Choke settings (% open)
Well 1	100%
Well 2	100%

Table 5.7 Optimal values of both methods in the two-well dry gas network.

	Approximate optimization	Exact optimization
Objective value	6.04	6.04
Manifold pressure (psia)	2017.4	2017.1
Well 1 flow rate (lbm/sec)	4.22	4.23
Well 2 flow rate (lbm/sec)	1.82	1.84

ear equations is likely to be poor and even relatively close to the linearization point. This leads to successive solutions being very close together which potentially slows the convergence. The case is depicted in Fig. 5.7, where the manifold pressure is linearized. In addition, the increased number of function evaluations can be explained by the fact that the exact optimization formulation has three nonlinear constraints while the approximation has only one.

Table 5.8 Computational performances of both methods in the two-well dry gas network.

	FE	IT
Exact optimization	180	30
Proposed algorithm	12	4

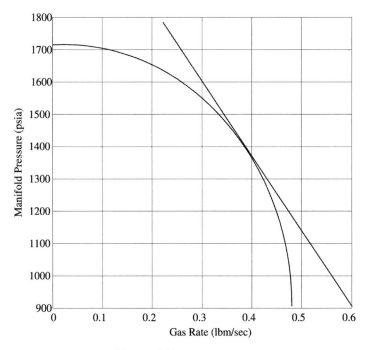

Fig. 5.7 Linearization of the manifold pressure.

The second example involves the pipeline network shown in Fig. 5.8. The network consists of four dry gas wells, four flow lines and the gas deliverability pressure was set equal to 1300 psia. The reservoir and well data are summarized in Table 5.9. The objective function was maximization of gas production and the resulting MINLP problem involved four nonlinear equality constraints, 20 binary variables. The optimality condition for the MILP subproblems was set equal to 0.0001% and was solved with CPLEX. The proposed method converged to the optimal solution after five iterations. The optimal point is presented in Table 5.10, where it is observed that well 4 is closed. The exact optimization formulation failed to converge since well 4 could not flow to the manifold pressure selected by the optimizer. It must be noted that the accuracy of the method has been shown in number of examples including naturally flowing and gas-lift wells.

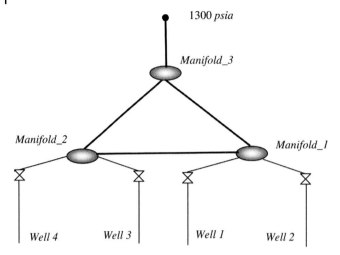

1300 *psia*

Manifold_3

Manifold_2

Manifold_1

Well 4   Well 3   Well 1   Well 2

Fig. 5.8 A four-well dry gas model.

Table 5.9 Reservoir, well, and separator data for the four-well dry gas model.

	Well 1	Well 2	Well 3	Well 4	Flow lines
Reservoir pressure (psia)	3660	3000	2500	2300	
α, b coefficients	793, 0.004	700, 0.003	650, 0.005	600, 0003	
Well vertical length (ft)	5000	5000	5000	5000	
Well horizontal length (ft)	2000	2000	2000	2000	8000
Diameter (in.)	2.5	2.5			
	Pressure (psia)				
Separator	1300				

Table 5.10 Optimal values of the four-well dry gas model.

	Well flow rate (lbm/s)		Pressure (psia)
Well 1	4.680	Manifold 1	1922.1
Well 2	2.446	Manifold 2	1817.5
Well 3	0.982	Manifold 3	1526.9
Well 4	0.0	Manifold 4	1300.0

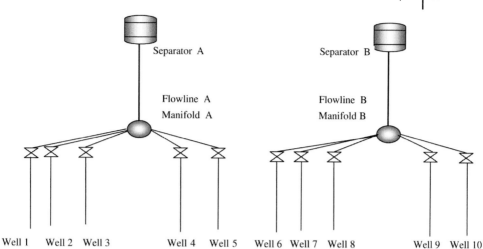

Fig. 5.9 Naturally flowing well example.

5.7
Examples

5.7.1
Naturally Flowing Wells

A field consisting of ten naturally flowing wells and two separators is analyzed. The pipeline network is depicted in Fig. 5.9 and the reservoir and well data are given in Table 5.11, while the operating pressure and capacity of each separator is given in Table 5.11. The objective function is maximization of oil production and the MILP subproblem involves 20 binary variables, 942 constraints, and 535 continuous variables. The problem is solved using CPLEX with optimality criterion equal to 0.0001%. The optimal fluid flow rates in the separators are given in Table 5.12. The results of Table 5.11 suggest that the gas capacity of separator A is active and the wells with the higher GOR are choked back as can be observed in Table 5.13 where the GOR and WOR of each well and the corresponding optimal choke settings are reported. Similarly, the water capacity of separator B is the bottleneck of the network as shown by Table 5.12. The wells with the highest WOR are operated with choke restrictions as can be observed from the results summarized in Table 5.13. Then the LP method proposed by Wang et al. [18] was applied to solve the naturally flowing well network. Their algorithm consists of the following steps:

- the well chokes are set fully open and the corresponding network is simulated without considering the separator capacity limits.
- The oil flow rates calculated from the previous step become upper bounds which along with separator handling limits and mass balances on the node constitute the LP formulation whose solution determines the optimal operating policy.

Table 5.11 Reservoir and well data for the naturally flowing wells example.

	Reservoir pressure (psia)	PI (stb/psia)	Horizontal (ft)	Vertical (ft)	Diameter (in.)
Well 1	1720	8.0	2000	3400	4.0
Well 2	1700	8.0	2000	3000	3.5
Well 3	1800	6.0	2000	3000	3.5
Well 4	2000	6.0	2500	4000	4.0
Well 5	2100	10.0	2500	4000	4.0
Well 6	1200	7.0	2500	4000	3.5
Well 7	1100	6.9	2500	4000	4.0
Well 8	1000	6.0	2000	3000	3.5
Well 9	1000	5.0	2000	3000	4.0
Well 10	1100	6.0	2500	4000	4.0
Flowline A				5000	5.0
Flowline B				5000	5.0

Table 5.12 Separator data and optimal separator flow rates.

	Sep_ A Capacity	Approx. method	LP method	Sep_ B Capacity	Optimal	Flow rate
Oil (barrel/day)	15000	12047.4	11989.1	10000	6738.6	6702.4
Water (barrel/day)	5000	2629.3	2446.8	2000	2000	2000
Gas (MMscf/day)	25000	25000	25000	18000	8561.8	8658
Pressure (psia)	600		300			

Table 5.13 Reservoir and optimal well choke settings of the naturally flowing well example.

	GOR (MMscf/stb)	WOR (stb/stb)	Choke settings (% open)
Well 1	2000	0.3	100
Well 2	2500	0.1	27.1
Well 3	2000	0.7	68.3
Well 4	3000	0.3	11.4
Well 5	2000	0.01	100
Well 6	1000	0.01	100
Well 7	1500	0.5	49.5
Well 8	2000	0.5	29.4
Well 9	1000	0.3	100
Well 10	1500	0.5	30.0

The results of the LP method are summarized in Table 5.12. The results of Table 5.12 suggest that the proposed method produces 94 more barrel of oil per day compared to the LP method. The reason is that setting the well chokes fully open does maximize oil production.

Table 5.14 Reservoir and well data for gas-lift well example.

	Reservoir pressure (psia)	PI (stb/psia)	Horizontal (ft)	Vertical (ft)	Diameter (in.)
Well 1	1920	13	1000	1000	4
Well 2	1900	9	1000	2000	4
Well 3	1800	6	2500	2700	4
Well 4	1720	10	3000	2000	4
Well 5	1600	5	2500	2000	4
Well 6	1820	3	2500	2500	4
Well 7	1420	4	2000	2000	4
Well 8	1320	2	2000	3000	4
Well 9	1450	3.5	2000	3000	4
Well 10	1550	5.5	1500	2000	4
Well 11	1720	10	2000	4000	4
Well 12	1720	10	2000	1500	4
Well 13	1520	8	3000	5000	4
Flowline A				5000	6.0
Flowline B				5000	6.0
	Pres (psia)	Oil capacity (stb/day)	Gas capacity (Mscf/day)	Water capacity (stb/day)	
Separator A	300	15 000	25 000	6000	
Separator B	300	15 000	15 000	4000	

Table 5.15 Comparison of the approximate and the *IGOR* heuristic in the gas lift example.

	Approx. method		IGOR method	
	Gas lift rate (Mscf/day)	Oil rate (stb/day)	Gas lift rate (Mscf/day)	Oil rate (stb/day)
Well 1	4000	2833.7	2500	2750.2
Well 2	3000	2651.1	3000	2734.0
Well 3	3000	1539.7	2530	1463.2
Well 4	1966	1979.0	2000	1987.2
Well 5	2750	1535.33	2500	1550.2
Well 6	631	948.4	1200	1153.1
Well 7	1500	869.6	1330	761.9
Well 8	750	353.2	650	240.87
Well 9	500	485.1	1082	710.9
Well 10	2250	1584.7	2082	1486.4
Well 11	663	1759.3	430	1613.2
Well 12	2250	2971.7	1910	2792.6
Well 13	500	393.6	560	238.8
Total		19904.4		19482.6

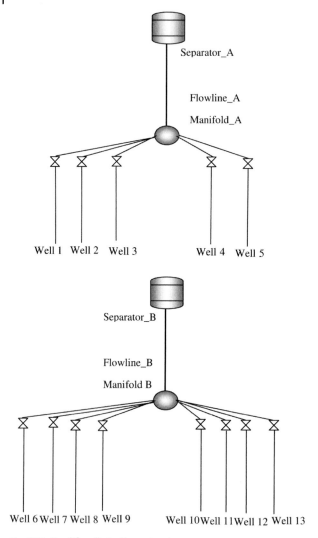

Fig. 5.10 Gas lift well pipeline network.

5.7.2
Gas-Lift Well Example

A field consisting of 13 gas-lift wells and two separators was analyzed. The pipeline network is depicted in Fig. 5.10 and the reservoir, well, and separator data are given in Table 5.14. The gas–oil relation for each gas-lift well was constructed by discretizing the manifold pressure between 300 to 800 psia every 100 psia and the gas injection rate at 0, 250, 500, 1000, 2000, 3000, 4000, 5000, 6000, 7000, and 8000 Mscf/day. The objective function was maximization of profit (Eq. (5.18)) where the price of the oil was 10$/barrel and the cost of gas compression was 0.1$/MMscf.

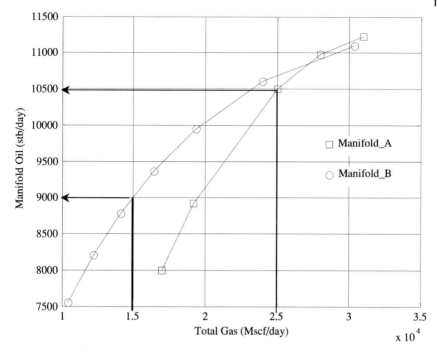

Fig. 5.11 Manifolds *IGOR* curves.

The MILP subproblem involved 348 binary variables, 912 continuous variables, and 1160 constraints and the optimality criterion was set equal to 0.0001%. The algorithm converges to the optimal solution after five iterations. The optimal well oil rate and gas injection rate for each well are summarized in Table 5.15. The results indicate that the field will produce a total oil rate of 19904.4 stb/day.

Then the problem is solved applying the *IGOR* heuristic rule as follows:

- A realistic manifold pressure is selected and the *IGOR* curve for each manifold is constructed. The pressure of each manifold was set equal to 575 psia and the manifold *IGOR* curve was constructed as follows. First the gas oil relation for each gas-lift well is constructed then by varying the value of *IGOR*, the manifold oil rate as a function of gas flow rate is plotted as shown in Fig. 5.11.
- The optimal *IGOR* is the one that maximizes oil production while satisfying separator gas capacity constraints. From Fig. 5.11 the optimal *IGOR* is calculated.
- From the optimal value of the *IGOR* the gas lift rates are calculated by fixing the *IGOR* of each well and the manifold pressure.
- Using the gas lift rates calculated from the previous step the network is simulated and the optimal well oil rates are determined.

The well oil rates resulting from application of the *IGOR* heuristic rule are summarized in Table 5.14, where a difference of 422 barrels of oil per day is observed in favor of the proposed method. The result can be explained by the fact that the

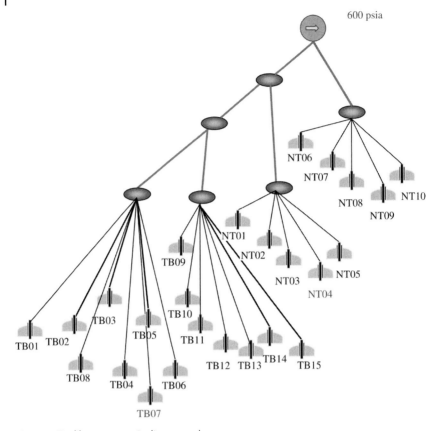

Fig. 5.12 Treelike structure pipeline network.

IGOR curves are constructed neglecting the interactions between wells sharing a common flow line since the manifold pressure was fixed to a rather arbitrary value.

5.7.3
Treelike Structure Pipeline Network

To evaluate the computational performance of the algorithm, the pipeline network shown in Fig. 5.12 was solved using the proposed method. The network consists of 25 wells where 15 of them are gas-lift wells (*TB01*, ..., *TB15*) and the remaining 10 wells are naturally flowing (*NT01*, ..., *NT10*). The delivery pressure was set equal to 600 psia. Two scenarios of the problem were solved. In the first scenario, there is an upper bound on the total gas production, while in the second there is an upper bound in the total water production. The objective in both cases was maximization of oil production and unlimited gas lift availability was assumed. The problem involved 612 binary variables, 1660 continuous variables, and 2133 constraints. The total oil production, the number of iterations, and the CPU time on an ultra 260 MHz are reported in Table 5.16. It must be noted that the cost of

Table 5.16 Computational statistics of treelike structure pipeline network.

Constraint	Oil production (stb/day)	Iterations	CPU (s)
$q_g < 110\ \text{MMscf/day}$	55 428	7	330
$q_w < 10\,000\ \text{stb/day}$	50 226	5	282

solving each MILP subproblem was between 3 and 5 sec while the rest of the time was consumed in function evaluations of the pipeline network. In addition, as can be seen from the results of the Table 5.16, the number of iteration required by the algorithm to converge was in the same order of magnitude as for the previous examples.

5.8
Conclusions

In this chapter, a formulation has been presented for the optimization of oil and gas production operations. The formulation simultaneously optimizes well rates and gas lift allocation, is able to handle flow interactions among wells, and can be applied to difficult situations where some wells are too weak to flow to the manifold or require a certain amount of gas lift to flow. The accuracy, efficiency, and robustness of the formulation has been established by comparison with an exact optimization formulation that was solved using an SQP method in a number of examples. The algorithm is applicable both to treelike structure pipeline networks and to pipeline networks with loops. Due to the ability of the algorithm to account for flow interactions it will always propose superior operating policies compared to the heuristic rules typically applied in practice. The proposed optimization method can be used for real-time production control since all the variables required for the construction of a well model can be measured and the discrete data can be directly incorporated in the formulation, a subject currently under further investigation.

Appendix A: Derivation of the Constraint (5.21)

Consistent initialization of the first stage of a gas-lift well involves the following system of nonlinear equations:

$$q_{o,i} = PI_i\left(P_{R,i} - P_i(L_0)\right) \tag{A.1}$$

$$q_{g,i} = GOR_i q_{o,i} + q_{g,i}^{inj} \tag{A.2}$$

$$q_{w,i} = WOR_i q_{o,i} \tag{A.3}$$

$$q_{L,i} = q_{o,i} + q_{w,i} \tag{A.4}$$

$$\frac{dP_i(L_o)}{dL} = f_P(P_i(L_0), q_{o,i}, q_{w,i}, q_{g,i}) \tag{A.5}$$

where PI_i, GOR_i, WOR_i are constant and known from a reservoir simulator. The unknowns of the above system are $q_{o,i}$, $P_i(L_0)$, $q_{g,i}^{inj}$, $q_{w,i}$, $q_{g,i}$, $q_{L,i}$, $\frac{dP_i(L_0)}{dL}$. Therefore, the first stage of a gas-lift well has two degrees of freedom. Consistent initialization of the second stage involves the solution of the following system of nonlinear equations:

$$q_{L,i} = f_c\left(d_i, P_i(L_{ch}^-), P_i(L_{ch}^+), y_i, GOR_i, WOR_i\right) \tag{A.6}$$

$$y_i = \max\left(y_c, \frac{P_i(L_{ch}^+)}{P_i(L_{ch}^-)}\right) \tag{A.7}$$

$$\frac{dP_i(L_{ch}^+)}{dL} = f_P\left(P_i(L_{ch}^+), q_{o,i}, q_{w,i}, q_{g,i}\right) \tag{A.8}$$

The unknowns of the system (A.6)–(A.8) are $P_i(L_{ch}^+)$, y_i, $\frac{dP_i(L_{ch}^+)}{dL}$. It must be noted that the choke is always fully open ($d_i = 1$), since reducing the choke diameter in a gas-lift well leads to an increased gas lift consumption for the same production of oil (Chia, Hussain, 1999). Actually, in a gas-lift well, the gas injection rate is used to control the well oil rate. In summary, a gas-lift well model has two degrees of freedom. One of them is the gas injection rate and the other is the manifold pressure in order to account for the interactions between wells sharing a common flowline. Therefore, the oil flow rate of a gas-lift well is given by

$$q_{o,i} = f\left(P_i(L_m), q_{g,i}^{inj}\right), \quad \forall i \in G \tag{A.9}$$

Appendix B:

To impose the adjacency condition in the SOS variables the following binary variables are introduced:

$$\lambda_{i,1} \leqslant y_{i,1} \tag{A.10}$$

$$\lambda_{i,j} \leqslant y_{i,j-1} + y_{i,j}, \quad j = 2, \dots, |J| \tag{A.11}$$

$$\lambda_{i,J} \leqslant y_{i,J-1} \tag{A.12}$$

$$\sum_j y_{i,j} = 1 \tag{A.13}$$

where $y_{i,j}$ are binary variables.

Appendix C: The Dry Gas Exact Optimization Formulation

The dry gas exact optimization formulation is as follows:

$$\max q_{g,m}$$

s.t.

$$P_m(L_s) = f(P_m, q_{g,m})$$

$$P_m(L_s) = P_s$$

$$\sum_i q_{g,i} = q_{g,m}$$

$$q_{g,m} \leqslant C$$

$$P_i(L_m) = g\left(q_{g,i}, P_i(L_{ch}^+)\right), \quad \forall i$$

$$P_{R,i}^2 - P_i^2(L_0) = a_i q_{g,i}^2 + b_i q_{g,i}, \quad \forall i$$

$$0 \leqslant q_{g,i} \leqslant q_{g,i}^U, \quad \forall i$$

where $q_{g,i}$ is the well gas flow rate and a_i, b_i are known as reservoir deliverability coefficients [51].

Nomenclature

$I = $ set of wells i
$N = $ set of naturally flowing wells i
$G = $ set of gas-lift wells i
$M = $ set of manifold m
$S = $ set of separators s
$d_i = $ diameter of choke i
$GOR_i = $ gas-oil ratio of well i
$q_{p,i} = $ flow rate of phase p in stock tank conditions
$q_{g,i}^{inj} = $ gas injection rate of gas-lift well i
$P_{R,i} = $ reservoir block pressure of well i
$P_i(L_0) = $ well bore pressure of well i
$P_m(L_m) = $ pressure of the manifold m
$P_s = $ separator pressure
$PI_i = $ productivity index of well i
$WOR_i = $ water-oil ratio
$y_i = $ pressure ratio of choke i

y_c = critical pressure ratio
w_o, w_g = weighting coefficient

Units

stb/day = barrels per day.
scf/day = standard cubic feet per day.

References

1 KANU, E. P., MACH, J., BROWN, K. B., Economic approach to oil production and gas allocation in continuous gas lift. *Journal of Petroleum Technology* 10 (**1981**), pp. 1887–1892.

2 FANG, W. Y., LO, K. K., A generalized well-management scheme for reservoir simulation. *SPE Reservoir Engineering* 5 (**1996**), pp. 116–120.

3 LITVAK, M. L., CLARK, A. J., FAIRCHILD, J. W., FOSSUM, M. P., MACDONALD, C. D., WOOD, A. R. O., Integration of Prudhoe bay surface pipeline network and full field reservoir models. *SPE 38885, SPE Annual Technical Conference and Exhibition, Texas, USA*, **1997**, *October 5–8*.

4 BEGGS, H. D., *Production Optimization*, OGCI Publications, Tulsa, **1991**.

5 STACKEL, A. W., BROWN, H. M., Predictive well management in reservoir simulation – a case study. SPE 7698, Presented at the 8th Symposium on Reservoir Simulation, Denver, Colorado, **1979**, February 1–2.

6 WALLACE, D. J., VAN SPRONSEN, E., A reservoir simulation model with platform production/injection constraints for development planning of volatile oil reservoirs. SPE 12261, Presented at the 58th Annual Technical Conference and Exhibition, San Francisco, CA, **1983**, October 5–8.

7 WEISS, J. L., MASINO, W. H., STARLEY, G. P., BOLLING, J. D., Large scale facility expansion evaluation at kuparuk river field. SPE 20046, Presented at the 60th California Regional Meeting, Ventura, CA, **1990**, April 4–6.

8 MATTAX, C. C., *Reservoir Simulation*. *SPE Monograph, Henry L. Doherty Series*, Richardson, TX, **1990**.

9 GeoQuest, *Eclipse 300 Technical Presentation 2000A*, GeoQuest, Schlumberger, **2000**.

10 Landmark, *VIP – Executive Technical Manual*, Landmark Graphics Corporation: **2001**.

11 REDDEN, J. D., SHERMAN, G. T., BLANN, J. R., Optimizing gas lift system. SPE 5150, Presented at 49th Annual Fall Meeting of the SPE of AIME, Houston, Texas, **1974**, October 6–9.

12 BARNES, D. A., HUMPHREY, K., MUELLENBERG, L., A production optimization system for Western Prudhoe bay field, Alaska. SPE 20653, Presented at the 65th Annual Technical Conference and Exhibition, New Orleans, LA, **1990**, September 23–26.

13 STOISITS, R. F., SCHERER, P. W., SCHMIDT, S. E., Gas optimization at the Kuparuk river field. SPE 28467, Presented at the 69th Annual Technical Conference and Exhibition, New Orleans, LA, 1994, September 25–28.

14 CARROLL, J. A. III, HORNE, R. N., Multivariate optimization of production system. *Journal of Petroleum Technology* 7 (**1992**), pp. 782–831.

15 FUJII H., HORNE, R., Multivariate optimization of networked production system. *SPE Production & Facilities* 8 (**1995**), pp. 165–171.

16 MARTINEZ, E. R., MORENO, J. W., MORENO, J. A., MAGGIOLO, R., Application of genetic algorithms on the distribution of gas lift injection. SPE 26993, Presented at the 69th Annual Technical Conference and Exhibition, New Orleans, LA, **1994**, September 25–28.

17 PALKE, M. R., HORNE, R. N., Nonlinear optimization of well production considering gas lift and phase behaviour. SPE 37428, Presented at the SPE Production Operations Symposium, Oklahoma City, Oklahoma, **1997**, March, 9–11.

18 WANG, P., LITVAK, M. L., AZIZ, K., Optimization production from mature fields. Presented at the 17th World Petroleum Congress, Rio De Janerio, Brazil, **2002**, September, 1–5.

19 DUTTA-ROY, K., KATTAPURAM, J., A new approach to gas lift allocation optimization. SPE 38333, Presented at the SPE Western Regional Meeting, Long Beach, California, **1997**, June 25–27.

20 HANDLEY-SCHACHLER, S., MCKIE, C., QUINTERO, N., A new mathematical technique for the optimization of oil & gas production systems. SPE 65161, Presented at the 2000 European Petroleum Conference, Paris, France, **2000**, October 24–25.

21 LO, K. K., STARLEY, G. P., HOLDEN, C. W., Application of linear programming to reservoir development evaluations. *SPE Reservoir Engineering* 2 (**1995**), pp. 52–58.

22 VAN DEN HEEVER, S. A., GROSSMANN, I. E., *Ind. Eng. Chem. Res.* 39 (**2000**), p. 1955.

23 VAN DEN HEEVER, S. A., GROSSMANN, I. E., VASANTHARAJAN, S., EDWARDS, K., *Comp. Chem. Eng.* 24 (**2000**), p. 1049.

24 VAN DEN HEEVER, S. A., GROSSMANN, I. E., VASANTHARAJAN, S., EDWARDS, K., *Ind. Eng. Chem. Res.* 40 (**2001**), p. 2857.

25 IERAPETRITOU, M. G., FLOUDAS, C. A., VASANTHARAJAN, S., CULLICK, A. S., *AIChE Journal* 45 (**1999**), p. 844.

26 ALARCON, G. A., TORRES, C. F., GOMEZ, L. E., Global optimization of gas allocation to a group of wells in artificial lift using nonlinear constrained programming. *Transactions of the ASME* 268 (**2002**), pp. 262–268.

27 CAMPONOGARA, E., NAKASHIMA, P. H. R., Optimal allocation of lift-gas rates under multiple facility constraints: a mixed-integer linear programming approach. *Transactions of the ASME* 128 (**2006**), pp. 280–289.

28 BANGERTH, W., KLIE, H., WHEELER, M. F., STOFFA, P. L., SEN, M. K., An optimization algorithm for the reservoirs oil well placement problem. *Computers & Geoscience* 10 (**2006**), pp. 303–319.

29 CAMPONOGARA, E., NAKASHIMA, P. H. R., Solving a gas-lift optimization problem by dynamic programming. *European Journal of Operational Research* 174 (**2006**), pp. 1220–1246.

30 CAMPONOGARA, E., NAKASHIMA, P. H. R., Optimizing gas-lift production of oil wells: piecewise linear formulation and computational analysis. *IEE Transactions* 38 (**2006**), pp. 173–182.

31 HERNANDEZ-BARRAGAN, V., ROMAN-VAZQUEZ, R., ROSALES-MARINES, L., GARCIA-SANCHEZ, F., A strategy for simulation and optimization of gas and oil production. *Computers and Chemical Engineering* 30 (**2005**), pp. 215–227.

32 KOSMIDIS, V. D., PERKINS, J. D., PISTIKOPOULOS, E. N., Optimization of well oil rate allocations in petroleum fields. *Ind. Eng. Chem. Res.* 43 (**2004**), pp. 3513–3527.

33 KOSMIDIS, V. D., Ph.D. Thesis, University of London, U.K., **2003**.

34 BRILL, J. P., MUKHERJEE, H., *Multiphase Flow in Wells. SPE Monograph, Henry L. Doherty Series*, Richardson, TX, **1999**.

35 HEWITT, G., *Handbook of Multiphase Systems*, HETSRONI, G. (ed.), Hemisphere, London, **1982**.

36 PEACEMAN, D. W., Interpretation of well-block pressure in numerical reservoir simulation. *SPE Journal* 253 (**1978**), pp. 183–194.

37 SACHDEVA, R., SCHMIDT, Z., BRILL, J. P., BLAIS, R. M., Two-phase flow through chokes. SPE 15657, Presented at the 61st Annual Technical Conference and Exhibition, New Orleans, LA, **1986**, October 5–8.

38 SAMSATLI, N. J., PAPAGEORGIOU, L. G., SHAH, N., Robustness metrics for dynamic optimization models under parameter uncertainty. *AIChE Journal* 44 (**1998**), pp. 1993–2006.

39 LITVAK, M. L., DARLOW, B. L., Surface network and well tubinghead pressure constraints in composition simulation. SPE 29125, Presented at the 13th Symposium on Reservoir Simulation, San Antonio, TX, **1995**, February 12–15.

40 VASSILIADIS, V. S., PANTELIDES, C. C., SARGENT, R. W. H., Solution of a class of multistage dynamic optimization problems 1. Problems without path constraints. *Indus-

trial & Engineering Chemistry Research 32 (**1994**), pp. 2111–2130.

41 BEALE, E. M. L., Branch and bound methods for numerical optimization of nonconvex functions, in: *COMPSAT 80: Proceedings in Computational Statistics*, vol. 11, Physica Verlag, Wien, **1980**.

42 WOLSEY, L. A., *Integer Programming*, Wiley, New York, **1998**.

43 LEA, J. F. JR., TIGHE, R. E., Gas well operation with liquid production. SPE 11583, Presented at the Production Operation Symposium, Oklahoma City, **1983**, February 27–March 1.

44 DURAN, M. A., Grossmann. I. E., A mixed integer nonlinear programming algorithm for process system synthesis. *AIChE Journal* 32 (**1986**), pp. 592–608.

45 GRIFFITH R. E., STEWART, R. A., A nonlinear programming technique for optimization of continuous processing systems. *Management Science* 7 (**1961**), pp. 379–392.

46 PALACIOS-GOMEZ, F., LASDON, L., ENGQUIST, M., Nonlinear optimization by successive linear programming. *Management Science* 28 (**1982**), pp. 1106–1118.

47 ZHANG, J., KIM, N., LASDON, L., An improved successive linear programming algorithm. *Management Science* 31 (**1985**), pp. 1312–1325.

48 BULLARD, L., BIEGLER, L. T., Iterated linear programming strategies for nonsmooth simulation: Continuous and mixed-integer approaches. *Computers and Chemical Engineering* 16 (**1992**), pp. 949–961.

49 BAZARAA, M. S., SHERALI, H. D., SHETTY, C. M., *Nonlinear Programming Theory and Applications*, 2nd edition, Wiley, New York, **1993**.

50 Mathwork, *Optimization Toolbox*, Mathworks, Massachusetts, **1999**.

51 LEE, J., WATTENBARGER, R. A., *Gas Reservoir Engineering. SPE Monograph, Henry L. Doherty Series*, Richardson, TX, **1996**.

6
Wind Turbines Modeling and Control
Konstantinos Kouramas and Efstratios N. Pistikopoulos

Keywords

wind turbines, wind energy, induction generator, reduced-order model, open-loop control problem, multiparametric controller (MPC)

6.1
Introduction

Wind turbines have been used for the production of wind energy in many applications worldwide [1]. Concentrations of wind turbines, which are called wind farms, have been used in the last few years, throughout the world, for the production of electrical energy. Their main advantage is that they can use wind power to produce electrical energy without any harmful byproducts and they constitute promising alternatives to fossil fuel-based electrical energy production such as gas, coal, etc. [1, 2].

The safe and productive operation of a wind farm requires the safe and productive operation of every wind turbine, thus posing many challenges for control. Nevertheless the issue of wind turbine control remains an open topic for research both in academia and industry [1, 3]. Classical control techniques such as PID controllers are used to regulate wind power. Previous work on control design assumed wind turbines operating under steady-state operating conditions without taking into account the dynamics of the wind and the wind turbines that demonstrate nonlinear characteristics [4]. Advanced control techniques are necessary to take care of the nonlinearities and performance constraints in order to guarantee improved performance, reliability, and safety [5].

An interesting characteristic of wind conversion systems is that the operating point is determined by the wind speed; it simply defines the available amount of energy that can be converted to electrical energy. Obviously the wind speed is governed by natural elements and cannot be controlled which is a characteristic that renders wind turbines different than most other process systems [6]. Thus regulating the produce power and maintaining constant power supply to the load or grid in the presence of changes in the wind speed are a challenging task.

Process Systems Engineering: Vol. 5 Energy Systems Engineering
Edited by Michael C. Georgiadis, Eustathios S. Kikkinides and Efstratios N. Pistikopoulos
Copyright © 2008 WILEY-VCH Verlag GmbH & Co. KGaA, Weinheim
ISBN: 978-3-527-31694-6

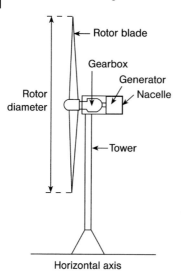

Fig. 6.1 Horizontal-axis wind turbine configuration [1].

Fig. 6.2 Wind conversion system [5].

In this work we offer a preliminary study of a wind turbine system and its characteristic and an attempt to design an advanced explicit multiparametric controller (explicit MPC) for this system that tries to maintain an active power level in the presence of wind speed variations [1]. Other aspects of the wind turbine control problem, such as reactive power regulation, etc., will not be considered here since this is a study, the first of its kind to our knowledge, for concept proving of the use of explicit multiparametric control into wind energy conversion systems. In the following section we will discuss the wind turbine system, its dynamic behavior according to a few simulation and the steps for the design of the explicit multiparametric controller.

6.2
Wind Turbine System Modeling

The wind turbine system considered here is shown schematically in Figs. 6.1 and 6.2. It is a horizontal axis wind turbine with a three-blade wind wheel, a high-speed asynchronous generator, and a gear box [1, 5]. Most wind turbine systems follow this basic design and horizontal-axis wind turbines currently dominate the market. Asynchronous machines are used due to construction simplicity, low operating costs, and investment and ability to perform under various operating conditions.

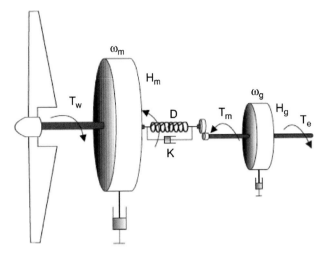

Fig. 6.3 Drive train system [5].

The main system consists of three parts: the rotor, the drive trains, and the electrical generator. The rotor consists of the blades and the hub on which the blades are attached. The drive train consists of a blade pitching mechanism, a hub with blades, a rotor shaft, and a gearbox that is connected to the generator (the drive train can be seen in more detail in Fig. 6.3). The generator is an alternating current (AC) asynchronous (induction) generator as we discussed previously. The kinetic energy of the wind is extracted by the rotor blades and is transformed into mechanical energy. The drive train then transfers the load and increases the rotational speed of the wind rotor to drive the electrical generator. Finally, the mechanical power produced in the wind rotor and the drive train is transformed into electrical by the generator.

In order to derive the model of the wind turbine one has to derive the models for each of the individual parts. This can be done by deriving the first-order principles of the mechanical and electrical parts of the wind turbine [1, 2, 5]. The models for each of the parts of the wind turbine will be analyzed in the following sections.

6.2.1
Modeling of the Rotor Aerodynamics

In order to describe the aerodynamics of the wind turbine rotor one will have to look into the conversion of the kinetic energy of the wind into mechanical energy. The total power available in the wind with constant speed, passing through the area A (m^3) swept by the blades of the wind turbine is given by [1, 7]

$$P_w = \frac{1}{2}\rho A v^3 \tag{6.1}$$

where ρ (kg/m^3) is the air density. The air density can be expressed as a function of the turbine elevation above sea level H,

$$\rho = \rho_0 - 1.194 \times 10^{-4} * H \tag{6.2}$$

where $\rho_0 = 1.225$ kg/m^3 is the air density at sea level.

A fraction of the total wind power is extracted by the blades of the wind turbine and converted into mechanical energy [1, 7]. The power extracted can be expressed as

$$P_{\text{BLADE}} = Cp(\lambda, \beta) P_w = Cp(\lambda, \beta) \frac{1}{2} \rho A v^3 \tag{6.3}$$

where $C_p(\lambda, \beta)$ is the coefficient of performance, λ is the tip speed ratio, and β is the blade pitch angle. The tip speed ratio λ is a variable that combines the effects of the rotational speed and the wind speed and is defined as the ratio of the speed of the tips of the blades to the wind speed:

$$\lambda = \frac{v_{\text{TIP}}}{v_{\text{WIND}}} = \frac{\omega R}{v} \tag{6.4}$$

where ω is the blade's angular velocity (rad/s) and R is the rotor radius (m). The coefficient of performance expresses the fraction of energy, extracted by the wind turbine, from the total wind energy that would have flowed through the area swept by the rotor if the wind turbine was not there. In other words, it expresses the fraction of the kinetic energy in the wind that can be converted into useful power. Its value varies with the wind speed. As the full data of $C_p(\lambda, \beta)$ are not always available, it is common to use the approximation of $C_p(\lambda, \beta)$ as a function of the tip speed ratio and the blade angle. A very accurate approximation is given by the following expression [7]:

$$Cp(\lambda, \beta) = c_1 \left(c_2 \frac{1}{\gamma} - c_3 \beta - c_4 \beta^x - c_5 \right) e^{-C_6 \frac{1}{\gamma}} \tag{6.5}$$

where γ is defined as

$$\frac{1}{\gamma} = \frac{1}{\lambda + 0.08\beta} - \frac{0.035}{1 + \beta^3} \tag{6.6}$$

For the wind turbine in this work the coefficients in (6.5) are given as $c_1 = 0.5$, $c_2 = 116$, $c_3 = 0.4$, $c_4 = 0$, $c_5 = 5$, and $c_6 = 21$. The parameter x does not have to be defined since $c_4 = 0$. The plot of the coefficient of performance versus λ for various values of the pitch angle β, is shown in Fig. 6.4.

The aerodynamic torque that occurs due to the rotation of the wind turbine rotor is given by

$$T_w = \frac{P_{\text{BLADE}}}{\omega_m} = \frac{\frac{1}{2} Cp(\lambda, \beta) \rho A v^3}{\omega_m} \tag{6.7}$$

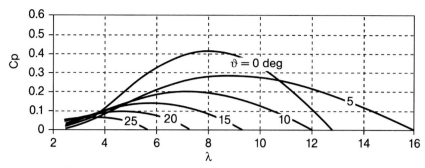

Fig. 6.4 Coefficient of performance $C_p(\lambda, \beta)$ [7].

Since $A = \pi R^2$, (6.7) can be rewritten as

$$T_w = \frac{\frac{1}{2}\pi C_p(\lambda, \beta)\rho R^2 v^3}{\omega_m} \tag{6.8}$$

The relations for the aerodynamic torque conclude the model of the rotor aerodynamics.

6.2.2
Drive Train Model

This section discusses the model of the drive train part of the wind turbine [1]. The drive train model presented here will include the inertia of both the turbine and the generator. In order to derive the model of the drive train it is assumed that both the hub with the blades (wind wheel) and the generator are represented as lumped masses (see Fig. 6.3) [7]. Then, the equation of motion of the induction generator is given as

$$H_g \frac{d\omega_g}{dt} = T_e + \frac{T_m}{n} \tag{6.9}$$

where ω_g is the generator rotor rotational speed, T_m is the mechanical torque, T_e is the electromagnetic torque in the generator, and H_g is the inertia constant. The motion of the windmill shaft is described by the following ordinary differential equation:

$$H_m \frac{d\omega_m}{dt} = T_w - T_m \tag{6.10}$$

where H_m is the inertia constant and T_w is the torque provided by the wind. The mechanical torque is given by

$$T_m = K\frac{\theta}{n} + D\frac{\omega_g - \omega_m}{n} \tag{6.11}$$

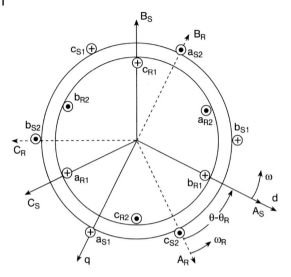

Fig. 6.5 The asynchronous generator [1].

where n is the gear ratio, ω_m is the speed of the turbine rotor, K and D are the drive stiffness and damping constants, and θ is the angle between the turbine rotor which is described by the following differential equation:

$$\frac{d\theta}{dt} = \omega_g - \omega_m \tag{6.12}$$

6.2.3
Modeling the Induction Generator

The electrical generator considered here, as discussed previously, is a three-phase AC induction generator. The generator has three-phase stator armature windings (A_S, B_S, C_S) and three-phase rotor windings (A_R, B_R, C_R) as shown in Fig. 6.5. The stator is the outer stationary part of the generator and the rotor is the inner rotating part. When an electrical current is supplied to the stator winding – the current in each winding is of the same amplitude but has the phase difference of one third of a period in comparison with the currents in the other windings – a rotating magnetic field is produced. The angular speed of the rotating field is called synchronous speed ω_s. The relative speed between the rotating magnetic field and the rotor induces a current in the rotor. As the mechanical torque drives the rotor to speeds beyond the asynchronous speed electrical energy is pumped up to the grid that the generator is usually connected.

The mathematical model of an induction generator is generally based on the following assumptions:

- The stator currents are positive when flowing toward the network.
- The real and reactive power are positive when fed to the grid.

- The stator and the rotor windings are placed sinusoidally along the air gap as far as the mutual effect with the rotor is concerned.
- The stator slots cause no appreciable variations of the stator inductances with rotor position.
- Magnetic hysteresis and saturation effects are negligible.
- The stator and rotor windings are symmetrical.
- The capacitance of all the windings can be neglected.

The above assumptions are standard assumptions for the modeling of induction machines and generators and the interested reader can seek for more details in [7, 8].

Generally, deriving an appropriate model for the induction generator is the most difficult and complicated task for the modeling of a wind turbine despite the fact that induction machines have been extensively studied in the literature. The detailed model of an induction generator [9] is a fifth-order model containing four electromagnetic state variables plus the generator speed and considers the electromagnetic transients both in the stator and the rotor. However, a more simplified model is usually used that does not take into account the stator transients and has only two electromagnetic state variables thus leading to a third-order model, the third state being the generator speed. Both these models are presented next.

6.2.3.1 The Fifth-Order Model

Usually the set of equations describing the asynchronous generator is transformed into another set of equations related to an arbitrary set reference frame – the so-called $0dq$ reference frame or synchronous reference frame [8, 10]. The frame consists of the direct (d) and quadrature (q) axis – the d-axis is aligned with the magnetic axis of the filed windings while the q-axis "leads" the d-axis by 90°. The stator and rotor voltages and the fluxes in the generator are both expressed in their components in the d- and q-axis.

The complete and detailed model of the induction generator consists of the following differential algebraic equations (DAE):

$$\phi_{ds} = X_s I_{ds} + X_m I_{dr} \tag{6.13}$$

$$\phi_{qs} = X_s I_{qs} + X_m I_{qr} \tag{6.14}$$

$$\phi_{dr} = X_r I_{dr} + X_m I_{ds} \tag{6.15}$$

$$\phi_{qr} = X_r I_{qr} + X_m I_{qs} \tag{6.16}$$

$$V_{ds} = -R_s I_{ds} + \omega_s \phi_{qs} - \frac{d\phi_{ds}}{dt} \tag{6.17}$$

$$V_{qs} = -R_s I_{qs} - \omega_s \phi_{ds} - \frac{d\phi_{qs}}{dt} \tag{6.18}$$

$$0 = -R_r I_{dr} + s\omega_s \phi_{qr} - \frac{d\phi_{dr}}{dt} \tag{6.19}$$

$$0 = -R_r I_{qr} - s\omega_s \phi_{dr} - \frac{d\phi_{qr}}{dt} \tag{6.20}$$

The voltages and fluxes, as we can see, have been transformed to their dq-axis components. The indexes s and r denote stator and rotor quantities, and d and q denote the d-axis and q-axis components of each variable, respectively. For example V_{dr} is the d-axis rotor voltage while I_{qs} is the d-axis stator current. The variables V, I, and φ are the voltage, current, and flux linkage of the stator and the rotor, respectively. Finally, X_s is the stator reactance, X_m is the mutual reactance, X_r is the rotor reactance, R_s is the stator resistance, R_r is the rotor resistance, ω_s is the synchronous speed, and s is the slip of the rotor which is given by

$$s = \frac{\omega_s - \omega_g}{\omega_s} \tag{6.21}$$

where ω_g is the generator rotational speed. Since in the generating mode $\omega_s \leqslant \omega_g$, the slip is always negative. The electrical torque is

$$T_e = \phi_{qr} I_{dr} - \phi_{dr} I_{qr} \tag{6.22}$$

and is negative for the generating mode. Finally, the active, reactive, and total power outputs are

$$P_{\text{active}} = V_{ds} I ds + V_{qs} I_{qs} \tag{6.23}$$

$$Q_{\text{reactive}} = V_{qs} I ds - V_{ds} I_{qs} \tag{6.24}$$

$$P = V_{ds} I ds + V_{qs} I_{qs} + V_{qs} I ds - V_{ds} I_{qs} \tag{6.25}$$

6.2.3.2 Reduced-Order Model

Equations (6.13)–(6.25) constitute a detailed model for the induction generator. Nevertheless, numerical integration of (6.13)–(6.25) requires a very small time step leading to increased computational times [11, 12]. Thus, a reduced-order model can be developed by neglecting the transients in the stators flux linkages $d\varphi_{ds}/dt$ and $d\varphi_{qs}/dt$ which is equivalent to assuming infinitely fast electromagnetic transients in the stator. This is a common when performing stability simulations [9]. By setting $d\varphi_{ds}/dt = 0$ and $d\varphi_{qs}/dt = 0$ in Eqs. (6.13)–(6.25) the following reduced-order model is obtained:

$$\phi_{ds} = X_s I_{ds} + X_m I_{dr} \tag{6.26}$$

$$\phi_{qs} = X_s I_{qs} + X_m I_{qr} \tag{6.27}$$

$$\phi_{dr} = X_r I_{dr} + X_m I_{ds} \tag{6.28}$$

$$\phi_{qr} = X_r I_{qr} + X_m I_{qs} \tag{6.29}$$

$$V_{ds} = -R_s I_{ds} + \omega_s \phi_{qs} \tag{6.30}$$

$$V_{qs} = -R_s I_{qs} - \omega_s \phi_{ds} \tag{6.31}$$

$$0 = -R_r I_{dr} + s\omega_s \phi_{qr} - \frac{d\phi_{dr}}{dt} \tag{6.32}$$

$$0 = -R_r I_{qr} - s\omega_s \phi_{dr} - \frac{d\phi_{qr}}{dt} \tag{6.33}$$

$$s = \frac{\omega_s - \omega_g}{\omega_s} \tag{6.34}$$

$$T_e = \phi_{qr} I_{dr} - \phi_{dr} I_{qr} \tag{6.35}$$

$$P_{\text{active}} = V_{ds} I ds + V_{qs} I_{qs} \tag{6.36}$$

$$Q_{\text{reactive}} = V_{qs} I ds - V_{ds} I_{qs} \tag{6.37}$$

$$P = V_{ds} I ds + V_{qs} I_{qs} + V_{qs} I ds - V_{ds} I_{qs} \tag{6.38}$$

The equations above with (6.9) constitute a reduced third-order model. This concludes the modeling of the wind turbine dynamics. The overall model of the wind turbine is obtained by combining the aerodynamic, the train drive, and induction generator models and consists of Eqs. (6.4)–(6.6), (6.8)–(6.12), and (6.26)–(6.38). This model will be used next to perform the simulation of the wind turbine performance and to design the explicit MPC controller. The wind turbine model parameters that will be used throughout this chapter are given in Table 6.1, and they were adapted from [1, 7, 9].

6.3
Dynamic Performance Simulation

6.3.1
Wind Turbine Dynamic Behavior

In this section the wind turbine model, presented above, is simulated for constant and variable wind speed operating conditions. The first case corresponds to ideal operating conditions when the speed of the wind is constant and hence the wind turbine operates under nominal conditions where no control is required. However, in reality wind speed is not constant and will vary with time as environmental conditions vary. It is obvious from (6.1) and (6.3) that if the speed of the wind doubles in value, then the power in the wind and the power extracted from the

Table 6.1 Wind turbine parameters.

Parameters	Values
Rotor radius, R	25
Air density, ρ	1.225
Aerodynamic coefficients c_1–c_6	$c_1 = 0.5$, $c_2 = 116$, $c_3 = 0.4$, $c_4 = 0$, $c_5 = 5$, $c_6 = 21$
Gear ratio, n	65.27
Damping, D	10^6
Stiffness, K	6×10^7
Rotor inertia, H_m	1.6×10^6
Generator inertia, H_g	35.184
Stator resistance, R_s	0.0121
Stator reactance, X_s	0.0742
Mutual reactance, X_m	2.7626
Rotor resistance, R_r	0.008
Rotor reactance, X_r	0.1761
Synchronous speed, ω_s	1

wind increases eight times. Thus, any fluctuation in the wind speed will result in fluctuations of the operating conditions of the wind turbine, the produced turbine torque and finally the electrical power produced.

Wind turbines reach their highest efficiency at wind speeds in the range of 10–15 m/s while their operating range is between 5 m/s and 25 m/s. Beyond these speeds the power of the wind turbine rotor has to be controlled to reduce the driving forces on the rotor blades as well as the load on the whole wind turbine structure [2, 13]. High winds occur only for short periods of time and have little influence in energy production; however if not controlled they will dominate the design and cost of the drive train and generator [3]. A controller is thus required to regulate these fluctuations under the operating conditions of the wind turbine, ensure control of the aerodynamic forces on the turbine rotor, limit power in high winds, and thus ensure the safe operation of the wind turbine.

A control policy to regulate the power output of the wind turbine under variable wind speed conditions is to manipulate the pitch angle of the turbine blades, which in the recent years has become a standard [2, 14]. Modern wind turbines are built with variable pitch angle blades. Pitch angle control can be employed to position the blades relative to the wind and limit the rotor speed by regulating the aerodynamic power flow. Normally, at high wind speeds above the range of 10–15 m/s the wind turbine will have to shut down to avoid damage; however with pitch angle control this is avoided by allowing the wind turbine to operate longer and safer [14]. On the other hand, in low or medium wind speeds the pitch angle can be changed to allow maximum conversion of the wind power. As can be seen from Fig. 6.4, when the wind speed changes, the pitch angle can be changed to adjust C_p to its optimal value and hence achieve maximum conversion of the wind power.

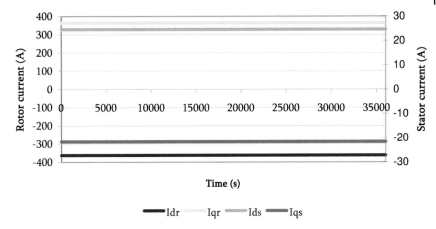

Fig. 6.6 Stator and rotor currents.

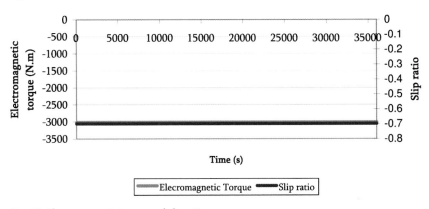

Fig. 6.7 Electromagnetic torque and slip ratio.

6.3.2
Simulations

The first simulation of the wind turbine system was set up to correspond to constant wind speed operations. The wind speed was set to the ideal value of $v = 10$ m/s in which the rated power of the wind turbine is obtained. The blade pitch angle was set to $\beta = 0$ thus assuming that the blades are fully exposed to the wind. The transmission system usually operates at voltages in the kilovolt (kV) range so V_{qs}, V_{ds} are assigned a value of 1000 kV. An operation of 10 h (36,000 s) is assumed. The d- and q-axis currents of the generator stator and rotor are shown in Fig. 6.6. The electromagnetic torque and generator slip are shown in Fig. 6.7 and they have negative values as expected. Finally, the generator output power is shown in Fig. 6.8.

A step change is then applied to the wind speed to observe the behavior of the wind turbine in varying wind speeds. It is assumed that the wind speed is initially 8 m/s and that after 10 h it changes to 10 m/s. The blade pitch angle and the voltage

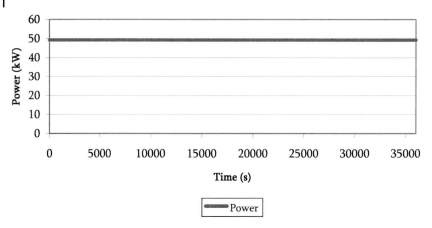

Fig. 6.8 Wind turbine output power.

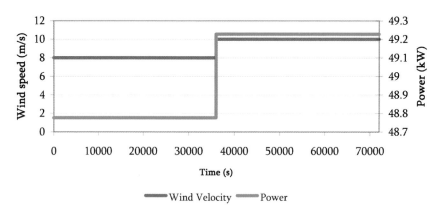

Fig. 6.9 Response of the wind turbine to wind speed step change.

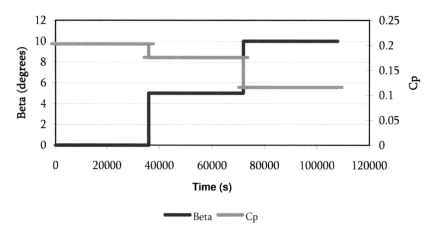

Fig. 6.10 Power coefficient versus pitch angle step change.

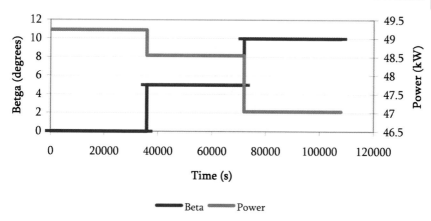

Fig. 6.11 Power output versus pitch angle step change.

supply are kept constant. The result is shown in Fig. 6.9, where it can be seen that the increase in the wind speed results in the increased output power. Similarly, the behavior of the wind turbine is examined for different adjustments of the blade pitch angle β. The wind speed is assumed constant to the ideal value of 10 m/s and two step changes are applied on β. The response of the turbine systems is shown in Fig. 6.10, where it is observed that an increase in β results in a decreased value for the coefficient of power C_p. Similarly, in Fig. 6.11 it is shown that the step increases in the pitch angle result in decreases in the output power. This is as expected since increases in β will result in excess wind flow to be shed through the blades reducing the converted power. Hence, when the wind speed is very high, the blade pitch angle can be increased to reduce the excess power.

6.4
Power Control

The objective in this section is to design an explicit MPC controller for the wind turbine system. MPC has become the standard for the control of complex, constrained, multivariable processes [15] and currently there is a lot of research focus on this control method. It is not in the scope of this chapter to provide a detailed analysis on MPC and the explicit MPC since these methods have been extensively researched in the past and discussed in many important publications. The interested reader can find more information in [15–17], and reference therein. However, we give a brief overview of the two control methods to enable a better understanding of the analysis that is going to follow. MPC is a control method in which the current control action is obtained by solving at each sampling instant a finite-horizon open-loop control problem, using the current state of the process as the initial condition [16]. The procedure is repeated at every sampling instant.

MPC has been successful for its ability to handle constrained, multivariable control problems even in the presence of uncertainty and disturbances [16, 18, 19].

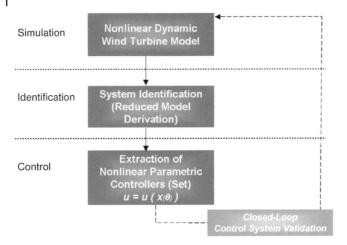

Fig. 6.12 Explicit multiparametric MPC design methodology.

However, its main drawback has been its dependence on online optimization – the finite-horizon open-loop control problem has to be solved at each sampling instant thus requiring solving an optimization problem online. This may limit the applicability of MPC to slow and small problems since the computational burden might be quite high [15, 17]. This drawback can be dealt by employing multiparametric programming methods [17–20] to solve the online optimization problem off-line and to obtain the optimal control action as (1) a function of the measured/available data (output, state, set points, etc.) of the process and (2) the critical regions (sets in the measured data space) where these functions are valid. The online implementation of the explicit multiparametric controller can be performed by simply measuring the current process data, identifying the critical region in which they belong, and calculating the control action by evaluating the function that corresponds to this critical region. Thus, explicit multiparametric control replaces online optimization with functions evaluations that could reduce the computational needs with subsequent cost reduction on the control implementation.

The methodology followed here to design the explicit multiparametric controller is shown in Fig. 6.12. In the first step of this procedure, the detailed nonlinear model of the system is developed and analyzed through simulations. This was the subject of the previous sections. In the second step a reduced-order model, which is suitable for control design, is obtained by model identification using the input–output data derived from the simulations. Finally, the controller is designed based on the derived reduced-order model and the problem specifications and is evaluated against the detailed nonlinear model of the system.

The detailed nonlinear model of the system was presented in Section 6.2 and consists of one manipulated variable (input) which is the blade pitch angle and one control variable (output) which is the output power. The aim is to adjust the pitch angle to regulate the changes in the output power due to the wind speed changes. Both the pitch angle and the power are constrained variables, the first due to limitations of the actuators moving the blades and the second due to the fact that no more power than the rated one, can be achieved. The lower bound for both the β and P is 0 while the upper bound for β is 90° and P is 50 kW. Finally, the set point which the power is required to follow is 47 kW.

An input–output linear ARX model was derived from the simulations in Section 6.2 [1]. A sampling time of 1 s was used for the data acquisition from the simulations which is also the sampling time used for the implementation of the controller. The derived ARX model is

$$y_{t+1} = y_t - 219.3u_t + 219.3u_{t-1} \tag{6.39}$$

$$y(t) = P(t) : \text{Power output}$$

$$u(t) = \beta(t) : \text{Pitch angle}$$

The above model is of relatively small degree and is prone to model error. However, one could derive models of higher degree to improve the approximation of the nonlinear model of the system. For simplicity and since this is the first attempt to design an explicit MPC controller for wind turbines, we will consider only this simple model approximation for the wind turbine.

The following MPC formulation was used for the power control problem and the design of the explicit MPC [1]:

$$J^* = \min_{\Delta u_t, \ldots, \Delta u_{t+N-1}} \sum_{i=t}^{t+N} Q(y_i - y_{\text{ref},i})^2 + \rho \sum_{i=t}^{t+N-1} R\Delta u_i^2 \tag{6.40}$$

$$\text{s.t.:} \ y_{i+1} = y_i - 219.3u_i + 219.3u_{i-1} \tag{6.41}$$

$$\Delta u_i = u_i - u_{i-1} \tag{6.42}$$

$$0 \leqslant u_i \leqslant 90° \tag{6.43}$$

$$0 \leqslant y_i \leqslant 50 \text{ kW} \tag{6.44}$$

$$Q = 50, \ R = 1, \ \rho = 10^{-3}, \ N = 5 \tag{6.45}$$

The above MPC problem is formulated to ensure that the tracking error of the power set points is minimized and the input and control constraints are satisfied. Instead of considering the input directly in the objective function, the increments Δu_t of the input are used to ensure smooth response of the controller to the output changes. In the traditional MPC implementation the optimization problem (6.40)–(6.45) will have to be solved online at each sampling instant. However, the optimization problem (6.40)–(6.45) is a multiparametric quadratic program (mp-QP) [17–20] where $\Delta u_t, \ldots, \Delta u_{t+N-1}$ are the optimization variables and $\theta_t = [y_t \ u_{t-1} \ y_{\text{ref}}]^T$ are the parameters. The methods of [18–20] can be used to solve (6.40)–(6.45) as an mp-QP problem. The parametric optimization software [21] was used to solve directly the problem (6.40)–(6.45).

The solution consists of 21 critical regions with 21 corresponding control laws. For example, the critical region 1 is expressed as the following set of linear inequalities:

$$
CR_1 : \begin{bmatrix}
1 & 0.0045579 & -0.0045579 \\
-1 & -0.0045579 & 0.0045579 \\
-2.4358\text{e}{-014} & 1 & -1 \\
2.4358\text{e}{-014} & -1 & 1 \\
-1 & 0 & 0 \\
1 & 0 & 0 \\
0 & -1 & 0 \\
0 & 1 & 0 \\
0 & 0 & -1 \\
0 & 0 & 1
\end{bmatrix} \cdot \theta_t \leqslant
\begin{bmatrix}
90 \\
0 \\
219.4 \\
219.4 \\
0 \\
90 \\
47{,}034 \\
2966 \\
47{,}034 \\
2966
\end{bmatrix}
$$

(6.46)

and its corresponding control law is

$$u_t = [1 \quad 0.0045579 \quad -0.0045579] \cdot \theta_t \tag{6.47}$$

The controller is then implemented as the following heuristic:

$$\text{If} \quad \theta_t \in CR_i \quad \text{then} \quad u_t = K_i \theta_t + c_i \tag{6.48}$$

where CR_i is the critical regions and K_i, c_i are the gain and off-set of the control law in this critical region. Obviously, for the critical region 1, $K_i = [1 \ 0.0045579 \ - 0.0045579]$ and $c_i = 0$. A 2D representation of the critical regions is shown in Fig. 6.13.

The controller is finally implemented and evaluated on the detailed nonlinear wind turbine. The implementation is as follows: (1) the process data θ is measured at each sampling instant, (2) the critical regions where θ is identified, (3) the heuristic (6.48) is applied to obtain the control action, and (4) the control action is applied to the system. In Figs. 6.14 and 6.15 the simulation of the controller implemented on the nonlinear model of the wind turbine is shown. The simulation was performed by assuming that the turbine initially operates at rated power and that at time zero an impulse change in the speed occurs. The controller then, as it can be seen from Fig. 6.15, increases the pitch angle to regulate the power back to its rated value.

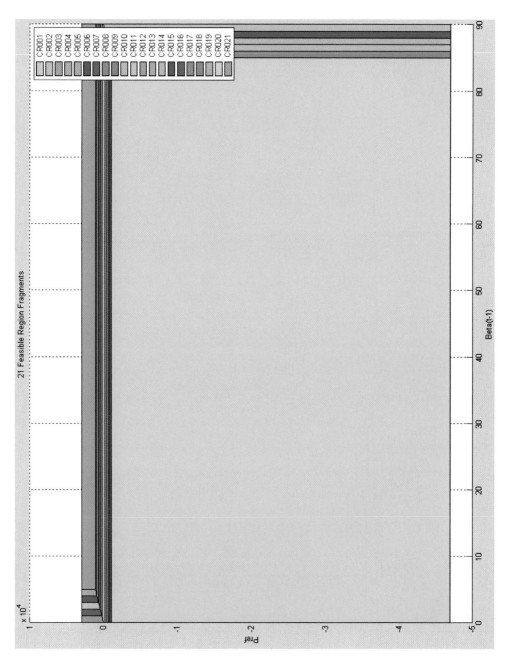

Fig. 6.13 Projection of the critical regions on the y_{ref}-β_{t-1} space.

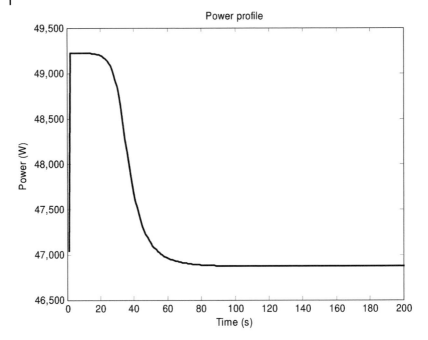

Fig. 6.14 Power output.

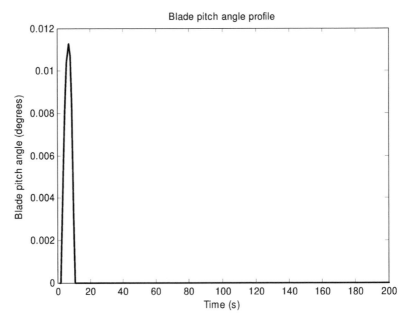

Fig. 6.15 Pitch angle.

6.5
Conclusions

The objective of this work was to discuss the performance, modeling, and power control of wind turbines as well as to design for the first time to our knowledge an explicit MPC controller for a wind turbine. A wind turbine with its detailed nonlinear, approximating nonlinear and approximating ARX model was presented. Simulation analysis was used to discuss the wind turbine's performance in the presence of wind speed variations as well as blade pitch angle variations. The latter was shown that can be manipulated to provide power regulation in the presence of wind speed disturbances. Finally, the steps for the design of an explicit MPC controller were shown and the resulting controller was evaluated against the actual nonlinear model of the system.

Acknowledgement

The help of Ms. J. Martinez is gratefully acknowledged and her help with data and analysis is greatly recognized.

References

1 MARTINEZ, J., Modelling and Control of Wind Turbines. MSc Thesis, Centre for Process Control Engineering, Imperial College London, **2007**.

2 ACKERMANN, T., *Wind Power in Power Systems*, Wiley, England, **2005**.

3 HARRISON, R., HAU, E., SNEL, H., *Larg Wind Turbines*, Wiley, England, **2000**.

4 BALAS, M., FINGERSH, L., JOHNSON, K., PAO, L., *IEEE Control Systems Magazine* 26 (**2006**), p. 70.

5 BOUKHEZZAR, B., SIGUERDIDJANE, H., in: *Proc. of the IEEE Conf. On Dec. and Contr.* (**2005**), p. 3456.

6 BONGERS, P., VAN BAARS, G., in: *Proc. of the IEEE Conf. on Dec. and Contr.* 2 (**1992**), p. 2454.

7 LUBOSNY, Z., *Wind Turbine Operations in Electric Power Systems*, Springer, Germany, **2003**.

8 JANGAMSHETTI, S., NAIK, R., in: *Proc. of the 5th World Wind Energy Conference*, **2006**.

9 MARTINS, M., PERDANA, A., LEDESMA, P., ANGEHOLM, E., CARLSON, O., *Renewable Energy* 32 (**2007**), p. 1301.

10 SLEMMON, G., *IEEE Transactions on Industry Application* 25 (**1989**), p. 1126.

11 EKANAYAKE, J., HOLDSWORTH, L., JENKINS, N., *Electrical Power Systems Research* 67 (**2007**), p. 207.

12 WASYNEZUK, O., YI-MIN, D., KRAUSE, P., *IEEE Transactions Power Apparatus Systems* 3 (**1985**), p. 598.

13 ACKERMANN, T., SODER, *Renewable and Sustainable Energy Reviews* 28 (**2000**), p. 315.

14 BUTTERFIELD, C., MULJADI, E., *IEEE Transactions on Industry Applications* 37 (**2001**), p. 240.

15 BEMPORAD, A., MORARI, M., DUA, V., PISTIKOPOULOS, E. N., *Automatica* 38 (**2002**), p. 3.

16 MAYNE, D. Q., RAWLINGS, J. B., RAO, C. V., SCOKAERT, P. O. M., *Automatica* 36 (**2000**), p. 789.

17 PISTIKOPOULOS, E. N., DUA, V., BOZINIS, N. A., BEMPORAD, A., MORARI, M., *Computers and Chemical Engineering* 26 (**2002**), p. 175.

18 PISTIKOPOULOS, E. N., GEORGIADIS, M., DUA, V., *Multi-Parametric Programming*, Wiley-VCH, Weinheim, **2007**.

19 PISTIKOPOULOS, E. N., GEORGIADIS, M., DUA, V., *Multi-Parametric Model-Based Control*, Wiley-VCH, Weinheim, **2007**.

20 Dua, V., Bozinis, N. A., Pistikopoulos, E. N., *Computers and Chemical Engineering* 26 (**2002**), p. 715.

21 Pistikopoulos, E. N., Bozinis, N. A., Dua, V., POP a Matlab implementation of parametric programming algorithms. Technical Report, Centre for Process Systems Engineering, Imperial College London, UK, **2002**.

22 Morari, M., Lee, J., *Computers and Chemical Engineering* 23 (**1999**), p. 667.

7
Stochastic Optimization of Investment Planning Problems in the Electric Power Industry

Daniel Kuhn, Panos Parpas, and Prof. Berç Rustem

Abstract

Decisions on whether to invest in new power system infrastructure can have far-reaching consequences. The timely expansion of generation and transmission capacities is crucial for the reliability of a power system and its ability to provide uninterrupted service under changing market conditions. We consider a local (e.g., regional or national) power system which is embedded into a deregulated electricity market. Assuming a probabilistic model for future electricity demand, fuel prices, equipment failures, and electricity spot prices, we formulate a capacity expansion problem which minimizes the sum of the costs for upgrading the local power system and the costs for operating the upgraded system over an extended planning horizon. The arising optimization problem represents a two-stage stochastic program with binary first-stage decisions. Solution of this problem relies on a specialized algorithm which constitutes a symbiosis of a regularized decomposition method and a branch-and-bound scheme.

Keywords
electric power industry, stochastic programming (SP), investment planning, capacity expansion planning, conditional value-at-risk (CVaR) constraint, mixed-integer linear program (MILP)

7.1
Introduction

We discuss the problem of capacity expansion and investment planning in a power system consisting of several generation units and transmission lines. On a conceptual level, this challenging management task can be described as follows. At an initial time point several investment decisions are taken. It is decided what types of new power plants and transmission lines should be built in which areas of the power system. Subsequently, the augmented power system is put into service during an extended operating period, that is, the output levels of all plants are dynam-

Process Systems Engineering: Vol. 5 Energy Systems Engineering
Edited by Michael C. Georgiadis, Eustathios S. Kikkinides and Efstratios N. Pistikopoulos
Copyright © 2008 WILEY-VCH Verlag GmbH & Co. KGaA, Weinheim
ISBN: 978-3-527-31694-6

ically optimized in response to random market conditions, which are gradually revealed. As far as power plants are concerned, there is usually a tradeoff between installation and operating costs. Expensive plants with low operating costs are used to cover the base load, whereas cheaper units with high operating costs are typically employed to cover the peak demand. In this paper, we propose a method to identify the optimal mix of power plants as well as the optimal transmission system infrastructure. The objective is to minimize installation and operating costs while meeting reliability requirements.

There is a vast number of papers that address various management problems in the electric power industry. Typical examples include the short-, medium-, and long-term scheduling of hydrothermal power systems [1, 2], optimal power flow dispatching [3], and electricity portfolio management [4]. Doege et al. [5] study a risk management problem, in which the decision maker uses a flexible (hydro) and a less flexible (nuclear) power plant to hedge an electricity portfolio. Their model is concerned with the short-term operation of the plant, and they also assume that the energy producer has access to a future market. Pritchard et al. [6] investigate the operation of hydropower plants and propose a dynamic programming algorithm which reconciles the different time scales of resource management and trading. Moreover, Fleten et al. [7] discuss hedging strategies for an energy producer who has access to a futures market, while Vehvilainen et al. [8] develop Monte Carlo techniques to optimize a portfolio of assets related to energy products. Capacity expansion planning has previously been investigated by Bloom [9], Leopoldino et al. [10], and Hobbs and Ji [11].

In this paper, we consider a regional electricity producer or a public utility which is responsible for covering the load demand in a designated area. We concentrate on the medium- to long-term decisions the energy provider needs to take. In addition, we try to capture the realities of deregulated markets, that is, we assume that the transmission grid within the utility's territory is connected to a superordinate grid on which spot market transactions can be effected. To keep our capacity expansion problem tractable, we do not model the structure of the superordinate grid but simply assume that there is a fictitious grid node representing the spot (pool) market. Investment decisions concerning new power plants and transmission lines are taken at time 0. The extended power system (including the new infrastructure installed at time 0) is then operated continuously over a time interval $[0, T]$. As we envisage long operating periods of several decades, it is reasonable to disregard temporal dependencies. Put differently, the operating decisions corresponding to different time points can be regarded as decoupled, which will allow us to formulate the capacity expansion problem as a two-stage stochastic program.

The following issues are important when addressing investment planning problems in liberalized markets:

- Before installing a new power plant, it is crucial to determine the required level of operational flexibility. When market prices are high, one should have excess capacity to sell energy on the market. When prices are low, on the other hand, it may be beneficial to shutdown certain plants and cover the regional load demand with cheap energy from the market. Nuclear power plants, for instance, are in-flexible but have low operating costs, whereas gas turbine plants are flexible but have high operating costs.

- With the maturing of electricity markets, it is becoming increasingly important to incorporate various financial instruments into one's power portfolio. For example, when a plant is taken off the network for maintenance, then the utility has to procure the missing energy from an alternative source, such as a forward contract or a swing option, see e.g., [12]. Since energy trading is becoming increasingly important, one has to design the transmission grid accordingly, that is, one has to ensure that market transactions can be processed without major bottlenecks.

- Given the high level of uncertainty of future electricity demand, spot prices, or the operational availability of power plants, it is clear that any reasonable energy planning model should exhibit some degree of robustness. This is especially true in consideration of the long planning horizons we have in mind. Robust models should ensure reliable operation of the power system under various challenging conditions, e.g., in the presence of outages and spiking spot prices, etc.

The model developed in this chapter is inspired by Leopoldino et al. [10], who study a related problem in a regulated environment. We extend this model to account for the new challenges faced by energy producers in liberalized markets. Furthermore, we elaborate on the underlying probabilistic model. In Section 7.2 we review some basic concepts of stochastic programming theory which will be used in Section 7.3 to formulate our capacity expansion problem as a two-stage stochastic program. This model accounts for market and technological uncertainties, i.e., it explicitly considers future electricity prices, production costs, and equipment outages as stochastic parameters. Section 7.4 outlines an approach to model the probability distribution of the underlying risk factors and also discusses the approximation of this distribution function by a simpler discrete one. In Section 7.5 we propose an algorithm to solve the model of Section 7.3 with the discretized distribution of Section 7.4. This algorithm combines a regularized Benders decomposition scheme with a branch-and-bound method.

7.2
Stochastic Programming Concepts

Problems of power system planning and operation are often addressed by using stochastic programming techniques. A survey of relevant work in this field has been published in [1], and therefore we do not review the multitude of existing energy models cast in a stochastic programming framework. However, in order to make this chapter as self-contained as possible, we give a brief introduction to stochastic programming with particular focus on the aspects that will arise in the later sections.

Stochastic programming (SP) is becoming an increasingly popular tool for addressing decision problems under uncertainty because of the flexible way uncertain phenomena can be modeled, and since real-world constraints can be imposed with relative ease. SP also injects robustness to the optimization process. It is not always possible to know the exact values of the problem data. Instead, we may only have estimations in the form of empirical or parametric probability distributions. The SP framework allows us to solve problems where the data of the problem are represented as random variables, yielding results that are robust to estimation errors.

The flexibility of SP does, however, come at a cost. Realistic models include many (or even an infinite number of) scenarios, and information about these scenarios is revealed sequentially at several (sometimes even an infinite number of) decision stages. To make a naive SP model computationally tractable, appropriate approximation schemes must be employed. Their aim is to reduce the number of scenarios and decision stages without sacrificing too much accuracy. After applying a suitable approximation scheme, we end up with a large-scale optimization problem with hundreds of thousands of variables and constraints. Models of this scale cannot be handled by general-purpose optimization algorithms. Thus, special-purpose algorithms attempt to take advantage of the specific structure of SP models. Decomposition methods that attempt to address these issues will be discussed later in this chapter.

7.2.1
Modeling Framework

A specific capacity expansion model will be elaborated in the next section. Here, we merely review some general concepts from SP that are necessary to formulate capacity expansion models in the electric power industry. Birge and Louveaux [13] provide a textbook introduction to SP that includes both modeling and algorithmic issues. To avoid technicalities, we discuss problems with fixed recourse [13, Chapter 3]. This class of problems is general enough for our purposes.

A linear two-stage stochastic programming problem is given by

$$\min_{x \in X} c^\top x + \mathcal{Q}(x)$$

$$\text{s.t. } Ax \leqslant b \tag{7.1a}$$

where the expectation functional $\mathcal{Q}(x) = \int_\Xi Q(x, \xi) P_\xi(\mathrm{d}\xi)$ is defined as the expectation of the recourse function

$$Q(x, \xi) = \min_{y \in Y} \left\{ q(\xi)^\top y \mid Wy \leqslant h(\xi) - T(\xi)x \right\}. \tag{7.1b}$$

The variable x represents the first-stage decision, ξ represents a random vector with distribution P_ξ supported on Ξ, and y represents the second-stage or "recourse" decision. The sets X and Y as well as all involved functions of the random vector are assumed to be sufficiently well behaved for the above problem to have a finite solution. Details on the existence of solutions can be found in [13, Chapter 3]. Problem (7.1) is to be interpreted as follows: in the first stage, we make a decision x that depends only on the distribution of ξ (but not on ξ). In the context of capacity expansion models, this first-stage decision determines the power system infrastructure. For example, we decide what type of power plants to construct in which areas and how to link the plants and load centers by transmission lines. The first-stage cost vector c associated with the construction of the system infrastructure is assumed to be deterministic. In the second stage, the random vector ξ is observed. In this paper, the components of ξ will represent fuel prices, electricity prices, load demand, and the availability of different components of the power system. Depending on the first-stage decision x and the realization of ξ, a second-stage decision y is selected. In the context of capacity expansion planning, the second-stage decision determines the power system operation, while $q(\xi)$ characterizes the stochastic operating costs of the system. Here, we consider the power system infrastructure as given and evaluate its operating costs as well as its reliability, i.e., we study how the system behaves in the presence of demand spikes and equipment failure.

Another modeling technique that can be used to formulate capacity expansion problems is the concept of chance constraints (or probabilistic constraints). In the recourse model (7.1), the second-stage constraints are required to hold in all scenarios. Sometimes, however, it may be more appropriate to enforce certain constraints only on a subset of scenarios whose probability is strictly smaller than 1. For example, we may wish to meet demand most of the time in most scenarios, but it may be too costly (or even impossible) to build a network that is resilient to all possible future states of the world. Chance constraints are formulated as

$$P_\xi \left(W y(\xi) \leqslant h(\xi) - T(\xi)x \right) \geqslant 1 - \alpha$$

where the parameter $\alpha \gtrsim 0$ can be interpreted as a confidence level. A detailed discussion of chance constraints is provided in [14]. It is worthwhile to mention that certain risk measures, such as the conditional value-at-risk (CVaR), can also be used to formulate similar constraints. The use of CVaR in energy planning models, for instance, is discussed in [5].

Having described different SP paradigms in abstract terms, we next elaborate a specific model for the capacity extension of an electric power system.

7.3
A Capacity Expansion Model

As discussed in Section 7.2.1, the proposed model has two stages. In the first stage, a given power system consisting of generation units and transmission lines is upgraded, that is, new generation and transmission infrastructure is added. In the second stage, which covers a continuum of decoupled time periods, the upgraded

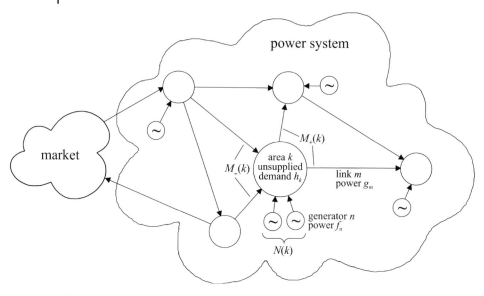

Fig. 7.1 Power system configuration.

system is put into operation. We now describe in detail how the two stages are modeled.

We assume that there are several regions $K = \{1, \ldots, \bar{k}\}$ that have to be supplied with electric energy (Fig. 7.1). An additional (fictitious) area with index 0 characterizes some electricity spot market, to which our system is connected. We consider several energy production units $N = \{1, \ldots, \bar{n}\}$ subdivided into a set of existing units $N_1 \subset N$ and a set of candidate units $N_2 \subset N$ that can be newly built. Similarly, we consider an entity $M = \{1, \ldots, \bar{m}\}$ of transmission lines (or network links) subdivided into existing transmission lines $M_1 \subset M$ and candidate transmission lines $M_2 \subset M$ that can be newly built. Each production unit is located in one of the regions $k \in K$, and the entity of production units located in area k is denoted by $N(k)$. Transmission lines constitute directed links between two different areas of the system (including area 0 representing the spot market). For each $k \in K \cup \{0\}$, we denote by $M_+(k)$ the set of links emanating from area k to some adjacent area. Analogously, we denote by $M_-(k)$ the set of links directed toward area k.

In the first stage, we decide which candidate energy production units and which candidate transmission lines should be built. If plant $n \in N_2$ is built, then a fixed cost c_n is incurred, and if transmission line $m \in M_2$ is built, then the amount d_m has to be payed. The decisions in the first stage are binary variables. Let the vector $u = (u_n)_{n \in N_2}$ be defined through

$$u_n = \begin{cases} 1 & \text{if production unit } n \text{ is built} \\ 0 & \text{otherwise} \end{cases}$$

while $v = (v_m)_{m \in M_2}$ is defined via

$$v_m = \begin{cases} 1 & \text{if transmission line } m \text{ is built} \\ 0 & \text{otherwise} \end{cases}$$

The first-stage decision problem thus reads

$$\min_{x=(u,v)} \sum_{n \in N_2} c_n u_n + \sum_{m \in M_2} d_m v_m + Q(x)$$

$$u_n \in \{0, 1\} \quad \forall n \in N_2$$

$$v_m \in \{0, 1\} \quad \forall m \in M_2 \tag{7.2a}$$

The expectation functional $Q(x)$ accounts for the expected operating costs conditional on the capacity expansion decisions $x = (u, v)$. The second-stage term is given by

$$Q(x) = \int_\Omega \int_0^T Q'(x, \xi(\omega, t)) \lambda(dt) P(d\omega)$$

where power system operation is assumed to extend over a time interval $[0, T]$, and λ represents a Borel measure on $[0, T]$ which may account for nontrivial time preferences and discounting. The mapping $\xi : [0, T] \times \Omega \rightarrow \Xi$ characterizes a measurable stochastic process on some probability space $(\Omega, \mathcal{F}, P)$ describing the dynamics of the operational risk factors. Concretely speaking, $\xi = (\alpha, \beta, \gamma, \delta, \epsilon, \pi)$ represents a combination of several subordinate stochastic processes:

(i) $\alpha = (\alpha_n)_{n \in N}$, where α_n characterizes stochastic outages of power plant n ($\alpha_n = 1$: plant n is fully functional; $\alpha_n = 0$: plant n is out of service);

(ii) $\beta = (\beta_m)_{m \in M}$, where β_m characterizes stochastic outages of transmission line m ($\beta_m = 1$: line m is fully functional; $\beta_m = 0$: line m is out of service);

(iii) $\gamma = (\gamma_n)_{n \in N}$, where γ_n denotes the cost of producing one unit of energy with power plant n (may be related to fuel prices);

(iv) $\delta = (\delta_k)_{k \in K}$, where δ_k represents the load demand in area k;

(v) $\epsilon = (\epsilon_k)_{k \in K}$, where ϵ_k denotes the cost of unsupplied demand[1] in area k;

(vi) π denotes the electricity spot price associated with area 0.

The integrand $Q'(x, \xi(\omega, t))$ stands for the stochastic cost rate function at time t, which depends only indirectly on time through its dependence on the process ξ.

---

1) It is difficult to develop a reasonable model for the dynamics of the process ϵ. Bental and Ravid [15], for instance, suggest a simple method for calculating the cost of industrial power cuts which is based on a revealed preference approach in the generators market. Moreover, Beenstock [16] and Caves et al. [17] propose different methods for estimating electricity outage costs by means of interacting with the consumers.

For the further argumentation, it is useful to consider ξ as a random vector defined on an augmented probability space $(\Omega', \mathcal{F}', P')$ with

$$\Omega' = [0, T] \times \Omega \quad \mathcal{F}' = \mathcal{B}([0, T]) \otimes \mathcal{F} \quad \text{and} \quad P' = \frac{1}{\lambda([0, T])} \lambda \otimes P$$

Without loss of generality, we may assume that Ξ represents the support of ξ. As usual, we let P_ξ denote the distribution function of the random vector ξ, implying

$$\mathcal{Q}(x) = \int_\Xi Q(x, \xi) P_\xi(d\xi)$$

where the recourse function $Q(x, \xi)$ is related to the cost rate function $Q'(x, \xi)$ through

$$Q(x, \xi) = Q'(x, \xi) \lambda([0, T])$$

The full cost rate function, in turn, is defined as the optimal value of the following parametric minimization problem:

$$Q'(x, \xi) = \min_{y=(f,g,h)} \sum_{n \in N} \gamma_n f_n - \sum_{m \in M_+(0)} \pi g_m + \sum_{m \in M_-(0)} \pi g_m + \sum_{k \in K} \epsilon_k h_k$$

$$\text{s.t.} \quad \sum_{n \in N(k)} f_n - \sum_{m \in M_+(k)} g_m + \sum_{m \in M_-(k)} g_m + h_k = \delta_k \quad \forall k \in K$$

$$0 \leqslant f_n \leqslant \alpha_n \bar{f}_n \quad \forall n \in N_1$$

$$0 \leqslant f_n \leqslant \alpha_n \bar{f}_n u_n \quad \forall n \in N_2$$

$$|g_m| \leqslant \beta_m \bar{g}_m \quad \forall m \in M_1$$

$$|g_m| \leqslant \beta_m \bar{g}_m v_m \quad \forall m \in M_2$$

$$0 \leqslant h_k \quad \forall k \in K \tag{7.2b}$$

The components of the second-stage decision vector $y = (f, g, h)$ are

(i) $f = (f_n)_{n \in N}$, where f_n denotes the rate of energy production in unit n;
(ii) $g = (g_m)_{m \in M}$, where g_m denotes the energy flow per time through transmission line m ($g_m > 0$ means that energy flows from the start to the end node of the link and vice versa);
(iii) $h = (h_k)_{k \in K}$, where h_k denotes the unsupplied load demand in area k.

Notice that y has units of energy per time. Thus, the objective of problem (7.2b) is to minimize the cost rate, that is, costs per time. The overall cost rate is given by the sum of the operating costs of all power plants, the costs for buying energy from the spot market, and the economic costs of unsupplied electricity. Any operation plan y must satisfy several constraints. First, we have to impose a power balance constraint for each region $k \in K$, that is, the sum of generated power, inflowing power, and unsupplied power must equal demanded power. Moreover, a set of capacity restrictions apply. The production of plant n may not exceed a given maximum power level $\bar{f}_n$ multiplied by the stochastic outage parameter α_n. For candidate power plants $n \in N_2$, the maximum power level is further multiplied by the first-stage decision u_n. This constraint ensures that $f_n \equiv 0$ if plant n has not been built, i.e., if $u_n = 0$. The fourth and fifth constraints impose similar capacity restrictions on the transmission lines, where $\bar{g}_m$ denotes the maximum power flow through line $m \in M$.

In order to guarantee that supply shortages are kept at a low level, one could extend the model (7.2) by appending a chance constraint of the form

$$P_\xi \left(\sum_{k \in K} h_k(\xi) \leqslant 0 \right) \geqslant 1 - \alpha \tag{7.3}$$

where α represents the percentage of P_ξ-scenarios in which at least some customers remain unsupplied. The drawback of chance constraints is that they frequently lead to nonconvex feasible sets, which severely complicates solution procedures. However, since the shortfall variables $h_k(\xi)$ are almost surely non-negative for all $k \in K$, we can rewrite (7.3) as an equivalent conditional value-at-risk (CVaR) constraint.

$$\mathrm{CVaR}_{1-\alpha} \left(-\sum_{k \in K} h_k(\xi) \right) \geqslant 0$$

As pointed out in [18, Section 4], CVaR-constraints of this type can conveniently be represented as linear constraints, and thus the linearity of the model (7.2) is preserved.

7.4
Probabilistic Model

In this section, we briefly outline how the vector-valued stochastic process ξ can be modeled.

The components α and β of ξ characterize operational failures of the power plants and the transmission lines, respectively. By convention, all possible realizations of (α, β) are elements of the finite state space $\Xi_{(\alpha,\beta)} = \{0, 1\}^{\bar{n}+\bar{m}}$. As equipment failures can hardly be foreseen, it is reasonable to assume that (α, β) follows a continuous-time Markov chain independent of the other risk factors. We further assume that the underlying transition rate matrix is constant. To ease the combinatorial state explosion in the presence of many plants and transmission lines, one can assume that never more than, say, three constituents of the power system infrastructure fail at the same time.

The production costs γ, the local electricity demand δ, the cost of unsupplied demand ϵ, and the electricity spot price π are clearly correlated. All of these processes should behave qualitatively like energy commodity prices. In the remainder of this section, we will therefore concentrate on the modeling of the electricity spot price process π. All components of γ, δ, and ϵ can be treated in a similar manner, i.e., they may be represented by other instances of the same process class. Our approach is inspired by [19, 20]. We propose a model in which the logarithmic spot price $\log(\pi)$ is determined by two risk factors X_1 and X_2 whose dynamics is governed by an affine jump diffusion process [21]. Concretely speaking, we assume that the two new risk-factors satisfy the system of stochastic differential equations

$$dX_{1,t} = -\kappa(X_{1,t} - \theta)\, dt + \sigma\, dB_t \tag{7.4a}$$

$$dX_{2,t} = 1_{\{X_{2,t}=0\}} J_t\, d\nu_{J,t} - X_{2,t}\, d\nu_{N,t} \tag{7.4b}$$

Here, B represents a standard Brownian motion, ν_J and ν_N are Poisson processes with arrival intensities λ_J and λ_N, respectively, and J constitutes a serially independent process of stochastic jump amplitudes with $J_t \sim \mathcal{N}(\mu_J, \sigma_J^2)$, $t \geq 0$. All sources of uncertainty are assumed to be mutually independent. This implies, in particular, that X_1 and X_2 are independent. Moreover, we assume that $\kappa, \sigma > 0$.

Below, dX_1 and dX_2 will be interpreted as the continuous and discontinuous parts of the instantaneous spot price return, respectively. There is evidence from many experimental studies that mean-reversion plays an important role in modeling electricity spot prices, see e.g., [20, 22]. Therefore, we require X_1 to follow an Ornstein–Uhlenbeck process with θ being the long-term mean, κ the mean-reversion rate, and σ some volatility parameter. If X_1 exceeds θ, then the drift term is negative, and the opposite occurs when X_1 drops below θ. The system (7.4) allows for analytical solution. First, the distribution of $X_{1,t}$ conditional on $X_{1,0}$ is given by

$$\mathcal{N}\left(e^{-\kappa t} X_{1,0} + \left(1 - e^{-\kappa t}\right)\theta, \; \frac{\sigma^2}{2\kappa}\left(1 - e^{-2\kappa t}\right)\right)$$

By independence, the second risk-factor X_2 can be treated separately. Note that this factor is needed to model the characteristic spot price spikes, i.e., short-lived upward jumps of electricity prices. By setting $\lambda_N \gg \lambda_J \gtrsim 0$, we generate sporadic spikes of short duration. Again, the conditional distribution of $X_{2,t}$ given $X_{2,0}$ can be calculated explicitly. It is given by

$$X_{2,t} \sim \begin{cases} \mathcal{N}(\mu_J, \sigma_J^2) & \text{with probability } \frac{-\lambda_N e^{(\lambda_N + \lambda_J)t} + \lambda_N}{\lambda_J + \lambda_N} \\ 0 & \text{with probability } \frac{\lambda_N e^{(\lambda_N + \lambda_J)t} + \lambda_J}{\lambda_J + \lambda_N} \end{cases}$$

if $X_{2,0} = 0$, and

$$X_{2,t} \sim \begin{cases} X_{2,0} & \text{with probability } e^{\lambda_J t} \\ \mathcal{N}(\mu_J, \sigma_J^2) & \text{with probability } \frac{-\lambda_J e^{(\lambda_N + \lambda_J)t} + \lambda_N}{\lambda_J + \lambda_N} - e^{\lambda_J t} \\ 0 & \text{with probability } \frac{\lambda_J e^{(\lambda_N + \lambda_J)t} + \lambda_J}{\lambda_J + \lambda_N} \end{cases}$$

if $X_{2,0} = J_0$. As already mentioned, the diffusive and spike risk factors are assumed to determine the spot price. If $E(\pi_t)$ denotes the expected spot price at time t, which accounts for all foreseeable seasonal fluctuations, then we can impose the following relationship between π_t and the risk factors X_1 and X_2:

$$\pi(t) = E(\pi_t) \frac{\exp(X_{1,t} + X_{2,t})}{E(\exp(X_{1,t} + X_{2,t}))}$$

We should emphasize once again that the processes γ, δ, and ϵ can be modeled in a similar manner by introducing additional risk factors X_3, X_4, etc. The latter also follow affine jump diffusion models of the type considered above. In general, we will assume that the diffusive risk factors are independent of the jump risk factors. However, the diffusive and spike factors may be mutually correlated among themselves, respectively.

Recall that we are ultimately interested in the distribution P_ξ, which is obtained when ξ is interpreted as a random vector on an augmented probability space. To this end, we select a "maturity" date T and a Borel measure λ on $[0, T]$. As power system expansion planning usually encompasses a very long planning horizon, it is reasonable to let T tend to infinity. Moreover, we select an absolutely continuous λ with density e^{-rt}. Thus, λ accounts for discounting. In order to generate a sample from the distribution P_ξ, we first generate a sample t from an exponential distribution with parameter r. Keeping t fixed, we then generate another sample from the distribution of the random vector ξ_t. This can be done by using the explicit solutions of the above stochastic differential equations. The resulting vector represents in fact a sample of the distribution P_ξ.

By construction, the marginal distribution (with respect to P_ξ) of the risk factors α and β is discrete. However, the other risk factors are continuously distributed. In principle, thus, the second stage problem (7.2b) must be solved for a continuum of ξ-values. Since this is impossible, one usually constructs a probability measure Q_ξ with finite support that approximates P_ξ in some sense. Then, problem (7.2) is solved with P_ξ replaced by Q_ξ. The selection of an appropriate discrete probability measure is referred to as *scenario generation* and represents a primary challenge in the field of stochastic programming. A survey and critical assessment of modern scenario generation techniques is provided in [23]. We use here the bounding approximation scheme developed in [24, 25]. This method exploits the structural properties of the model (7.2) to construct Q_ξ in such a way that the approximate problem associated with Q_ξ provides an upper (or lower) bound on the original problem associated with P_ξ. By calculating bounds bracketing the true optimal value, we can therefore obtain a deterministic error estimate.

7.5
Decomposition Algorithms

Once the capacity expansion problem (7.2) has been discretized, then it becomes a large scale mixed-integer linear program (MILP). An important characteristic of the proposed model formulation is that only the first stage decisions constitute integer variables. Moreover, once the first stage decisions have been fixed, then the second-stage decision variables belonging to different scenarios become independent of each other, and we can solve the resulting scenario subproblems very efficiently. We generalize a decomposition algorithm due to Parpas and Rustem [26] to take into account the integer variables. The method is similar to the standard L-shaped algorithm for integer variables (see for example [13]), except that we use a proximal term which improves the numerical performance. Before we delve into decomposition algorithms, we introduce some terminology that will be used in the next section.

The recourse model described in Section 7.2.1 naturally accounts for *nonanticipativity*. This means that decisions are only based on past and present observations but not on future ones. There are two ways to capture this idea, namely compact and split-view formulations (see, e.g., [27]).

The compact variable formulation maps the problem directly to a tree structure, that is, a *scenario tree*; see Fig. 7.2(a).

The root of the tree represents the first decision stage at which the decision maker has no information about ξ. As we move down the scenario tree, information about ξ is gradually revealed. Each level of the tree represents a different de-

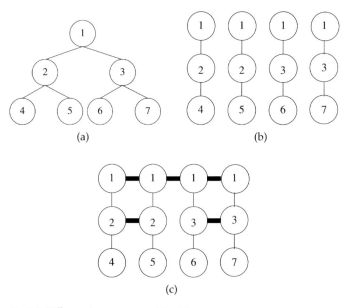

Fig. 7.2 Different views on nonanticipativity.

cision stage. Benders decomposition, to be introduced in the next section, exploits this structure and decomposes the large scale problem into several subproblems, each corresponding to a node in the tree.

In a split-variable formulation, we artificially increase the number of decision variables. Concretely speaking, in each decision stage we introduce a separate decision variable for each scenario. This has the advantage that the large-scale problem decomposes into $|\operatorname{supp} Q_\xi|$ simple subproblems, where $|\operatorname{supp} Q_\xi|$ is the number of scenarios under the discrete probability measure Q_ξ. Since the first-stage decision may now be scenario-dependent, this approach relaxes nonanticipativity, see Fig. 7.2(b). To restore nonanticipativity, new constraints are introduced that link the subproblems. These constraints force decisions corresponding to undistinguishable scenarios to be equal, which is usually achieved by using some penalty function (see Fig. 7.2(c)).

The properties of our model suggest that a tree representation is more appropriate, and we therefore adopt this formulation in the proposed solution algorithm.

7.5.1
Nested Benders Decomposition

Benders decomposition was first proposed by Benders [28], and has been applied to SP by Van Slyke and Wets [29]. In the context of SP, Benders decomposition is usually referred to as the *L-shaped method*, which hints to the structure of the underlying constraint matrix. The extension to nonlinear convex mathematical programs is due to Geoffrion [30], while the extension to general convex stochastic programs goes back to Birge and Rosa [31]. The algorithm has also been applied to multi-stage problems in a parallel environment by Birge et. al. [32]. More recent studies are due to Nielsen and Zenios [33]. Furthermore, Louveaux [34] has studied the algorithm in the convex quadratic case. Recall that the model (7.2) reduces to a simple two-stage problem since the operational decisions belonging to different time points in the interval $[0, T]$ are decoupled. Our algorithm essentially coincides with the regularized decomposition algorithm by Ruszczyński [35], combined with a branch-and-bound scheme. The branch-and-bound framework is thoroughly explained in [36], while numerical experiments can be found in [37, 38]. The basic idea is to approximate the first-stage problem (7.1) by

$$\min_{x \in X, \theta \in \mathbb{R}} \quad c(x) + \theta \tag{7.5a}$$

$$\text{s.t.} \quad Ax \leqslant b \tag{7.5b}$$

$$Dx \geqslant d \tag{7.5c}$$

$$Ex + \theta \geqslant e \tag{7.5d}$$

Note that the expected recourse function $Q(x)$ has been eliminated, while a new decision variable θ and two types of new constraints have been added. The constraints (7.5c) are known as feasibility cuts determining the effective domain $\{x|Q(x) < +\infty\}$ of the expected recourse function. The constraints (7.5d), on the other hand, are called optimality cuts and represent outer linearizations of Q on its effective domain. The derivation of these cuts can be found in [26, 35]. Note also that the integer constraints $x \in X$ will be relaxed during the solution procedure. The branch-and-bound algorithm creates a list of unexplored nodes within a branch-and-bound tree, each of which corresponds to a polyhedral subset $\tilde{X}$ of the convex hull of X. We denote this list by $\mathcal{G}(i)$, where i represents the iteration counter.

Our solution algorithm can now be described as follows:

Step 0 Set the iteration counter to $i = 0$.

Step 1 Select a node $\tilde{X}$ from $\mathcal{G}(i)$. If none exists, stop.

Step 2 Set $i := i + 1$, and solve the current relaxed problem,[2] i.e., problem (7.5) with X replaced by $\tilde{X}$. If the current problem has no feasible solution, or if its optimal value is below the current best solution, then fathom the current node and go to Step 1.

Step 3 If any second-stage constraint is violated, add a feasibility cut; otherwise, add optimality cuts (details of this step can be found in [26]).

Step 4 If an integer constraint is violated, create two new nodes according to the branch-and-bound procedure. Append the two nodes to $\mathcal{G}(i)$.

Step 5 If the current solution is optimal and below the current best solution, then fathom the current node. Return to Step 1.

Acknowledgements

The authors wish to acknowledge support from EPSRC grants GR/T02560/01 and EP/C513584/1.

References

1 WALLACE, S. W., FLETEN, S.-E., Stochastic programming models in energy. In: *Stochastic Programming*, Vol. 10 of *Handbooks Oper. Res. Management Sci.*, Elsevier, Amsterdam, **2003**, pp. 637–677.

2 GRÖWE-KUSKA, N., RÖMISCH, W., Stochastic unit commitment in hydrothermal power production planning. In: *Applications of Stochastic Programming*, Vol. 5 of *MPS/SIAM Ser. Optim.*, SIAM, Philadelphia, PA, **2005**, pp. 633–653.

3 NOGALES, F. J., PRIETO, F. J., CONEJO, A. J., A decomposition methodology applied to the multi-area optimal power flow prob-

lem. *Annals of Operations Research* 120 (**2003**), pp. 99–116. OR Models for Energy Policy, Planning and Management, Part I.

4 EICHHORN, A., RÖMISCH, W., WEGNER, I., Mean-risk optimization of electricity portfolios using multiperiod polyhedral risk measures. In: *IEEE St. Petersburg PowerTech Proceedings*, **2005**.

5 DOEGE, J., SCHILTKNECHT, P., LÜTHI, H.-J., Risk management of power portfolios and valuation of flexibility. *OR Spektrum* 28(2) (**2006**), pp. 267–287.

6 PRITCHARD, G., PHILPOTT, A. B., NEAME, P. J., Hydroelectric reservoir optimization

---

2) Notice that the relaxed problem constitutes a linear program.

in a pool market. *Mathematical Programming, Ser. A* 103(3) (**2005**), pp. 445–461.

7 FLETEN, S.-E., WALLACE, S. W., ZIEMBA, W. T., Hedging electricity portfolios via stochastic programming. In: *Decision Making Under Uncertainty*, Vol. 128 of *IMA Vol. Math. Appl.*, Springer, New York, **2002**, pp. 71–93.

8 VEHVILAINEN, I., KEPPO, J., Managing electricity market price risk. *European Journal of Operational Research* 16(1) (**2003**), pp. 136–147.

9 BLOOM, J., Solving an electricity generating capacity expansion planning problem by generalized Benders decomposition. *Operations Research* 31(1) (**1983**), pp. 84–100.

10 LEOPOLDINO, C. M. A., PEREIRA, M. V. F., PINTO, L. M. V., RIBEIRO, C. C., A constraint generation scheme to probabilistic linear problems with an application to power system expansion planning. *Annals of Operations Research* 50 (**1994**), pp. 367–385. Applications of Combinatorial Optimization.

11 HOBBS, B. F., JI, Y., Stochastic programming-based bounding of expected production costs for multiarea electric power system. *Operations Research* 47(6) (**1999**), pp. 836–848.

12 JAILLET, P., RONN, E., TOMPAIDIS, S., Valuation of commodity based swing options. *Manage. Science* 50(7) (**2004**), pp. 909–921.

13 BIRGE, J. R., LOUVEAUX, F., *Introduction to Stochastic Programming*, Springer, New York, **1997**.

14 PRÉKOPA, A., *Stochastic Programming*, Vol. 324 of Mathematics and its Applications, Kluwer Academic Publishers Group, Dordrecht, **1995**.

15 BENTAL, B., RAVID, S. A., A simple method for evaluating the marginal cost of unsupplied electricity. *Bell Journal of Economics* 13(1) (**1982**), pp. 249–253.

16 BEENSTOCK, M., Generators and the Cost of Electricity Outages. In: *Energy Economics* (**1991**).

17 CAVES, D., HERRIGES, J., WINDLE, R., The cost of electric power interruptions in the industrial sector. *Land Economics* 68 (1) (**1992**), pp. 49–61.

18 ROCKAFELLAR, R., URYASEV, S., Conditional value-at-risk for general loss distributions.

Journal of Banking & Finance 26 (**2002**), pp. 1443–1471.

19 DENG, S., JOHNSON, B., SOGOMONIAN, A., Exotic electricity options and the valuation of electricity generation and transmission assets. *Decision Support Systems* 30(3) (**2001**), pp. 383–392.

20 BERGER, M., KLAR, B., MÜLLER, A., SCHINDLMAYR, G., A spot market model for pricing derivatives in electricity markets. *Quantitative Finance* 4(1) (**2004**), pp. 109–122.

21 DUFFIE, D., PAN, J., SINGLETON, K., Transform analysis and asset pricing for affine jump-diffusions. *Econometrica* 68(6) (**2000**), pp. 1343–1376.

22 PILIPOVIC, D., *Energy Risk: Valuing and Managing Energy Derivatives*, McGraw-Hill, New York, **1998**.

23 DUPAČOVÁ, J., CONSIGLI, G., WALLACE, S. W., Scenarios for multistage stochastic programs. *Annals of Operations Research* 100 (**2000**), pp. 25–53 (2001). Research in Stochastic Programming (Vancouver, BC, 1998).

24 KUHN, D., Aggregation and discretization in multistage stochastic programming. *Mathematical Programming Ser. A* 113(1) (**2008**), pp. 61–94.

25 FRAUENDORFER, K., Barycentric scenario trees in convex multistage stochastic programming. *Mathematical Programming* 75(2) (**1996**), pp. 277–294.

26 PARPAS, P., RUSTEM, B., Computational assessment of nested benders and augmented Lagrangian decomposition for mean-variance multistage stochastic problems. *INFORMS Journal on Computing* 19(2) (**2007**), pp. 239–247.

27 ROCKAFELLAR, T., WETS, R., Scenarios and policy aggregation in optimization under uncertainty. *Mathematics of Operations Research* 16(1) (**1991**), pp. 119–147.

28 BENDERS, J., Partitioning procedures for solving mixed-variables problems. *Numerische Mathematik* 4 (**1962**), pp. 238–252.

29 VAN SLYKE, R., WETS, R. J.-B., L-shaped linear programs with applications to control and stochastic programming. *SIAM Journal on Applied Mathematics* 17 (**1969**), pp. 638–663.

30 GEOFFRION, A., Generalized Benders decomposition. *Journal of Optimization The-*

ory and Applications 10(4) (**1972**), pp. 237–260.

31 BIRGE, J. R., ROSA, C. H., Parallel decomposition of large-scale stochastic nonlinear programs. *Annals of Operations Research* 64 (**1996**), pp. 39–65.

32 BIRGE, J. R., DONOHUE, C. J., HOLMES, D. F., SVINTSITSKI, O. G., A parallel implementation of the nested decomposition algorithm for multistage stochastic linear programs. *Mathematical Programming* 75 (2) (**1996**), pp. 327–352.

33 NIELSEN, S. S., ZENIOS, S. A., Scalable parallel Benders decomposition for stochastic linear programming. *Parallel Computing* 23(8) (**July 1997**), pp. 1069–1088.

34 LOUVEAUX, F., Piecewise convex programs. *Mathematical Programming* 15 (**1978**), pp. 53–62.

35 RUSZCZYŃSKI, A., A regularized decomposition method for minimizing a sum of polyhedral functions. *Mathematical Programming* 35 (**1986**), pp. 309–333.

36 FLOUDAS, C., *Deterministic Global Optimization*, Vol. 37 of *Nonconvex Optimization and its Applications*, Kluwer Academic Publishers, Dordrecht, **2000**. Theory, Methods and Applications.

37 ADJIMAN, C., ANDROULAKIS, I., FLOUDAS, C., Global optimization of mixed-integer nonlinear problems. *AICHE Journal* 46 (9) (**2000**), pp. 1769–1797.

38 ADJIMAN, C., ANDROULAKIS, I., FLOUDAS, C., Global optimization of minlp problems in process synthesis and design. *Computers and Chemical Engineering* 21 (**1997**), pp. 455–450.

8
Integrated Design of CO_2 Capture Processes from Natural Gas

Frances E. Pereira, Emmanuel Keskes, Amparo Galindo, George Jackson and Claire S. Adjiman

Keywords

natural gas, Ryan–Holmes process, CO_2 capture, flowsheet, process model, total capital investment (TCI), operating costs (OC)

8.1
Introduction

The increasing importance of natural gas as a source of energy poses difficult gas separation design challenges, as the streams recovered from gas fields are at high pressures (typically about 10 MPa) and can contain a high proportion of CO_2 (up to 70%). These natural gas streams also have very high flowrates, in the region of 100–1000 million standard cubic feet per day (MMSCFD). This may give rise to 70–700 MMSCFD of CO_2, comparable with the CO_2 production of a typical power station. In addition, as the implementation of the Kyoto protocol would require the capture of large quantities of CO_2, its injection into depleted, or near-depleted, reservoirs for enhanced oil/gas recovery operations will become increasingly frequent. This is likely to result in gas streams that are even richer in CO_2, and that have an increasing CO_2 concentration over the life of the gas field. Conventional CO_2 separation techniques used in the industry are usually restricted to low CO_2 content or low-pressure feeds. The dynamic nature of the natural gas composition also creates an additional challenge of designing a flexible separation process that is capable of dealing with these highly variable inlet conditions. As a consequence, there exists a need for the development of a robust, flexible, and economically viable process for the removal of CO_2 from natural gas streams.

In this chapter, we introduce such a new process for CO_2 capture from natural gas, and then describe an integrated strategy for its design and optimization. The proposed process strips the natural gas stream of CO_2, through physical absorption into an *n*-alkane solvent. The CO_2 concentration in the outlet sale gas stream is reduced to below 3%, the requirement of standard commercial grade gas. The design methodology involves simultaneous optimization of the solvent chain length and process conditions, and is undertaken for various process flowsheets.

Process Systems Engineering: Vol. 5 Energy Systems Engineering
Edited by Michael C. Georgiadis, Eustathios S. Kikkinides and Efstratios N. Pistikopoulos
Copyright © 2008 WILEY-VCH Verlag GmbH & Co. KGaA, Weinheim
ISBN: 978-3-527-31694-6

8.1.1
Choice of Separation Technique

The techniques used in the gas separation industry include adsorption onto solid substrates, chemical absorption, gas permeation, and physical absorption [1]. Adsorption is economical for purification, typically reducing the CO_2 content from 3% down to 0.5%. Chemical absorption has been used successfully for low-pressure gas streams containing between 3% and 25% of CO_2. Such processes have very high selectivities, of >99%, but involve large solvent regeneration costs, which hamper their application to higher CO_2 contents. The absorption is limited by the stoichiometry of the chemical reaction (with an amine to CO_2 ratio of 2:1 for monoethanolamine) so that the use of this process for CO_2-rich gas streams will lead to high solvent circulation flowrates and high energy requirements. Gas permeation techniques are compact and flexible, adapting easily to changes in CO_2 content. However, reliability is a concern, especially because natural gas contaminants can lead to the deterioration of the membrane.

Physical absorption can also be used successfully. Its main advantage is the potentially greater absorption limits of physical absorption solvents with respect to CO_2, as compared to chemical absorption solvents. This difference is due to the nature of the absorption limits in both cases; the physical limit is thermodynamic, whereas the chemical limit is stoichiometric. At high CO_2 partial pressure, the CO_2 loading capacity of the solvent has the potential to be higher for a physical solvent than for a chemical solvent. Hence, physical absorption processes are particularly appropriate for the treatment of CO_2-rich gas streams. The commercial viability of physical absorption processes for CO_2 capture with n-alkane solvents is exemplified by the successful application of the Ryan–Holmes cryogenic separation process [2] to natural gas treatment.

8.1.2
Choice of Solvent for Physical Absorption

The choice of solvent is one of the key decision variables impacting on the performance and economics of a physical absorption process. Many solvents have been used for the absorption of CO_2 and CH_4, including various formulations of tributyl phosphate, polycarbonate, methylcyanoacetate, and n-formyl morpholine [3]. Unfortunately, there are major drawbacks with all of these for practical operations; the solvents are not easily disposable and could be involved in side reactions with other natural gas constituents. A more suitable solvent, which does not react and can easily be handled in an oil and gas environment, is n-butane. This is used in the afore-mentioned Ryan–Holmes process [2]. The process has a satisfactory CO_2/CH_4 separation factor, but operation at low temperatures is very energetically demanding. Like n-butane, other alkanes, such as n-decane, are known to absorb CO_2 preferentially to CH_4 [4–7]. The use of higher alkanes, or alkane *blends*, may provide a promising route toward adapting the Ryan–Holmes process to the temperatures and pressures typical of gas fields.

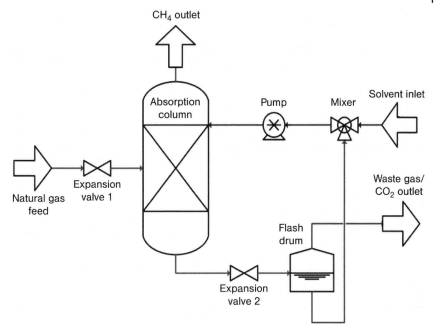

Fig. 8.1 Flowsheet A: Basic process flowsheet for the removal of CO_2 from natural gas through physical absorption into an n-alkane solvent.

Often, the choice of solvent is made prior to the design of the flowsheet structure and operating conditions. However, these decisions are closely linked, and the best process performance can only be achieved by considering them simultaneously. Several methods for integrated process and solvent design have been proposed [8–10], but they usually rely on thermodynamic models that are only applicable at low pressures. In this chapter, we therefore seek to demonstrate how the incorporation of advanced thermodynamic methods into computer-aided process design may be used to tackle the design of a solvent and a process for high-pressure separation, in an integrated manner.

8.1.3
A Basic Physical Absorption Process Flowsheet

A starting point for the design of an optimal flowsheet is the consideration of the units required for the basic functioning of the absorption process. This flowsheet is shown in Fig. 8.1, and contains the following unit operations:

- Absorption column
- Flash drum
- Expansion valve
- Mixer
- Pump

The process operates as follows: The natural gas stream is expanded, to ensure a constant flowrate, and to allow control of the pressure in the absorption column. It then enters the absorption column, where it comes into contact with an n-alkane solvent. The solvent absorbs CO_2 preferentially to CH_4 and the gaseous stream is consequently stripped of CO_2. The CO_2-enriched solvent is then expanded into a flash drum to form a CO_2-rich vapor phase and CO_2-lean liquid phase. The regenerated solvent is recycled to a mixer where new solvent is added, and the CO_2-rich vapor phase leaves the process for disposal, or further use. The cleaned gas stream, or 'sale gas', is sent for retail. For simplicity, the natural gas stream is modeled as containing only CH_4 and CO_2.

Although this basic flowsheet is sufficient for the job of separating the CO_2 and CH_4, the design may be improved, so as to maximize the process efficiency. The performance of the process depends on many factors. These include the thermodynamic behaviour of the solvent + CH_4 + CO_2 mixture, which is highly influenced by process conditions. In addition, the general process layout will affect both the degree of separation and also the process efficiency. In particular, the number of absorption/desorption stages plays an important role in the purity of the two outlet streams. Finally, the physical design of the absorber/desorber, and the flow patterns within them, will have an effect, although this will not be investigated here. The extent of the absorption of CO_2 into the solvent is partially controlled by adjusting the temperature and pressure in the absorber. The pressure inside the absorber may be controlled through the expansion valve at the inlet, provided the required absorber pressure is below that of the inlet stream. The specifics of modeling the individual process units are described in the next section.

8.2
Development of a Process Model

The process model has a modular construction, allowing units to be easily added and removed. This facilitates the assessment of both the technical and economic aspects of various flowsheets. The process model performs simultaneous mass, energy, and phase equilibria calculations. Steady-state mass and energy balances for all the units in the flowsheet have been implemented, and thermodynamic equilibrium is assumed at each stage. The main assumptions made in each unit model are summarized in Table 8.1.

Within the model one is able to identify whether one or two phases are present in any of the units. This is particularly important for the compressors, where vapor phase conditions should be maintained. Two-phase equilibrium is specified via the equality of pressures and chemical potentials of all components in the two phases. The crystallization of the hydrocarbon solvent and regions of liquid–liquid immiscibility, and three-phase separation, are avoided by constraining the process conditions.

The following physical properties are required to develop the process model: enthalpy, entropy, heat capacity, chemical potential, and pressure. These properties

Table 8.1 Main assumptions used in the construction of unit models.

Unit	Assumptions
Entire process	Perfect mixing, thermodynamic equilibrium, conservation of mass and energy balances.
Flash drum	Liquid and vapor outlet streams have qualities 1 and 0, respectively, i.e., complete separation of vapor and liquid.
Mixer	Perfect mixing.
Absorption column	Each stage modeled as flash drum. Thermodynamic equilibrium reached at every stage.
Expansion valve	Isenthalpic expansion.
Compressor	Isentropic compression.
Pump	As compressor.
Heat exchanger	Heat exchange fluid: water/steam. Counter current shell and tube configuration. Heat transfer coefficients from Douglas [11].

should be calculated using an equation of state, given the high-pressure conditions and nonideality of the mixture. The SAFT–VR equation of state [12, 13] has been chosen for its ability to represent phase behavior accurately over a wide range of conditions. In particular it can treat homologous series of n-alkane + CO_2 mixtures, in a predictive way [14, 15]. The SAFT–VR equation of state is described in the next section.

Various correlations were developed for this model, either to allow the n-alkane solvent chain length to be treated as a noninteger, continuous variable, or to improve on the accuracy of literature correlations. Specifically, correlations were developed for absorption column tray efficiency, n-alkane critical temperature and pressure, n-alkane melting temperature, unit pressure cost factors and overall installation factors (OIF). Details of these correlations are given in Keskes' thesis [16]. The absorption column tray efficiency correlation requires a knowledge of the solvent viscosity, as does the calculation of the pump power. The viscosity of the solvent *blend* is obtained from the correlation of Ducoulombier et al. [17].

In general, the costing of the separation process is split into total capital investment (TCI) and operating costs (OC). Apart from the correlations developed in this work, the TCI is calculated through literature correlations with the size or power of specific units, quoted by Perry et al. [18] or by Douglas [11]. It not only includes the purchased cost of the units, but also a variety of associated charges, such as installation costs and insurance. The OC include such contributions as labor, utilities and maintenance. A full description of the unit models and cost estimation methodology may be found in [16].

8.2.1
SAFT–VR Model for CO_2/CH_4/n-Alkane Mixtures

In the SAFT–VR equation of state, nonassociating molecular species, as is the case here, are treated as follows. Each component, i, is described by two molecular size

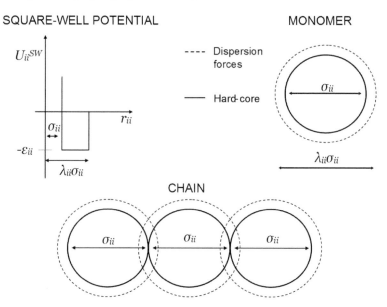

Fig. 8.2 The square-well potential (U_{ii}^{SW}) of the interactions between monomer segments in SAFT–VR, a monomer segment and a chain of three monomer segments. Each molecule, i, is represented as a chain of m_i identical, tangentially connected spheres, each with a diameter σ_{ii}. These monomer segments interact through a square-well potential with a depth ϵ_{ii} and a range λ_{ii}.

parameters; the number of spherical segments in a molecule (m_i), and the hard-core diameter of the segment (σ_{ii}), and also two energy parameters; the depth (ϵ_{ii}) and range (λ_{ii}) of the square-well shaped dispersive interactions. The physical interpretation of these four parameters is shown in Fig. 8.2.

Intermolecular parameters for CO_2 and CH_4 have previously been optimized to describe coexistence data up to 90% of the critical temperature [14, 19]; this usually leads to an over prediction of the critical pressure and temperature. Such a problem is prevalent with equations of state, and is due to a discrepancy between the real experimental and theoretical critical exponents. As the region near the critical point of these two compounds is of interest here, the size and energy parameters σ_{ii} and ϵ_{ii} have been rescaled to reproduce the critical temperature and pressure (see also [14] for more detail). The resulting parameters are presented in Table 8.2.

The SAFT–VR equation allows a complete family of solvents to be considered in design. For the n-alkane solvent used in the process, a correlation that gives the SAFT–VR parameters as a function of the molecular weight MW_a, in $g\,mol^{-1}$, of n-alkane a has been used [19]. This is described by Eqs. (8.1)–(8.5):

$$MW_a/(g\,mol^{-1}) = 14 \times N_a + 2 \tag{8.1}$$

$$m_a = 0.02376 \times MW_a/(g\,mol^{-1}) + 0.6188 \tag{8.2}$$

Table 8.2 SAFT–VR square-well segment parameters for CH_4 and CO_2. (For component i;
m_i = number of monomer segments in chain,
σ_{ii} = diameter of each segment, in Å, ϵ_{ii} = depth of the intermolecular potential between 2 segments and
λ_{ii} = range of the intermolecular potential between 2 segments, k_B = the Boltzmann constant.)

Component i	m_i	σ_{ii} (Å)	ϵ_{ii}/k_B (K)	λ_{ii}
CH_4	1.0	4.0576	156.50	1.4479
CO_2	2.0	3.1364	168.89	1.5157

$$m_a(\sigma_{aa}/\text{Å})^3 = 1.53212 \times MW_a/(\text{g mol}^{-1}) + 30.753 \tag{8.3}$$

$$m_a(\epsilon_{aa}/k_B)/\text{K} = 5.46587 \times MW_a/(\text{g mol}^{-1}) + 194.263 \tag{8.4}$$

$$m_a\lambda_{aa} = 0.04024 \times MW_a/(\text{g mol}^{-1}) + 0.6570 \tag{8.5}$$

where N_a is the carbon number of the n-alkane, a.

Standard mixing rules can be used to model mixtures of different molecules [13]. These mixing rules invariably involve the determination of binary parameters, to describe the interactions between species i and j. The following relations have been used to calculate these parameters:

$$\sigma_{ij} = \frac{\sigma_{ii} + \sigma_{jj}}{2} \tag{8.6}$$

$$\epsilon_{ij} = \left(1 - k_{ij}^\epsilon\right)\sqrt{\epsilon_{ii}\epsilon_{jj}} \tag{8.7}$$

$$\lambda_{ij} = \frac{\sigma_{ii}\lambda_{ii} + \sigma_{jj}\lambda_{jj}}{\sigma_{ii} + \sigma_{jj}} \tag{8.8}$$

where k_{ij}^ϵ is an additional parameter that captures the deviation of the unlike interaction energy from the geometric mean.

The k_{ij}^ϵ interaction parameter has been estimated based on isothermal vapor–liquid equilibrium data over a wide range of temperatures and pressures for each of the three relevant binary mixtures (CO_2/CH_4, CO_2/n-decane, and CH_4/n-decane). For CO_2/CH_4 mixtures, several sources of experimental data were used [20–32]. n-decane (C_{10}) has been used as a representative compound in the n-alkane series, as large sets of experimental data are available. For CO_2/C_{10} and CH_4/C_{10} mixtures, experimental data were selected at pressures below 10 MPa and temperatures below 477 K; this corresponds to the operating range of the process. For the CO_2/C_{10} mixture, experimental data were obtained from the sets in [33–39], and for the CH_4/C_{10} mixture from those in [40–44]. Each k_{ij}^ϵ was estimated by using the maximum likelihood objective function for the total pressure and the vapor mole fraction, as a function of liquid mole fraction and temperature. A constant variance

Table 8.3 Errors in prediction of pressure and vapor mole fraction, using the estimated SAFT–VR parameters (AAPE, absolute average percentage error; AAD, absolute average deviation.)

Mixture	AAPE% for the pressure	AAD (mol) for the vapor mole fraction
CH_4/CO_2	0.021	0.0236
CO_2/C_{10}	0.099	0.0010
CH_4/C_{10}	0.076	0.0027

model was assumed for the error on the experimental data. All the experimental data available in [45] were used, namely 111 points for CH_4/C_{10} and 85 points for CO_2/C_{10}, both over 15 temperature values and 312 points over 24 temperatures for CH_4/CO_2. The errors between experiment and prediction, for the pressure and the vapor mole fraction, can be found in Table 8.3.

The following interaction parameters were obtained:

$$k^\epsilon_{CH_4,C_{10}} = -0.053006, \quad k^\epsilon_{CH_4,CO_2} = +0.036798, \quad \text{and}$$

$$k^\epsilon_{CO_2,C_{10}} = +0.089642.$$

Since the *n*-alkanes belong to a homologous series, we assume that the binary interaction parameters between CH_4 (or CO_2) and *any* *n*-alkane solvent are the same as those between CH_4 (or CO_2) and *n*-decane. Furthermore, *n*-alkane *i*/*n*-alkane *j* interactions are described by Eqs. (8.6), (8.7), and (8.8), with $k^\epsilon_{ij} = 0$. To a first approximation, this allows for *n*-alkane solvent *blends* to be represented by treating the *blend* as a single species with a molecular weight that is the average of the *blend*.

One key assumption of the thermodynamic model is that ternary phase behavior can be captured by using binary interaction parameters. To test the adequacy of a pair interaction theory in describing the ternary system, we compare predicted tie-lines for mixtures of CO_2, CH_4, and *n*-decane with experimental data [4]. The SAFT–VR thermodynamic description is in very good agreement with the data, even at pressures above 10 MPa, as shown in Fig. 8.3.

The first part of this chapter has focused on describing the development of the thermodynamic description and process models for the absorption process. In the following sections, the optimization of the process model and the impact of this optimization on the process performance are discussed.

8.3
Optimization Methodology

In the physical absorption process, there are three interconnected optimization areas, that each impact on process efficiency. These three areas are as follows:

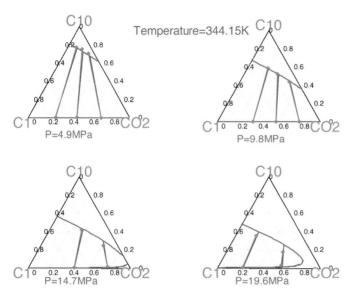

Fig. 8.3 Vapor–liquid equilibrium of the ternary CO_2 + CH_4 + n-decane mixture. Three data sets obtained at different pressures of 4.9 MPa, 9.8 MPa, 14.7 MPa, and 19.6 MPa, and a temperature of 344.15 K, are shown. The experimental points are shown as diamonds (joined by the tie-lines) [4], and the SAFT–VR calculations as the continuous curves.

1. The process flowsheet
2. The process conditions
3. The absorption solvent

The conditions and composition of the inlet stream also impact heavily on the process performance, but this is treated as an input parameter, rather than a design variable. Optimization of the process flowsheet includes changes to the layout of the existing units in the basic flowsheet, and also the addition of further units, such as heat exchangers and compressors, or the duplication of existing units, such as the absorber. The impact of different flowsheets on process performance is examined, although the flowsheet itself is not optimized in this work. The process conditions include the temperature and pressure of each unit, and also the solvent flowrate. In many cases, and in the case of the process being examined , there are a number of similar solvents that are all suitable absorption media (this is discussed in Section 8.1.2). Multiple solvents, or solvent *blends*, must therefore be screened to determine that which performs the best. For solvents to be compared on a fair basis, the process conditions must be optimized for each candidate solvent/*blend*.

The basic absorption flowsheet, shown in Fig. 8.1, is used as a starting point for the flowsheet layout. Different flowsheets are then screened, during which the process conditions and solvent chain length (or solvent molecular weight (MW_a), cf. Eqs. (8.1)–(8.5)) are optimized for the given flowsheet. The highest achievable value of the objective function for the flowsheet in question is calculated. This procedure is demonstrated, for a limited choice of prototype flowsheets, in Section 8.3.2. The optimization problem solved for the individual flowsheets is described in the next section.

8.3.1
Formulation of the Design Problem

For a feed of fixed flowrate and composition, and a fixed number of trays in each absorber ($N_T = 10$), the vector of design variables, d, for the gas separation process consists of the solvent flowrate (FS), the absorber pressures (PA_0, PA_1, ..., PA_{NA}, where NA is the number of absorption columns), the flash unit pressures (PF_0, PF_1, ..., PF_{NF}, where NF is the number of flash units), the heat exchanger areas (AHE_1, AHE_2, ..., AHE_{NHE}, where NHE is the number of heat exchangers), the flowrates of heat exchange media (FHE_1, FHE_2, ..., FHE_{NHE}) and the carbon number of the alkane solvent (N_a). Only n-alkanes with a carbon number between 7 and 14 are considered, because smaller n-alkanes are likely to be too volatile, while larger n-alkanes would be too viscous and could solidify at the thermodynamic conditions of interest.

In addition, there are a number of constraints on the operating conditions of the process. These include: a maximum column height (HC_0, HC_1, ..., HC_{NA}) of 50 m; a maximum column cross-sectional area (AC_0, AC_1, ..., AC_{NA}) of 30 m^2; a maximum solvent viscosity in the columns (denoted by vector μ_S) of 100 cP; a gas stream temperature into the absorption columns ($TC_{IN,1}$, $TC_{IN,2}$, ..., $TC_{IN,NA}$) of between 273.15 K and 500 K; a temperature constraint to maintain all process temperatures (T, a vector) above the melting temperature of the alkane chain length in question (T_{SOLID}(n-alkane)), to avoid solidification of the solvent around the process; a minimum 5K temperature difference between hot and cold heat exchanger streams (DT_1, DT_2, ..., DT_{NHE}) [11]; a minimum purity of the CH$_4$ outlet stream ($x_{CH_4OUTLET}$) of 97% (i.e., a CH$_4$ mole fraction of 0.97), as mentioned in Section 8.1; and finally, a minimum volume difference between liquid and vapor phases (DV, a vector), to avoid convergence to the trivial solution in phase equilibrium calculations, and to enforce the formation of 2 phases, where necessary, around the process.

The general optimization problem is formulated as follows:

$$\max_{d} \quad f(d, x)$$

$$\text{s.t.} \quad h(d, x) = 0$$

$$FS \geqslant 0$$

$$PA_0 \geqslant PA_1 \geqslant \cdots PA_{NA} \geqslant 0$$

$$PF_0 \geqslant PF_1 \geqslant \cdots PF_{NF} \geqslant 0$$

$$AHE_1, AHE_2, \ldots, AHE_{NHE} \geqslant 0$$

$$FHE_1, FHE_2, \ldots, FHE_{NHE} \geqslant 0$$

$$14 \geqslant N_a \geqslant 7$$

$$50 \geqslant (HC_0, HC_1, \ldots, HC_{NA} \geqslant 0$$

$$30 \geqslant AC_0, AC_1, \ldots, AC_{NA} \geqslant 0$$

$$100 \geqslant \mu_S \geqslant 0$$

$$500 \geqslant TC_{IN,1}, TC_{IN,2}, \ldots, TC_{IN,NA} \geqslant 273.15$$

$$T \geqslant T_{SOLID}(n\text{-alkane}) + 5$$

$$DT_1, DT_2, \ldots, DT_{NHE} \geqslant 5$$

$$x_{CH_4 OUTLET} \geqslant 0.97$$

$$DV \geqslant 10^{-4} \tag{8.9}$$

where f is a performance measure, h are the model equations, which include material and energy balances, thermodynamic relations, and the solvent structure–parameter relations, i.e., Eqs. (8.1)–(8.5), and x is a vector of process state variables and SAFT–VR solvent parameters. Variable N_a is a continuous variable as it represents the average carbon chain length of the solvent *blend*. Vector inequalities are understood component-wise. This problem is therefore a nonlinear programming (NLP) problem, which can be solved with standard solvers in the chosen modeling platform, gPROMS [46]. The profit-based objective function in this optimization has the form $f = gas\_sales - separation\_costs$. Here, the variable *gas_sales* is simply the net present value at the end of the plant life, of the quantity of methane leaving in the clean gas outlet multiplied by the price of sale gas input into the model. The variable *separation_costs* represents the net present value of the combined *OC* and *TCI*.

8.3.2
Process Optimization

A feed of 1 kmol s^{-1}, consisting of 30 mol% CO_2 and 70 mol% CH_4, at 7.96 MPa and 301 K is considered [47]. A target outlet pressure of 0.1 MPa is set for the CO_2 stream and the sale gas price is set as 10 USD/million BTU. The plant life is considered to be 15 years, and since there is uncertainty surrounding both future interest rates, and future natural gas prices, changes in the two are deemed to offset each other. Three potential flowsheets are optimized for process conditions and alkane

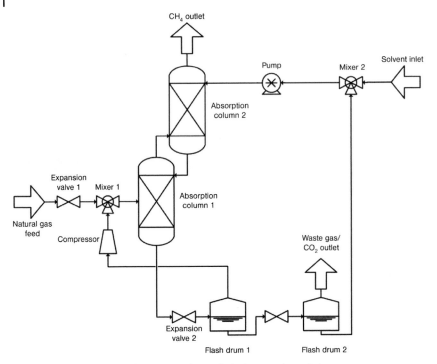

Fig. 8.4 Flowsheet B: Extra absorption column, flash drum, and recycle loop.

chain length (N_a), according to the problem as presented in Eq. (8.9). The process model equations are implemented in gPROMS [46]. The SAFT–VR calculations are implemented in Fortran90 and accessed via a Foreign Object interface [48].

The first flowsheet (A) is the most basic physical absorption layout, shown in Fig. 8.1. The second flowsheet (B) contains an extra absorber and flash drum, working on a hypothesis that since each absorber is limited to only ten stages, this may not be enough for an efficient separation, and that similarly, additional desorption stages may lead to improved separation. A recycle loop is also added from the first flash drum to the first column inlet, to reduce the loss of methane in the CO₂ outlet stream. This flowsheet is shown in Fig. 8.4. The third flowsheet (C) is the same as the second, except that a sea-water cooler is added before the first absorption column, since sea-water at low temperatures is readily available in many off-shore applications. The absorption is favored by high pressure and low temperature, therefore some reduction in absorber temperature may improve performance. This flowsheet is shown in Fig. 8.5.

8.4
Optimal Designs

The best objective function (profit) values obtained for each of the three flowsheets are shown in Table 8.4. Flowsheet C yields the highest profit of the three flowsheets

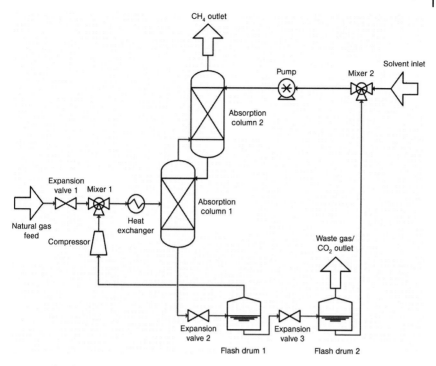

Fig. 8.5 Flowsheet C: Extra absorption column, flash drum, recycle loop, and cooler before absorption column.

investigated. It is interesting to note the large increase in the objective function between flowsheets B and C. The addition of a cooler before the first absorption column has a considerable impact on the objective function. This is due to the absorption of the CO_2 into the solvent being favored by low temperatures and high pressures. In process B, the best temperature/pressure compromise must be found simply through expansion of the inlet stream, which reduces both the temperature and the pressure. The inclusion of the cooler in process C allows the temperature in the absorption columns to be reduced, while simultaneously maintaining the high pressure of the natural gas stream. The absorption is consequently improved, since the column is at both low temperature and high pressure. This is demonstrated by the actual figures obtained for the absorption column pressures and temperatures, in both flowsheets; absorber 1 in flowsheet B has a pressure of 3.59 MPa and a bottom tray temperature of 311.06 K, whereas for flowsheet C these figures are 6.12 MPa and 302.00 K.

The improved absorption increases the revenue from natural gas sales. The TCI and OC also increase for flowsheet C, mainly due to increased recycle rates from the first flash drum, and the higher process pressures. However, the improvement in the separation performance is such that the increase in clean gas production (of around 30%) more than offsets these expenditures. The most promising flowsheet,

Table 8.4 The costs and profit associated with each flowsheet, for a CO$_2$ inlet condition of 30%. (All quantities are the net present value after a process life of 15 years, in USD million.)

Flowsheet	Capital costs (TCI)	Operating costs (OC)	Revenue from gas sales	Value of objective
A	53.5	80.9	1425.7	1291.3
B	119.1	156.0	1783.9	1508.8
C	189.6	337.1	2544.5	2017.8

Table 8.5 Values of the optimized design variables for flowsheet C and the concentration of CH$_4$ in the CO$_2$ outlet stream, over a range of CO$_2$ inlet concentrations.

% CO$_2$ in inlet stream	10%	30%	50%	70%
Average solvent chain length	13.35	12.67	13.20	13.05
Solvent flowrate (mol s^{-1})	1364.69	1669.39	1542.11	1351.4
Coolant flowrate (mol s^{-1})	9.54	10.01	7.41	4.66
Cooler area (m^2)	8596.80	7763.97	2522.29	366.85
Absorber 1 pressure (MPa)	7.33	6.12	6.42	7.39
Tank 1 pressure (MPa)	0.86	1.41	2.37	3.37
% CH$_4$ in CO$_2$ outlet stream	41.80	22.69	16.45	10.12

in terms of profitability, is therefore flowsheet C, and this will be the subject of the following, more detailed, discussion.

It has been demonstrated that the process remains profitable, under all the constraints included in Eq. (8.9), and achieves the CH$_4$ purity target of 97%. However, there is a further process requirement: flexibility. The properties of a natural gas stream do not remain constant throughout the life of the gas field and consequently the process must remain effective and profitable under varying input conditions. To assess the process capability in this respect, flowsheet C is optimized over a range of inlet compositions, from CO$_2$ inlet concentrations of 10% to 70%. The results of these optimizations, including the values of each of the optimized design variables, are presented in Table 8.5. The values in Table 8.5 are plotted in Fig. 8.6, and the profit and costs associated with the process over the different CO$_2$ inlet concentrations are presented in Table 8.6.

Some interesting trends may be observed in Fig. 8.6. The solvent recirculation rate has a maximum at around 30% CO$_2$ concentration in the input stream. In terms of the effectiveness of the separation this seems an unlikely trend, since the quantity of CO$_2$ to be absorbed in the solvent is increasing. This trend must therefore be due to the profit-based objective function. After a certain point, it becomes economically favorable to lose an increasing proportion of CH$_4$ to the waste gas stream, rather than increase recycle rates to allow for a clean separation. The 97% purity requirement of the CH$_4$ stream is still met, but since the purity of the CO$_2$ outlet stream and the quantity of CH$_4$ lost are not considered in the optimiza-

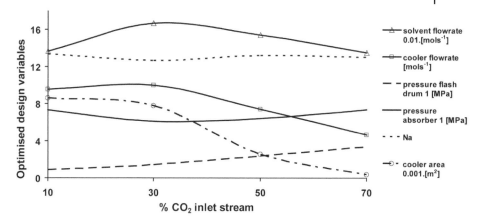

Fig. 8.6 The optimized design variables for flowsheet C over a range of CO_2 inlet concentrations.

Table 8.6 Costs and profit for flowsheet C over a range of CO_2 inlet concentrations. (All quantities are the net present value after a process life of 15 years, in USD million.)

% CO_2 in inlet stream	10%	30%	50%	70%
Revenue from gas sales	3489.7	2554.5	1664.5	914.5
Capital costs (TCI)	234.0	189.6	166.6	156.4
Operating costs (OC)	397.1	337.1	282.6	247.5
Profit	2858.7	2017.8	1215.3	510.6

tion, savings are made on reduced separation costs. This trend is reflected in the decreasing capital and operating costs with CO_2 inlet concentration, observed in Table 8.6. The cooler area and coolant flowrates are in line with the solvent recirculation rate, and support the trend of reduced costs and separation quality at high natural gas CO_2 concentrations. In general, the decreasing profits with CO_2 inlet concentration, presented in Table 8.6, are explained by a reduced fraction of the revenue-creating CH_4 in the total volume of gas being processed.

The optimal solvent chain length is relatively constant, at around a carbon number of 13, which corresponds to the diesel fraction. Solvents with higher values of N_a absorb larger quantities of CO_2. Therefore, if the energy costs of pumping a viscous solvent, and also the solidification constraints, did not contribute importantly to the objective function, one would expect to obtain the largest chain length permissible (i.e., 14 for this optimization). The fact that the optimal solvent chain length is consistently below this indicates the energetic favorability of conducting the absorption at lower temperatures, but with a shorter alkane chain lengths.

The pressure of the first flash drum increases monotonically with increasing CO_2 concentration. This is explained by the altered phase equilibrium associated with the changing composition of the stream leaving the bottom of absorption column 1, which becomes increasingly CO_2 rich. The pressure in flash drum 1 must be set so as to optimize the CH_4 recycled to the column, while minimising the total volume of gas recycled. The optimal pressure will therefore be that which favors a phase split in flash drum 1 that retains a maximum amount of CO_2 in the liquid phase, while recycling the greatest possible quantity of CH_4 to the absorption column.

The optimization of this flowsheet includes many conflicting demands, such as a high CH_4 outlet stream purity, and a profit-based objective function. This demonstrates the need for integrated optimization methodologies, since it would be impossible to find the optimal combination of operating strategy and solvent *blend* without simultaneous optimization of both. The optimizations performed here allowed the equipment sizes to vary over different CO_2 concentrations, the sizes being functions of factors such as flowrate. In practice, the equipment sizes could not alter with CO_2 inlet concentration. Consequently, the process would either have to be built with the "worst-case scenario" in equipment sizing, which would accommodate the inlet conditions requiring the greatest volumetric throughput, else there could be some retrofitting after a threshold point, as the natural gas composition changed over the life of the field. Systematic techniques for the design of flexible processes (e.g., [49]) could be applied to identify an optimal design over the life cycle of the field.

8.5
Conclusions

In this chapter, we have presented a process for the capture of CO_2 from natural gas streams though physical absorption into an *n*-alkane solvent. A methodology for the optimization of the solvent *blend* and operating conditions has also been presented, in the context of three different process flowsheets. The model indicates that this process is commercially viable, since profit through gas sales far outweighs capital and operating costs, over all CO_2 inlet concentrations studied. The flowsheet layout was not optimized in this study, however this would be possible, reformulated as a mixed-integer optimization problem. The process model could also be extended to investigate nonequilibrium situations, or to the screening of solvents outside the *n*-alkane family. A group contribution equation of state, such as SAFT-γ [50], would be particularly valuable in modeling these solvents.

This work has also demonstrated the benefits of employing a complex equation of state within a process modeling and optimization framework. The use of SAFT–VR facilitates the accurate prediction of phase behavior, even under highly non-ideal, high-pressure conditions. Since a physical absorption process is centered around this phase behavior, its accurate prediction is a crucial part of any valid process model.

Acknowledgements

E.K. and F.E.P. are grateful to Schlumberger Cambridge Research and the Engineering and Physical Sciences Research Council (EPSRC) of the UK, respectively, for PhD studentships. Additional funding from the EPSRC (grants GR/T17595, GR/N35991, GR/R09497, and EP/E016340), the Joint Research Equipment Initiative (JREI) (GR/M94427), and the Royal Society-Wolfson Foundation refurbishment grant is also acknowledged.

References

1 Kohl, A. L., Nielsen, R. B., *Gas Purification*, 5th Edition, Gulf, Houston, **1997**.
2 Holmes, A. S., Ryan, J. M., US Patent No. 4 318 723-A, **1979**.
3 Newman, S. A., *Acid and Sour Gas Treating Processes*, Gulf, Houston, **1985**.
4 Dunyushkin, I. I., Skripka, V. G., Nenartovich, T. L., Phase Equilibria in the Systems Carbon Dioxide – *n*-Butane – *n*-Decane and Carbon Dioxide – Methane – *n*-Decane, Imperial College London, VINITI issue, 2180-77, **1977**.
5 Orr, F. M., Silva, M. K., *Journal of the Society of Petroleum Engineers* 23(2) (**1983**), pp. 272–280.
6 Orr, F. M., Johns, R. T., Dindoruk, B., *SPE Reservoir Engineering* 8(2) (**1993**), pp. 135–142.
7 Blunt, M., Fayers, F. J., Orr, F. M., *Energy Conversion and Management* 34(9–11) (**1993**), pp. 1197–1204.
8 Eden, M. R., Jorgensen, S. B., Gani R., El-Hawagi, M. M., *Chem. Eng. Proc.* 43 (**2004**), pp. 595–608.
9 Pistikopoulos, E. N., Stefanis, S. K., *Comp. Chem. Eng.* 22 (**1998**), pp. 717–733.
10 Marcoulaki, E. C., Kokossis, A. C., *Comp. Chem. Eng.* 22 (**1998**), pp. S11–S18.
11 Douglas, J. M., *Conceptual Design of Chemical Processes*, McGraw-Hill, London, **1988**.
12 Gil-Villegas, A., Galindo, A., Whitehead, P. J., Mills, S. J., Jackson, G., Burgess, A. N., *J. Chem. Phys.* 106 (**1997**), p. 4168.
13 Galindo, A., Davies, L. A., Gil-Villegas, A., Jackson, G., *Mol. Phys.* 93 (**1998**), p. 241.
14 Galindo, A., Blas, F. J., *J. Phys. Chem. B* 106 (**2002**), p. 4503.
15 Blas, F. J., Galindo, A., *Fluid Phase Equilb.* 194 (**2002**), pp. 501–509.
16 Keskes, E. K., Integrated Process and Solvent Design for CO_2 Removal from Natural Gas, PhD Thesis, Imperial College London, **2007**.
17 Ducoulombier, D., Zhou, H., Boned, C., Peyrelasse, J., Saint-Guirons, H., Xans, P. J., *Phys. Chem.* 90 (**1986**), pp. 1692–1700.
18 Perry, R. H., Green, D. W., Maloney, J. O., *Perry's Chemical Engineers' Handbook*, 7th Edition, McGraw-Hill, London, **1997**.
19 Paricaud, P., Galindo, A., Jackson, G., *Ind. Eng. Chem. Res.* 43 (**2004**), p. 6871.
20 Donnelly, H. G., Katz, D. L., *Ind. Eng. Chem.* 46 (**1954**), p. 511.
21 Neumann, A., Walch, W., *Chem. Eng. Tech.* 40 (**1968**), p. 241.
22 Al-Sahhaf, T. A., Kidnay, A. J., Sloan, E. D., *Ind. Eng. Chem. Fundamentals* 22 (**1983**), p. 372.
23 Knapp, H., Yang, X., Zhang, Z., *Fluid Phase Equilib.* 54 (**1990**), p. 1.
24 Wei, M. S. W., Brown, T. S., Kidnay, A. J., Sloan, E. D., *J. Chem. Eng. Data* 40 (**1995**), p. 726.
25 Webster, L. A., Kidnay, A. J., *J. Chem. Eng. Data* 46 (**2001**), p. 759.
26 Davalos, J., Anderson, W. R., Phelps, R. E., Kidnay, A. J., *J. Chem. Eng. Data* 21 (**1976**), p. 81.
27 Somait, F. A., Kidnay, A. J., *J. Chem. Eng. Data* 23 (**1978**), p. 301.
28 Toriumi, T., Kaminishi, G., *Report Research Laboratory Asahi Glass Foundation* 14 (**1968**), p. 67.
29 Xu, N., Dong, J., Wang, Y., Shi, J., *Fluid Phase Equilib.* 81 (**1992**), p. 175.

30 Bian, B., PhD Thesis, University of Nanjing, **1992**.

31 Xu, N., Dong, J., Wang, Y., Shi, J., *Hua Hsueh Kung Yeh Yu Kung Cheng* 43 (**1992**), p. 640.

32 Bian, B., Wang, Y., Shi, J., *Fluid Phase Equilib.* 90 (**1993**), p. 177.

33 Reamer, H. H., Sage, B. H., *J. Chem. Eng. Data* 8 (**1963**), p. 508.

34 Iwai, Y., Hosotani, N., Morotomi, T., Koga, Y., Arai, Y., *J. Chem. Eng. Data* 39 (**1994**), p. 900.

35 Inomata, H., Tuchiya, K., Arai, K., Saito, S., *J. Chem. Eng. Jpn.* 19 (**1986**), p. 386.

36 Jennings, D. W., Schucker, R. C., *J. Chem. Eng. Data* 41 (**1996**), p. 831.

37 Chou, G. F., Forbert, R. R., Prausnitz, J. M., *J. Chem. Eng. Data* 35 (**1990**), p. 26.

38 Nagarajan, N., Robinson, R. L., *J. Chem. Eng. Data* 31 (**1986**), p. 168.

39 Sebastian, H. M., Simnick, J. J., Lin, H. M., Chao, K. C., *J. Chem. Eng. Data* 25 (**1980**), p. 138.

40 Koonce, K. T., Kobayashi, R., *J. Chem. Eng. Data* 9 (**1964**), p. 490.

41 Reamer, H. H., Olds, R. H., Sage, B. H., Lacey, W. N., *Ind. Eng. Chem.* 34 (**1942**), p. 1526.

42 Lavender, H. M., Sage, B. H., Lacey, W. N., *Oil Gas J.* 39 (**1940**), p. 48.

43 Bett, K. E., Juren, B., Reynolds, R. G., *Proc. Symp. on Physical Properties of Liquids and Gases for Plant and Process Design*, HMSO, London, **1968**, pp. A53–A76.

44 Lin, H. M., Sebastian, H. M., Simnick, J. J., Chao, K. C., *J. Chem. Eng. Data* 24 (**1979**), p. 146.

45 Fletcher, D. A., McMeeking, R. F., Parkin, D., The United Kingdom Chemical Database Service. *J. Chem. Inf. Comput. Sci.* 36 (**1996**), pp. 746–749.

46 Process Systems Enterprise, gPROMS v3.0.3, www.psenterprise.com, **2006**.

47 Anderson, C. L., Siahaan, A., Case Study: Membrane CO₂, Removal from Natural Gas, Grissik Gas Glant, Sumatra, Indonesia, Air Liquide and Conoco-Philips Report, **2005**.

48 Kakalis, N. M. P., Kakhu, A. I., Pantelides, C. C., *Proc. 6th International Conference on Foundations of Computer Aided Process Design*, CACHE Publications, Michigan, **2004**, p. 537.

49 Mohideen, M. J., Perkins, J. D., Pistikopoulos, E. N., Optimal Design of Dynamic Systems under Uncertainty. *AIChE J.* 42 (**1996**), pp. 2251–2272.

50 Lymperiadis, A., Adjiman, C. S., Galindo, A., Jackson, G., *J. Chem. Phys.* 127 (**2007**), p. 234903.

9
Energy Efficiency and Carbon Footprint Reduction
Jiří Jaromír Klemeš, Igor Bulatov and Simon John Perry

Keywords
energy-saving technologies, life-cycle assessment (LCA), pinch technology, dividing wall distillation technology, heat integration flowsheeting simulation, Stirling engine, fluid catalytic cracking (FCC)

9.1
Introduction

The increase in the world population that has occurred, especially in the last 20 years, has placed increasing pressure on the demands of world society, and most especially that of industrial and agricultural production. Industrial production has increased sharply in countries that are now demanding the goods and services that have been available to countries that have been industrialized at an earlier date. The increase has required a large and continuous supply of energy delivered principally from natural resources, mainly in the form of fossil fuels, such as coal, oil, and natural gas.

The accelerating development of these countries which also have large populations, such as China and India, has resulted in increased demands on agricultural production and processing, which in turn has resulted in further increases in energy demands. These increases in energy demands are not without cost. There have been sharp increases in costs of all forms of energy.

There have also been costly side effects. The amount of energy-related emissions of CO_2, NO_x, SO_x, dust, black carbon, and waste of combustion processes [1] has increased to such a level that there is now an unpredictable effect on the world's climate. Therefore it is becoming increasingly important to ensure that the production/processing industry takes advantage of recent developments in energy efficiency thereby contributing to the reduction in energy required by society, and also significantly reducing the amount of waste that is being produced.

Energy efficiency is generally defined as the effectiveness with which energy resources are used for useful purpose. Within many industries, and within society generally, a large proportion of energy that is being consumed is in the form of heating and cooling requirements. Consequently there is much emphasis on ther-

Process Systems Engineering: Vol. 5 Energy Systems Engineering
Edited by Michael C. Georgiadis, Eustathios S. Kikkinides and Efstratios N. Pistikopoulos
Copyright © 2008 WILEY-VCH Verlag GmbH & Co. KGaA, Weinheim
ISBN: 978-3-527-31694-6

mal efficiency, which measures the efficiency of energy conversion systems such as office and home heating and cooling systems, process heaters, chilling and refrigeration systems, steam and hot water systems, engines, and power generators.

The efficiency of energy use also varies considerably from process to process despite the availability of information and the help available to improve the design and operation of the energy processing systems. There are many reasons why energy is not being used effectively. The efficiency can be limited by the initial design of the energy system or the system that is consuming energy to deliver products, by mechanical, chemical, or other physical parameters, or by the age and design of equipment employed within the overall system. In some cases, operating and maintenance practices contribute to lower than optimum efficiency. As energy is required by so many different systems, it is clear that increasing the efficiency of energy use and thereby reduce the amount of energy that is consumed (and related emissions produced) could result in substantial economic and social benefits to the world economy [2].

However, there are some specific features in various industrial, power-generating, and domestic sectors which make the optimization of energy efficiency and total cost reduction more difficult when compared to other traditional processing industries such as oil refining, where there is a continuous mass production concentrated in a few locations and which offer an obvious potential for large energy saving [3]. In the food processing industry, which is frequently distributed over very large areas and often produces during specific and limited time periods, for example in the case of campaigns in the sugar industry, producing energy efficient and reduced wastage systems can become extremely complex. An additional complication is that the industry relies heavily on small producers and processors, which results in nonoptimal equipment and systems due to the relatively small sizes involved. Other industries, such as pulp and paper producers, have been found to consume large amount of energy and create large amounts of both gaseous and water-based emissions, or if properly designed and operated have been found to produce virtually zero emissions and water-based effluents.

The domestic sector is responsible for around 30% to 50% of all energy use, principally related to heating, cooling, and lighting of buildings. This sector requires an integrated design methodology for the efficient heating and cooling of individual buildings and complexes. Such a methodology would be required to include the design basis for combined heat and power systems, refrigeration, air conditioning and heating with pump-based systems. The methodology would be equally applicable for single family houses as well as large building complexes and meet a major challenge in the design of heating and cooling systems, namely the complexity of energy and power integration. The efficient use of available heating and cooling resources for serving buildings of various sizes and designations can significantly reduce energy consumption and emissions. Although some specific and highly efficient energy efficiency methods such as heat integration (or pinch technology) have been practiced for several decades and have been very successful in large industrial applications, they have only recently been applied to improving the energy efficiency of buildings and building complexes.

There is significant scope for the application of these types of methodology in the context of rising energy prices and the requirement to significantly reduce energy related emissions. Methodologies for increasing energy efficiency can also be used to integrate renewable energy sources such as biomass, solar PV, and solar heating into the combined heating and cooling cycles.

This chapter starts by providing an overview of the energy efficiency problem. After a general overview of energy-saving techniques that are currently available, relatively simple but effective screening and scoping approaches are reviewed, including energy auditing, benchmarking, and good housekeeping. The next section covers in more detail related energy-saving techniques, such as balancing and flow-sheeting simulation, and specifically an integrated approach – heat integration. The chapter continues with energy efficiency optimization and emerging energy-saving technologies. Here very recent information resulting from an advanced study selected and co-financed by the European Community is presented.

A significant part of the chapter is devoted to the increasingly significant issue of emissions and carbon footprint assessment and minimization. A "carbon footprint" (CFP) is defined by the UK Parliamentary Office for Science and Technology – POST [4], as the total amount of CO_2 and the other greenhouse gases emitted over the full life cycle of a process or product. There have been numerous studies, e.g., Albrecht [5] or Fiaschi and Carta [6], stressing the carbon neutrality of renewable sources of energy. However, even renewable energy sources make some contribution to the overall carbon footprint, which is frequently not accounted for in assessment studies. The assessments of various renewable technologies applicable in the domestic sector can assist in making the correct selection and consequently reduce the release of CO_2. In the same manner the carbon footprint should also be incorporated into the life-cycle assessment (LCA) (see, e.g., [7]).

The chapter concludes with a number of industrial applications: an energy efficiency retrofit study of from a large energy-consuming petrochemicals industry, followed by a pulp and paper plant energy minimizations study and several examples from the food industry. Case studies related to the domestic sector energy are demonstrated as well. These cover a hospital and a total site comprising a sugar plant and a nearby town.

9.2
Overview of Energy-Saving Techniques

The task of saving energy, more particularly at a time of rising energy costs, is a task that has to be considered seriously by all communities and industries. However, the driving force in society is still dominated by the economics of individual situations, and no section of society worldwide can be expected to save energy at any cost. Energy-saving measures have to be considered within the context of issues such as legislative measures, environmental factors, and consumer pressure.

The simplest and most obvious technique that can be employed involves energy auditing and good housekeeping measures. In many cases even these simple measures have not always been fully understood and completed in sufficient detail. In order to undertake a worthwhile energy audit correct and proper measurements should be available. This has been closely related to energy consumption monitoring over specific periods of time as in many cases energy demand is not constant, but fluctuates considerably.

They are many examples of recommended techniques available from utility companies, such as SEMPRA ENERGY [8], governmental agencies, such as the US Department of Housing and Urban Development [9], and international agencies, such as the International Energy Agency [10].

The improvements in energy efficiency that are desired by society often have to be achieved by more complex means, often associated with improved design and operation. It is of paramount importance that all energy-related processes, from small-scale domestic systems to large-scale industrial systems, operate with maximum efficiency and minimum energy input. Additionally these systems should ensure that they operate as much as possible with low-value inputs or wastes, such as process outputs, e.g., off-gases and hot water waste [11].

To ensure that these designs are as efficient as best practice allows, optimization methods are frequently employed for process/plant/building grassroots design, retrofit, and control and intelligent support systems. One technology that has been practiced for over 20 years, and has a strong reputation for improving energy efficiency through better design, is pinch technology [18]. Although it is widely practiced, the technology has been continuously developed and expanded Klemeš et al. [19], Smith [20]. Further details of this application and how it has been successfully applied in various sectors will be described later in this chapter.

The efficiency of energy systems can also be considerably improved by making use of renewable energy sources, such as biofuels, wind, and water power, all of which can significantly reduce the generation of greenhouse gases. Further, the implementation of heat and power systems [11], rather than individual power and individual heat systems, can also considerably improve the overall efficiency of energy systems. Other fast improving technologies, such as heat pumps, compact heat exchangers, intensified technologies, and fuel cells, can also improve the overall situation. These new intensified and advanced process systems and methods are underway internationally. These are not yet fully commercialized but will become available for the future. They include [11]:

- Advanced gas turbines for both utility and industrial applications, including co-generation.
- Fuel cells are electrochemical devices, which may be fuelled by hydrogen, methane, or other organic fuels. These systems produce heat in addition to electrical power, and are suited for use in cogeneration systems. It is estimated that there is typically a full fuel cycle CO_2 saving of around 25% relative to a gas turbine. However further advances are required to be made before there is full and economical practical application. Some advances were made recently to integrate CHP and fuel cells [12].

- Dividing wall distillation technology [14]. The technology can be applied to the separation of three components (or groups of components) in a mixture. In the past this would have required two columns, with heating and cooling provided for each column. The dividing wall technology combines the separation process into a single vessel, with energy savings of about 30% and capital savings of 25% [16].
- Compact heat exchangers are generally made from thin metallic plates rather than tubes. The plates form complex and small flow passages that result in large heat transfer surface area per unit volume. Multistream versions of these exchangers can incorporate 12 or more streams. Compact heat exchangers can produce energy savings as well as reduced capital and installation costs. In a case study at a UK refinery, potential capital savings amounted to between 69% and 84% [17].

Finally, cogeneration is finding increasing application in most sectors of the economy. For example, many oil refineries buy power as an alternative to onsite generation or to supplement onsite generation (typically 50% is purchased electricity). Cogeneration offers an opportunity to reduce net greenhouse gas emissions from the combined power grid/refinery system through more efficient combustion compared with many existing power generation technologies, and through the use of recovered heat.

9.3
Screening and Scoping: Energy Audit, Benchmarking, and Good Housekeeping

Screening and scoping tools have had a considerable effect over the years in reducing the energy and effluent treatment costs and consequently reducing plant profit margins. For example, in the food and drink industry, energy audits performed on various food and drink processes have resulted in cost savings of 15 to 30% with attractive returns on investment [21, 22]. As profit margins are generally relatively low in this industrial sector, efficient management of energy can help increase net profit margins, while reducing environmental impacts.

The Carbon Trust [23] has suggested the following steps for improvement in energy use and consequently the improvement in energy efficiency:

Energy Efficiency – Good Housekeeping

- Staff attitude and awareness
- Locating leakages of heat
- Preventive maintenance
- Insulation
- Heating/cooling/lighting justification
- Fouling
- Monitoring and control

Energy Efficiency – Energy Audits

- The examination of records of energy cost and consumption
- Production of an energy balance sheet
- Provision of good quality data on energy consumption and costs
- The ongoing collection, analysis, and review of energy information are important areas
- A "benchmark" of energy consumption against other organizations or accepted standards

Energy Efficiency Environmental Management System ISO 14001

- Management system is a network of interrelated elements.
- Elements include responsibilities, authorities, relationships, functions, processes, procedures, practices, and resources.
- A management system is establishing policies and objectives and developing ways of applying policies and achieving these objectives.

Energy Efficiency – Responsible Use

Typical energy management activities include:

- Energy procurement
- Metering and billing
- Performance measurement
- Policy development
- Assigning energy management responsibility
- Energy surveying and auditing
- Training and education
- Capital project management

The above steps have been found to be central to any resource and waste management program (UK Government's Energy Efficiency Best Practice programme – EEBPp [24]).

Depending on the required depth of the study, the process to be analyzed and the size of the plant, energy screening, and scoping audits can fall into two main categories. First, a simple cursory audit could be employed, identifying only obvious saving opportunities. Second, a more comprehensive audit may be used, which provides extremely detailed information and analysis, involving one or several specialists. In identifying opportunities to reduce plant energy consumption, three types of audits are most often used, either individually or in combination:

- the walk-through audit, to provide a quick snapshot of certain opportunities,
- the detailed audit, to conduct an in-depth analysis of specific components, and
- the process integration audit, which analyzes the plant as a whole and takes a systematic look at all processing steps and their interconnections.

These audits may be performed by plant personnel, by external experts or by teaming-up in-house resources with outside experts. Although there are many possible options and levels of detail, typical activities include [21]:

- determination of the production base-case and the reference period;
- collection of energy total consumption and cost (information usually available from fuel, electricity invoices),
- development of a process flow chart with material, energy inputs and outputs for the main processing steps,
- for the largest consumers, collection of energy data from the plant-metering devices, control systems, and process flow diagrams (if current operating conditions are close to design data).

Specific measurements can be performed using portable instruments (flow rate, temperature, humidity, etc.):

- determination of the overall plant steam, refrigeration, compressed air, and hot water production,
- interviews of key personnel and operators.

9.4
Energy-Saving Analysis: Balancing and Flowsheeting Simulation

Balancing reconciliation and flowsheeting simulation tools are frequently used for energy-saving analysis and have become one of the main tools in the engineers' toolbox. These tools help engineers to develop complete mass and energy models based on measurements and/or design values and mathematical models. Consequently these simulation tools play an important role in the technical and economic decision-making activity related to the planning and/or design stage of processes under development and to the operation of actual existing equipment.

A number of computer-based systems have been developed over the years in order to assist the process engineer in the energy and mass balance calculations procedures. However, due to the ongoing cost of development, only a limited number have continued in the market, and they have only been secured by a substantial number of sales. An early overview has been presented by Klemeš [25].

The balancing, data validation, and reconciliation technology is made up of a set of procedures which have been incorporated into a software tool. Process data reconciliation has become the main method for monitoring and optimizing industrial processes as well as for component diagnosis, condition-based maintenance, and online calibration of instrumentation. According to Heyen and Kalitventzeff [26], its main goal is to

- detect and correct deviations and errors of measurement data so that these satisfy all balance constraints,

- exploit the structure and the knowledge of the process system together with the measurement data to compute unmeasured data wherever it is possible, in particular the key performance indicators (KPI), and
- determine the postprocessing accuracy of measured and unmeasured data including KPIs.

A comprehensive computer-based system, DEBIL, which included balancing, flowsheeting simulation, and optimization, was developed in 1970s and 1980s [13], and has been further developed by BELISIM into the balancing and reconciliation tool VALI, the latest version of which is VALI III [27]. The application of the tool for various energy efficiency tasks have been published elsewhere, including the efficiency of nuclear power plants [28], and regenerative heat exchangers [29].

The online encyclopedia, Wikipedia, [30] presents a comprehensive list of software tools which are available for the simulation of material and energy balances of chemical processing plants:

a) ASCEND& Aspen HYSYS by Aspen Technology
b) ASSETT and D-SPICE by Kongsberg Process Simulation
c) CHEMCAD
d) COCO simulator
e) Design II by Winsim
f) EcosimPro
g) Environment for Modelling, Simulation and Optimisation (EMSO)
h) Dymola
i) GIBBSim
j) gPROMS by PSE Ltd
k) OLGA by SPT Group (Scandpower)
l) Omega by Yokogawa
m) OpenModelica
n) Petro-SIM
o) ProMax
p) SimSci-Esscor DYNSIM & PRO/II by Invensys
q) SysCAD
r) UniSim Design & Shadow Plant by Honeywell
s) VMGSim

Wikipedia [30] also provides a comprehensive list of software tools that can be used for dynamic process simulation:

A. Aspen Dynamics & Aspen HYSYS Dynamics by Aspen Technology
B. gPROMS by PSE Ltd
C. OLGA by SPT Group (Scandpower)
D. Omega Equatran by Yokogawa
E. SimSci-Esscor DYNSIM by Invensys
F. UniSim Design & Shadow Plant by Honeywell

and also those available for operator training:

A1. HYSYS Dynamics by Aspen Technology
B1. OLGA by SPT Group (Scandpower)
C1. Omega Equatran by Yokogawa
D1. SimSci-Esscor DYNSIM by Invensys
E1. UniSim Design (based on a HYSYS version) & Shadow Plant by Honeywell

In many cases there is considerable overlap in products as many simulators contain facilities for steady-state simulation, dynamic simulation, and operator training.

ASPEN PLUS™ has become one of the most widely used simulators, followed by HYSYS which was acquired in 2002 by Aspen Technology, Inc.

A growing share of the marked has been claimed by gPROMS, developed by PSE Ltd., London, UK. There is comprehensive information available on the web (www.psenterprise.com) as well as examples of various energy-related applications (www.bioenergynoe.com/?_id=243 and www.psenterprise.com/gproms/applications/fuelcell/index.html).

9.5
Integrated Approach: Heat Integration

Process integration methodology provides the design foundation for combined heat and power systems, refrigeration, air conditioning, and heating with pump systems. It is equally applicable for single-family houses and large building complexes as well as for big industrial sites such as oil refineries with petrochemicals production and power stations. The technology answers one of the major challenges in the design of heating and cooling systems, namely the complexity of energy and power integration. The efficient use of available heating and cooling resources for serving complex systems of various sizes and designations can significantly reduce energy consumption and emissions.

There is significant scope for the application of this technology, as energy prices continue to rise and energy-related emissions need to be reduced. This methodology can also be used to integrate renewable energy sources such as biomass, solar PV, and solar heating into the combined heating and cooling cycles.

Since 1995 the energy consumption of the EC member countries have risen by 11%, to the value of 1637 Mt of oil equivalent [31]. This increase in energy consumption contrasts with the population of the EC member states which is growing at a much slower rate, approximately 0.4%/year [31].

Generally, much of society's attention is directed toward reducing energy consumption related to industrial activities. However, the overall share of total energy consumption by industry is declining in most countries. For example, in the UK domestic energy consumption has risen from 35.6 Mt of oil equivalent to 48.5 Mt in the period from 1971 to 2001, an increase of 36%; despite that energy efficiency increases [32].

The increase in energy consumption per household however was considerably less, at 5%, over the same period. This difference reflects a substantial increase in energy efficiency in individual dwellings. However, pressure on the housing market caused by reduction in family size and family composition has resulted in a proposal in the UK to significantly increase the number of houses built over the next 10 years to 200,000/year [33].

These changes are likely to have serious implications for energy consumption and greenhouse gas emissions. The increase in energy consumption from individuals [32], dwellings, and buildings has been generally met by the burning of carbon-based fuels, or in some cases, by the increase in nuclear power. The proposition of renewable energy making a large contribution to energy consumption has so far failed to materialize, and the contribution toward total domestic energy use is still small (in the UK 1.0 Mt of oil equivalent compared to a total use of 47.0 Mt of oil equivalent).

Process integration technology (or heat integration/pinch technology) has been extensively used in the processing and power generating industry over the last 30 years and was developed and pioneered by the Department of Process Integration, UMIST (now the Centre for Process Integration, CEAS, The University of Manchester) in the late 1980s and 1990s.

The publications covering main issues include Linnhoff et al. [35]; Linnhoff and Vredeveld [18], Klemeš et al. [19], Smith [20] and [36].

A second edition of the Linnhoff et al. book was recently published by Kemp [38]. A specific food industry overview of the technology and its application was presented by Klemeš and Perry [39].

Process integration technology analyzes the potential of exchanging heat between the heat sources of systems or processes that use energy and the heat sinks via the use of heat exchangers. The ultimate goal of the technology is to make more efficient use of the energy in the system and to reduce the amount of external heating and cooling requirements. Coupled with this essential targeting technology, a systematic design procedure has been developed to provide the final energy reduction design of the system.

The technology, as well as looking at individual energy using systems or processes, also has the potential to reduce energy demand and emissions on a large-scale industrial sites which are made up of a large number of individual processing units and an integrated utility system. The technology, simultaneously, can maximize the production of cogeneration shaft power. The method has further been developed to specify the source of heating and cooling required, e.g., steam, hot water, and cooling water [19].

Further examples and details heat integration methodology are provided by Linnhoff et al., 1992 and [35], Shenoy [37], Smith [36], and Kemp [38].

The process integration or heat integration methodology is based on the analysis and understanding of heat exchange between process streams through the use of a temperature–enthalpy diagram. The specific steps for drawing the curves in this diagram are presented in Figs. 9.1–9.3.

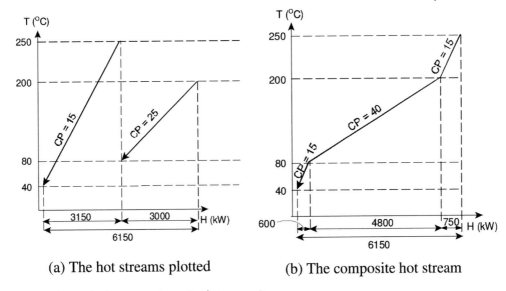

(a) The hot streams plotted

(b) The composite hot stream

Fig. 9.1 Composing hot streams to create a hot composite curve (after [34]). (a) The hot streams plotted separately. (b) The composite hot stream.

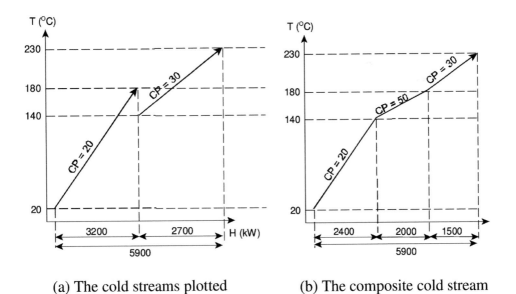

(a) The cold streams plotted

(b) The composite cold stream

Fig. 9.2 Composing cold streams to create a cold composite curve [34]. (a) The cold streams plotted separately. (b) The composite cold stream.

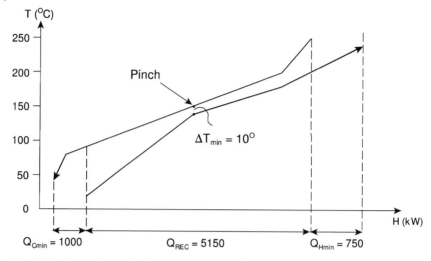

Fig. 9.3 Plotting the hot and cold composite curves (after [34]).

Table 9.1 Sources (hot) and sink (cold) streams [39].

Stream	Type	Supply temperature T_S (°C)	Target temperature T_T (°C)	ΔH (kW)	Heat capacity flowrate CP (kW/°C)
Fresh water	Cold	20	180	3200	20
Hot product 1	Hot	250	40	−3150	15
Juice circulation	Cold	140	230	2700	30
Hot product 2	Hot	200	80	−3000	25

Initially the methodology first identifies sources of heat (termed hot streams) and sinks of heat (termed cold streams) in the process flowsheet. Table 9.1 presents a simple example.

These sources of heat can be combined to construct the composite hot stream (Fig. 9.1) and the sinks of heat can similarly be combined to construct the composite cold stream (Fig. 9.2).

The second stage in the process is to select a minimum permissible temperature approach between the hot and cold streams, ΔT_{min}. The optimum value of this ΔT_{min} results from an economical assessment and tradeoff between the capital and operating costs (mainly for energy usage) of the process.

The chosen value of ΔT_{min} provides the information for the location of these curves on the temperature–enthalpy diagram and further depicts the allowable temperature difference for heat exchange in the overall process. In this example a ΔT_{min} of 10 °C was selected for simplicity.

If ΔT_{min} is large then this implies that energy use is higher than that when a lower value of ΔT_{min} is chosen. This in turn means that energy costs for the overall process heat exchange will be higher but that capital costs, in the form of heat exchanger area, will be lower. If ΔT_{min} is lowered, then the heat exchanger system is allowed to recover more energy, but at the expense of additional capital which is required to pay for the increased heat transfer area. The selection of the correct value of ΔT_{min} is discussed in greater detail elsewhere, for example Taal et al. [40]; Smith [36] and Donnelly et al. [41].

If the resultant composite curves (CC) are plotted in the same graphical space (Fig. 9.3), then the maximum heat recovery in the process and minimum use values of hot and cold utilities can be obtained. These derived values are commonly known as targets and form the basis of a future design scenario. For this case, and using a ΔT_{min} value of 10 °C, the minimum hot utility requirement is 750 kW and minimum cold utility requirement is 1000 kW.

The graphical representation of the heat sources and heat sinks in the form of the composite curves, and the incorporation of minimum temperature for heat exchange in the form of ΔT_{min} as shown in Fig. 9.3, can also be used to determine the position of the pinch. This is the point of closest approach between the hot and cold streams on the composite curves graph. The pinch in the composite curves has also provided an alternative name, which is commonly used for the heat integration process, namely pinch technology.

Using the targets obtained from the composite curves, and making use of the temperature location of the pinch, has allowed the development and implementation of different design methods, which have been applied both for grass roots designs (Linnhoff and Vredeveld, 1998) and, more importantly, for the retrofit of existing plants [3, 42, 43].

The process integration/heat integration/pinch technology methodologies are supported by process integration software which provides both design and retrofit support and automated design [44, 45].

In the above example we have identified that a target amount of external heating and cooling is required by the process in order to achieve the energy balance at the chosen minimum approach temperature. We can also identify easily from this graphical representation the temperature at which the heating and cooling sources have to be supplied. In reality there are more than one hot and one cold utility that could potentially provide these heating and cooling requirements. In these situations we should use a systematic methodology that allows the evaluation of the potential candidate or group of candidates based on cost (Fig. 9.4).

The Grand Composite Curve (GCC), and the derived Balanced Grand Composite Curve (BGCC), and Balanced Composite Curves (BCC) have been developed to systematically analyze the placement in terms of temperature and the amount of heating and cooling requirement at these chosen temperatures. An example of selection of utilities and its placement is shown in Fig. 9.5.

The GCC has also been found to be a useful tool for targeting the cooling requirements in subambient processes (such as in gas processing processes or domestic air conditioning systems) which require some form of compression refrigeration or chilling. The Grand Composite Curve analysis has also been extended to provide information on the shaftwork requirements of these low-temperature processing systems [46]. For this additional analysis the temperature axis of the composite curves

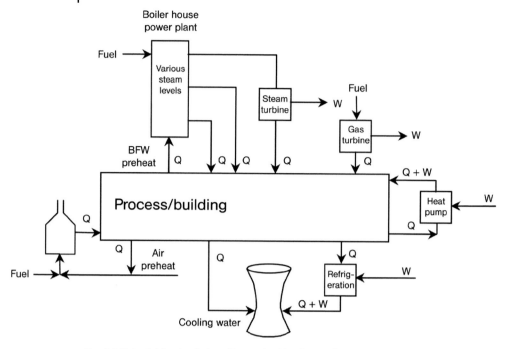

Fig. 9.4 Potential for the choice of hot and cold utilities (after [34]).

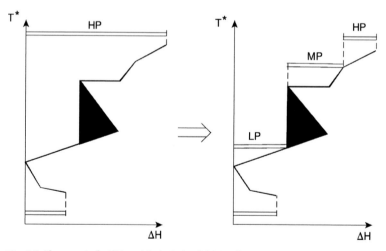

Fig. 9.5 Placement of utilities with the help of GCC (after [34]).

and the grand composite curve is replaced by a Carnot factor axis $\eta_c = (1 - T_O/T)$. The transformed graphics are shown in Fig. 9.6.

The graphs show the transformation from the standard composite curves, to the exergy composite curves, and finally the exergy grand composite curve. These new diagrams show a shaded area between the hot and cold composite curves, in the

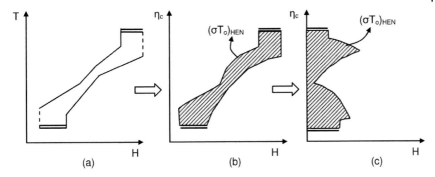

Fig. 9.6 (a) Composite curves and resulting, (b) exergy composite curves and (c) exergy grand composite curves (after [46]).

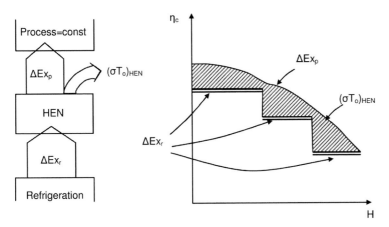

Fig. 9.7 A representation of the amount of ideal work lost in the process of transferring heat (after [46]).

case of the exergy composite curve, which is proportional to the amount of ideal work lost in the process of transferring heat. The shaded area of the exergy grand composite also shows the same. This exergy loss takes place in the heat exchanger network. A schematic diagram of the shaftwork targeting procedure can be seen in Fig. 9.7.

The flow diagram starts with the refrigeration system. This supplies exergy (ΔEXr) to the heat exchanger network. The next stage of the flow diagram shows the flow of exergy from the heat exchanger network to the process (ΔEXp). This in turn provides the required cooling of the process streams. Part of the exergy supplied by the refrigeration system is lost in the transfer process which is represented on the diagram by $\sigma T_0 HEN$. Finally, making use of the exergy efficiency, the overall shaftwork requirements of the system can be calculated.

Linnhoff and Dhole [46] tested this methodology against a case with known shaftwork requirements and found that the method was accurate to less than 2%. The determination of the value for the exergy efficiency for the system can be problem-

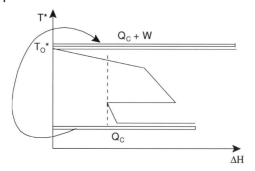

Fig. 9.8 The grand composite curve with cooling provided by a single refrigeration level and heat rejected to the process (after [34]).

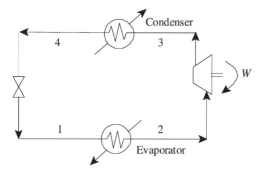

Fig. 9.9 Schematic of a simple compression refrigeration system (after [34]).

atic. However, Dhole [47] found that for a given refrigeration fluid and for a range of temperatures, the exergy efficiency remained essentially constant.

A different approach to targeting for shaftwork requirements in low-temperature processes has been presented by Lee [48], Lee et al. [49], and Smith [36]. Their approach makes use of the original temperature/enthalpy representation of the grand composite curves. They couple this with a procedure for calculating the refrigeration power requirements of the refrigeration system. This coupled procedure, starting with the grand composite curve, as shown in Fig. 9.8, is used to calculate the power requirements of a simple compression refrigeration system as shown in Fig. 9.9.

The section 1–2 in Fig. 9.9 represents the evaporation of the refrigerant, which provides the cooling for the process. This cooling load is derived from the grand composite curve along with the temperature of the required cooling. The refrigerant is vaporized in the heat exchanger, after which the vapor enters the compressor. The compressor increases both the pressure and the temperature of the vapor.

The next stage in the cycle involves the condensation of the vapor in the condenser, which is represented on the grand composite curve (Fig. 9.8) by the heat provided to the process and the heat load. The final stage sees the refrigerant exiting the condenser and being expanded to start the cycle again. Smith [36] stated that the procedure was able to estimate the refrigeration power requirements prior to the completed design, assess the performance of the whole process before detailed design is performed, allow alternative designs to be assessed rapidly, and to be able to accurately calculate energy and capital costs of the refrigeration system design.

The method makes use of the physical properties of the refrigeration fluid and is fully derived in the original text [36]. A more detailed description has been provided elsewhere [50].

Heat integration (or pinch technology) has been used for 20 years in industry throughout the world to increase energy efficiency of any processing plant that has heating or cooling requirements, and also has a requirement for power to provide electricity or directly drive machinery. Energy savings of over 30% have been recorded, and the methodologies developed have been incorporated into the design offices of all major producing companies. The same methodologies and design rules can also be applied in buildings or their complexes.

9.6
Energy Efficiency Optimization

There are a number of methods available for the optimization of energy efficiency. In the last section optimization of the design process, both for grassroot processes and for existing (retrofit) processes, was discussed. In this section we look at the second step which involves the optimal operation of the process system, namely Energy Management and Control. The third and final step is optimal maintenance and production planning.

9.6.1
Energy Management and Control

This topic has received attention over a period of many years [51–53]. A comprehensive source of information has been published in the Energy Management Reference Library edited by Turner [54].

Specific applications to oil refineries and petrochemicals have also been well described by Worrell and Galitsky [55]. The major findings from this report have shown that required improvements in energy efficiency need to be approached from several directions. First there is a need for a strong, corporate-wide energy management program that is fully supported by management and all other employees. Second, it is required that equipment, such as boilers, compressors, and pumps, which are common to most processing plants, should be carefully reviewed to ensure that they are being operated as efficiently as possible, and also reviewed in the light of the introducing new more efficient replacements. Thirdly, techniques are required to be adopted that allow fine-tuning or optimization of operations to produce additional savings.

9.6.2

Energy Management Systems (EMS) and Programs

The introduction of a company-wide supported energy management program has been frequently demonstrated as the ideal and most cost-effective way to bring about energy efficiency improvements. The introduction of such a program produces a culture of improvement that is supported by all employees and allows the introduction of carefully formulated guidelines which allow for the efficient management of energy use within the company or group of companies. Without this program and set of guidelines, then energy management is likely to be haphazard and produces a sense of confusion in the workforce. Areas of confusion are quite often linked to a lack of properly organized communication systems which are needed to implement a program of energy efficiency changes. Considering the opportunities that usually exist in the improvement of energy efficiency in many energy-using processes, it is surprising that there are still barriers that prevent the adoption of wide-scale improvements in energy management within organizations.

Worrell and Galitsky [55] with the U.S. EPA, through ENERGY STAR, have worked with many of the leading industrial manufacturers to identify the basic aspects of an effective energy management program [56].

The major elements are depicted in Fig. 9.10.

In order to set up an effective energy management program, there is a need to establish an energy director who can oversee and manage the supported program. This can be followed by the appointment of an energy team, which can then establish the required procedures needed for such actions to assess performance, regularly review the energy information that is available and assess the need of additional data collection, assess any technical requirements that are needed, and carry out some form of benchmarking. Using this type of management system, an organization is able to develop an initial baseline of the current performance related to energy use and set goals for improvement [57].

If current performance is correctly established, then the setting goals is obviously much easier, and this in turn allows the implementation plan that can provide the necessary support to achieve these goals. The success of this performance improvement cannot be made unless all employees of the organization are informed and involved. There is usually a need for staff training that can be carried out inhouse, or outside experts can also be brought in for this. The level of this support helps to cement the idea that improvement in energy efficiency is a concern for all members of staff. Communication to staff on how the improved performance is progressing is vitally important. The progression of performance toward the initial goals needs to be assessed and communicated, and any best practices that have been identified should be implemented as widely as possible.

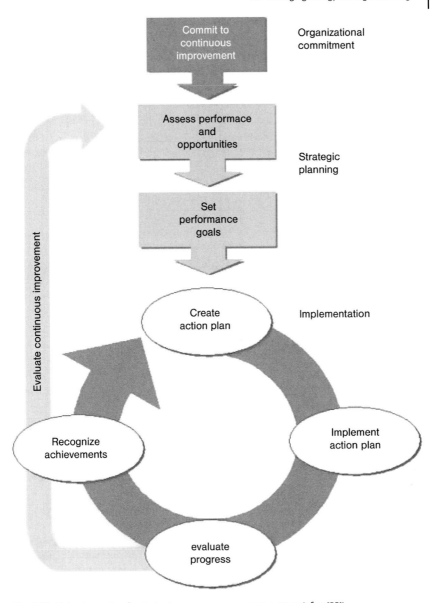

Fig. 9.10 Main elements of a strategic energy management system (after [55]).

9.7
Emerging Energy-Saving Technologies

The lead time for the development of a new energy technology, from the initial idea to the commercial application, can take many years. The reduction of this lead time has been the main objective of the EC DGTREN, who has funded two related recent projects, EMINENT and EMINENT2 (Early Market Introduction of New Energy Technologies). These projects were implemented in order to identify and accelerate the introduction and implementation of leading-edge European technology in the field of energy saving into the market place [58].

The principal features of the projects included the production of a software tool and an integrated database of new technologies and sectoral energy supplies and demands. The software tool has the capability to analyze the potential impact of new, underdeveloped energy technologies in different sectors emerging from different countries. In addition, the software tool has been used to perform case studies which have been used to illustrate the new technologies.

Many novel energy-related concepts are formulated and developed in leading research centers and laboratories throughout the industrial world. If and when these are implemented, they could lead to significant economic and social benefits. The problem is in identifying them, evaluating their practicality, and speeding up the time of getting them into the market.

During the first project launched by EC DG TREN, an expert network for the systematic evaluation and rapid dissemination of early stage technologies (ESTs) was established [58]. A software tool was developed enabling the evaluation of ESTs within different national and economic contexts as no similar tool existed. The tool provides a rapid appraisal of the geographic potential and a reduction in market lead time of the promising ESTs.

The aim of EMINENT tool is to evaluate the market potential of energy-related ESTs in various energy supply chains, and their performance in terms of (i) CO_2 emissions, (ii) costs of energy supply, (iii) use of primary fossil energy, (iv) in different subsectors of society. Technology developers and financial supporters are frequently not aware of the application potential and the market attractiveness across countries and society sectors. The EMINENT project consequently provides insight into the future market attractiveness and can accelerate the development of technologies which benefit research and development efforts as they can be targeted more effectively.

The EMINENT tool which evaluates given ESTs makes use two databases:

- National energy infrastructures, which contains information regarding the number of consumers per sector, type of demand, typical quality of the energy required, and the consumption and installed capacity per end-user;
- ESTs and other already commercial technologies, which include key information on new energy technologies currently under development, and proven energy technologies are available and in use.

9.7.1
Methodology – the Description of the Algorithm

As the availability, price, and geographical conditions of primary energy resources differ significantly worldwide, there is a requirement to evaluate the impact of ESTs within a national energy supply system [59], see Fig. 9.11.

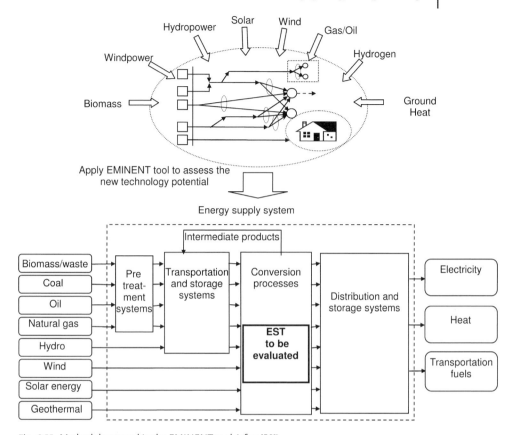

Fig. 9.11 Methodology used in the EMINENT tool (after [59]).

The algorithm for designing and evaluating energy supply chains [59] consists of different steps:

Step 1: Composing energy supply chains
For each individual energy demand to be covered, different combinations of maximum five energy technologies are proposed to form various energy supply chains that could potentially supply the energy requested at a specific demand. Linkages are only done on the basis of energy type; no checks are yet done. For each individual chain, steps 2–6 are performed.

Step 2: Dimensioning of final energy technology (eventually EST) and validating technology linkages
Starting from the characteristics of one single final energy demand, a check is performed on various characteristics of the suggested previous technology (which could be the EST) to validate that it can indeed supply the requested energy. This validation contains checks on

1. *Characteristics of the energy demand*
   Examples are temperature and pressure levels in the case of heating or cooling. In case this check fails, the whole energy supply chain will not be feasible.
2. *Requested energy supply capacity*
   It is evaluated if (eventually multiple units of) the technology can exactly deliver the requested energy.
   The proposed number of units np to be operated in parallel is calculated as the ratio of requested capacity Pdem to the maximum unit size of the supplying technology in which the technology is available (Pout, max) and rounded to the above integer:

   $$np = Pdem/Pout, max +1.$$

3. *The corresponding output capacity is established by the ratio capacity as requested by the demand*

   $$Pout = Pdem/np$$

4. *Capacity matching*
   In the case of electricity, it is assumed that a match can always be made, as mismatch in demand and supply can be counterbalanced by the public grid.
   For other energy forms, the suggested capacity should be within the available capacity range:

   $$Pout, min < Pout < Pout, max$$

   Then the option is possible and the number of units and output capacity is known. The dimensioning is then ready.
   If the suggested output capacity is however smaller than the minimum capacity available (Pout < Pout, min) but by using multiple units in parallel (np > 1), then it could be considered if the output capacity could be suitable if one unit less is applied:

   $$if\ np > 1,\ then\ np = np - 1$$

   Repeat the procedure at 3. If np is however already equal to 1, then it implies that there is always too much energy generated, even with the smallest technology and the whole chain can therefore be considered as impossible.

After the above checks on the suitability of the final energy technology to deliver final energy demand, the energy inputs and outputs of the final energy technology are dimensioned based on the given energy conversion matrix.

Step 3: Dimensioning of other energy technologies (including EST) and validating technology linkages
In case the proposed energy chain comprises multiple individual energy technologies, Step 2 is repeated for all other previous technologies using the demand of subsequent until all technologies in the chain are dimensioned. If ready, go to Step 4.

Step 4: Availability check of energy resources

After dimensioning the first technology in an energy supply chain under Step 3, the availability of sufficient appropriate energy resources is checked. This is for example important in the case of biomass fuels that are available in limited quantities.

If resource availability is sufficient, the whole design of the energy chain is finalized. Continue to Step 7.

If resource availability is however insufficient, then dimensioning of the energy chain should be done starting with a limited resource. This is done under Step 5.

Step 5: Redimensioning of first energy technology (eventually EST) and validating technology linkages

Starting from the energy resource availability, the input of the first energy technology is dimensioned. A validation checks on temperature levels, etc. have already been carried out successfully before; this validation only contains checks related to dimensioning of equipment.

It is evaluated if (eventually multiple units of) the technology can consume the requested energy. This is done as follows:

1. The ideal number of units to be operated in parallel is evaluated by first evaluating the ratio of the available resource capacity to the maximum supply capacity.
2. The proposed number of units np to be operated in parallel is calculated as the ratio of the requested capacity Pres to the maximum unit size of the supplying technology in which the technology is available (Pin, max) and rounded to the above integer:

$$np = Pres/Pin, max + 1$$

3. The corresponding output capacity is established by the ratio capacity as requested by the demand:

$$Pin = Pres/np$$

4. If the suggested capacity is within the available capacity range

$$Pin, min < Pin < Pin, max$$

then the option is possible and the number of units and input capacity is known. The dimensioning is then ready.
5. If the suggested input capacity Pin is however smaller than the minimum capacity available (Pin < Pin, min) but by using multiple units in parallel (np > 1), then it could be considered if the output capacity could be suitable if one unit less is applied:

$$if\ np > 1,\ then\ np = np - 1$$

Repeat the procedure at 3. If np is however already equal to 1, then it implies that there is always too much energy consumed compared to the resource, even with the smallest technology and the whole chain can therefore be considered as impossible.

After the above checks on the suitability of the first energy technology to consume whatever limited resource is available deliver, the energy inputs and outputs of the first energy technology are dimensioned based on the given energy conversion matrix. Continue to Step 6.

Step 6: Dimensioning of other energy technologies (including EST) and validating technology linkages

In case the proposed energy chain comprises multiple individual energy technologies, Step 5 is repeated for subsequent technologies, using the energy output of previous technologies, until all technologies in the chain are dimensioned.

Step 7: Feeding side inputs and using side outputs

After dimensioning and validating the main energy supply chain with Steps 1–6, it is considered if other energy inputs of technologies used can be met using outputs of the same or other energy technologies in the chain. If not, energy needs to be purchased from other sources available. Similarly, it is examined if energy outputs can be made useful to cover other demands in the same sector, or energy inputs in the same energy supply chain.

Step 8: Evaluating energy chains

After dimensioning the complete energy supply chains, all energy flows are known and the total chain efficiency, CO_2 emission, and use of renewable energy are assessed.

9.7.2
EMINENT Tool Description

The EMINENT software tool consists of integrated resource manager, demand manager, EST manager, databases on resources, demand, and the analysis tool (Fig. 9.12):

Figure 9.13 shows main components of EMINENT and their interface.

Resource manager modifies, enters, and selects data on resources in a country (electricity, fuels, geothermal, hydro, ocean tidal, wave and wind energy).

Demand manager describes energy demands per subsector in a country, modifies and enters new data, and selects data for the technology assessment.

Technology manager contains key data for existing technologies and ESTs.
User input:

- The sectoral energy demands to which EST applied is to be evaluated.
- Other peripheral technologies to establish full energy supply chains.
- Resources that may feed the full energy supply chains with the EST.

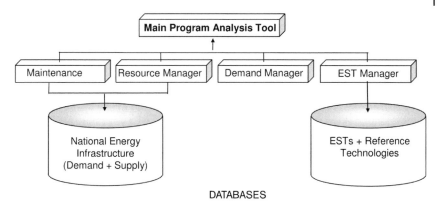

Fig. 9.12 The EMINENT software tool.

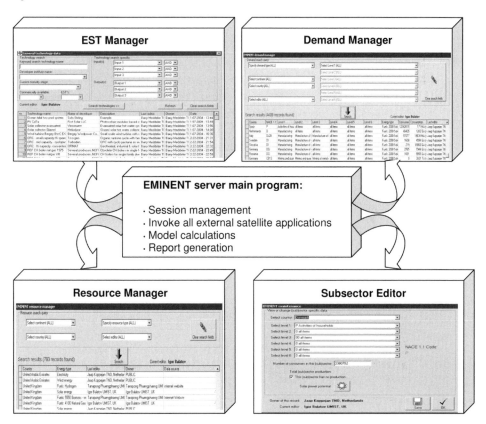

Fig. 9.13 EMINENT software components.

Output:

- aggregate numbers,
- application potential of ESTs per (sub)sector,

- annual costs of energy delivery per consumer and per (sub)sector,
- annual CO_2 emission.

Performance indicators:

- chain efficiency,
- primary fossil energy usage,
- CO_2 emission per MWh,
- costs of delivered energy (€/MWh).

Several case studies have been analyzed using the EMINENT tool (Table 9.2).

Most of ESTs analyzed still have to improve to achieve the cost levels of the existing technologies. Some of the ESTs e.g., Zero Emission Norwegian Gas (ZENG), Molten Carbonate Fuel Cell (MCFC) could become competitive with relatively small additional efforts aimed at reduction of costs.

Various promising future trends have been located with the help of the EMINENT tool and supported by the policy makers

Among them in the imminent term:

- Diverse energy system (supplies and management and control of demand),
- Market-based grid with large power stations (incl wind farms of different types),
- Local distributed generation (biomass, waste, wind),
- Micro generation (CHP, fuel cells, photovoltaic),
- Diverse energy system (supplies and management and control of demand),
- Market-based grid with large power stations (incl wind farms of different types),
- Local distributed generation (biomass, waste, wind),
- Micro generation (CHP, fuel cells, photovoltaic),
- New homes with near carbon zero emissions,
- Gas will form a large part of the energy mix,
- Coal fired generation lesser role or combined with carbon capture and storage,
- Mixed fuels – coal with biomass, natural gas with hydrogen.

In the longer term:

- Nuclear power stations (subject to later decision),
- Fuel cells,
- Hydrogen generated primarily by nonelectricity,
- Low carbon biofuels,
- Nuclear fusion.

9.8
Emissions and Carbon Footprint

Energy use and prices continue to rise with the emissions of CO_2. Energy efficiency methods have been applied across economic sectors, and efficiency gains and energy use per manufactured unit have fallen, particularly in relation to the chemical

Table 9.2 ESTs analyzed using the EMINENT tool.

Name of EST	Brief description	Maturity
Power cycle with fuel cells (FC)	EMINENT shows that with power production cost 58.29 $/MWh vs. market power price 55.00 $/MWh an MCFC operating standalone cannot compete if the power is the only product to sell. The system may become more competitive if the fuel prices rise or/and waste heat generated by the MCFC system is used. Some deeper process integration analysis carried out using other tools [44] shows that a FC combined cycle. The heat exchanger network synthesized used as a steam boiler and a steam turbine added forming a Rankine cycle [78].	Lab.
Biomass/waste fluidized bed reactor	Fluidized bed reactor has been developed for biomass and/or waste gasification [78].	Lab.
Autothermal reforming (ATR) with CO_2- and H_2-separating membrane reactors	EMINENT analysis shows that this technology is still considerably less cost effective than the existing reference technology: energy supply costs per unite delivered energy are €/y 52.5 vs. €/y 29.3 of the reference technology. Thus the EMINENT tool shows the scope for improvement for the ATR. So far, the ATR concept is commercially viable. The gap between capture cost and EUA price level to be closed – e.g. by technology improvements [68].	Pilot
Zero Emission Norwegian Gas (ZENG)	A process with closed cycle, the only products of which are water and 100% captured CO_2. Analysis shows that this technology is still considerably less cost effective than the existing reference technology: Energy supply costs per unite delivered energy are €/y 42.2 vs. €/y 29.3 of the reference technology. The EMINENT tool shows the scope for improvement for the ATR. So far, the ATR concept is commercially viable. The gap between capture cost and EU Allowance (EUA) price level to be closed – e.g. by technology improvements. Concerted R&D action, industrial involvement and incentives are required including the public money, new market opportunities for CO_2 e.g., Enhanced Oil Recovery (EOR)/Enhanced Gas Recovery (EGR), clarification of carbon storage as a policy measure under the Kyoto treaty [68].	Demo
Electricity transport by ship	An EST envisaging electricity transport by ship from Iceland to the Netherlands is based on the Redox fuel cells concept. Two ships of different capacities and a reference technology – cable to Iceland – have been analyzed. That electricity transport by ship may be feasible in future, the crucial factor being energy density. Further R&D is required on Redox flow systems and high energy density electrolytes [68].	Paper idea

processing industry. Residential, work place, leisure, and service sectors still however use large amount of energy and consequently produce large emissions of CO_2 despite efficiency gains.

The use of renewables in meeting these demands is still comparatively small, and if the carbon footprint is used as a measure of CO_2 reduction, some apparently zero or low carbon emission technologies have a measurable effect on overall CO_2 production.

Successful methodologies and design strategies used in the processing industry for integrating energy systems and increasing efficiency can be applied to this nonindustry sector, which allows integration of renewable energies on a scale that allows demands to be satisfied and carbon footprints to be minimized.

It is likely to have serious implications for energy consumption and consequently greenhouse gas emissions if carbon-based energy sources are not replaced more rapidly by noncarbon or renewable sources of energy. For example, renewables only contribute, in the UK, 1.0 Mt of oil equivalent toward total domestic energy use compared to a total use of 47.0 Mt of oil equivalent.

Although the release of CO_2 from carbon-based fuels has become the prime driving force to reduce energy use, the simple measure of CO_2 emissions related to fuel use is only part of the potential production of CO_2 from prom production processes/products and/or buildings. A "carbon footprint" (CFP) is defined [4] as the total amount of CO_2 and the other greenhouse gases emitted over the full life cycle of a process or product.

There have been numerous studies (e.g., [5, 6]) stressing the carbon neutrality of renewable sources of energy. However, renewables make some contribution to the CFP.

The assessments of various renewable technologies can assist in making the correct selection and consequently reduce the release of CO_2. Electricity and heating/cooling needs can be partially satisfied by the use of renewable energy sources related to low- and zero-carbon technologies on the local scale.

Such energy sources include community CHP (combined heat and power), micro-CHP (at the household level), heat pumps, biomass, photovoltaics (PV), solar heating and cooling, wind turbines, etc. Numerous examples exist of how this can be achieved with individual dwellings. Remaining electrical and heating/cooling duties have to be met by either local sources (for example boilers) using carbon-based fuels, such as gas, or from the national grid.

What has not been investigated is the possibility of integrating local renewable energy sources with district renewable energy sources, supplemented by carbon-based systems at times of high demand or low production. This paper will attempt to provide an overall framework for such an integration based on existing studies of integrated energy systems for industrial scale processes.

9.8.1
Technologies and Their CFPs

9.8.1.1 Micro-CHP

Different consumers, typically buildings, have different energy requirements but a basic common requirement is the ability to satisfy building electricity and heat demands. Technology determines the specific electricity/heat production ratio and it cannot be significantly adjusted during operation. But to comply with the building energy demands it should be comparable with the building-specific ratio of required electricity to required heat. The temperature levels required should also fit the heat production available. As the minimum required temperatures in building applications vary between 40 °C and 80 °C, an output temperature of approximately 100 °C should be designed for a CHP system. A special issue is the cooperation with a utility company to deal with a shortage or surplus of energy. The electricity and heat demand could have varying load conditions which become crucial with for smaller size buildings. The energy demand profile of a building is characterized periodically by unpredictability. A short response time is demanded which modern CHP systems fail to provide en masse [60]. In some cases (e.g., hospitals), reliability is a very significant requirement [61].

Alanne and Saari [63] presented the following eight criteria to evaluate the applicability of small-scale CHP technologies in buildings: electrical efficiency; length of life cycle of a technology; space demanded to install a technology in a building; emissions; flexibility of control (in practice, response time); availability of fuel in the short term; level of noise generated (loudness); cost. As in many works the carbon footprint is not considered and should be explicitly added together with the life cycle analysis (LCA) of the technology.

Several competing technologies in the microscale can be used for combined heat and power generation: reciprocating engines, microturbines (electric power below 250 kW), Stirling engines, and fuel cells.

9.8.1.2 Reciprocating Engines

A typical reciprocating engine (diesel, gas, or multiple fuel) and a generator linked to the engine are quite efficient in producing electricity and have a large power range and a versatile assortment of fuels (natural gas or diesel, oil as fuels, biodiesel, and possibly regenerative biomass). The applicability of gas engines is at its best in backup systems, whereas diesel engines are recommended for continuous use. However the relatively noisy operation makes them unattractive for residential applications. The moving parts of the engine require regular maintenance which further adds to the carbon footprint. CO_2 and SO_2 emissions are dependent on the type of fuel used. CHP based on reciprocating engines are more applicable to larger buildings with less peaked electricity and heat consumption profiles [63].

9.8.1.3 Stirling Engines

The Stirling engine is a reciprocating engine with its cylinder closed and combustion taking place outside of the cylinder. The piston moves in the cylinder because of compression or expansion of the working gas (helium, hydrogen) due to the alternating heating and cooling of the cylinder by means of external combustion. Stirling engines are characterized by rather low emissions (especially NO_x) and lower noise levels.

External combustion also causes a decreased maintenance level which reduces the carbon footprint. Due to its external combustion, various fuels are also suitable, including biomass. However this type of CHP has rather low electrical efficiency, about 25–30% when natural gas is used as a fuel. When solid fuels (e.g., biomass) are used, the efficiency can be as low as 15%. The total efficiency is not significantly lower than that of other CHP applications. Stirling engines are very applicable to residential buildings, especially because the electricity/heat ratio is suitable. Their low efficiency, however, supports their use as backup power supplies rather than the one in continuous use [63]. Stirling engines are now considered as being beyond the demonstration phase and entering the market in the near future.

9.8.1.4 Fuel Cells

A fuel cell produces electricity electrochemically, by combining hydrogen and atmospheric oxygen. If the fuel is not available as pure hydrogen, it can be released from various fuels by means of a reformation process. There are many factors that make fuel cells beneficial. The most frequently mentioned benefit is its electrical efficiency, which can be as high as 45–55% [63]. Reformation of fuel decreases the efficiency, but an efficiency of approximately 40% is still achieved. On the other hand, the higher the temperature the more efficient the fuel cell will be. In addition, the electrical efficiency of fuel cells is both quite immune to load changes and not power range dependent.

Another benefit is its very low emission rate. If pure hydrogen is used, the only emission is water. If reformation is used, CO_2 and a minimal amount of oxides of sulfur and nitrogen are formed, depending on the fuel. Other benefits are noiselessness, reliability, modularity, and rapid adaptability to load changes. There are still problems to be solved by further development of the technology. The most important drawback is the investment cost. Fuel cells are more demanding than other technologies in respect of fuel production, storage, and transportation.

At the present technology stage, fuel cell plant costs can be even three times higher compared to those of reciprocating engines. Another problem is that fuel cells are more demanding than other technologies in respect of fuel production, storage, and transportation.

There may be significant environmental impacts associated with the manufacture of the fuel cells, the magnitude varying with the type of fuel cell. It is therefore critically important to carry out a full life-cycle assessment of different schemes in order to minimize overall environmental costs.

Peacock and Newborough [64] assessed the carbon footprint of some of the CHP technologies, namely the Stirling engine and fuel cells applied for a single dwelling in Central England. They underlined the significance of the control strategies for micro-CHP in buildings. Under the optimal control strategy, the Stirling engine yields daily savings of > 2.5 kg CO_2 in winter and less than 1 kg CO_2 in summer.

Table 9.3 Carbon performance of micro-CHP systems [64].

Micro-CHP	Annual reduction in CO_2 emissions, $kgCO_2$ (%)
Stirling engine – unrestricted thermal surplus	−145 (+3%)
Stirling engine – restricted thermal surplus	+574 (−10%)
Stirling engine – restricted thermal surplus and part-load capacity	+512 (−9%)
Fuel cell (1 kW)	+892 (−16%)
Fuel cell (3 kW)	+2247 (−40%)

The reduced thermal output of the 1 kW fuel cell system causes significantly less seasonal variation and yields daily savings of > 3 kg CO_2 in winter and > 2 kg CO_2 in summer.

Overall, micro-CHP systems can offer considerable CO_2 emissions reduction compared to a condensing boiler and network electricity. For a single house, both the economic and carbon rationale for deployment improve with the production of electrical output for instantaneous use in the house or for export. In case no thermal surplus is produced, the increased utilization of a micro-CHP system will yield further CO_2 and energy-cost savings. The deployment of Stirling engine technology with appropriate control has the potential to reduce the carbon footprint in the order of 0.5 t CO_2. Substantially greater savings are achievable if higher efficiency prime movers become available, whereupon the deployment of systems of greater electrical output appears desirable especially at low penetration levels (Table 9.3).

Micro-prime movers of high efficiency and acceptable cost are not yet available, but such micro-CHP systems would offer the greatest savings potential (e.g., a 1 kW system with an electrical efficiency of 50% savings 0.9 t CO_2/year for the considered dwelling). An identical system of 3 kW output is predicted to yield an annual saving of almost 2.3 t CO_2/year. This increase in savings further questions the deployment of micro-CHP systems: is it more beneficial to deploy micro-CHP systems of higher capacities in a smaller number of homes or lower efficiency micro-CHP systems in a larger number of homes? The overall evaluation of the CHP system carbon footprint requires a complete LCA, including the provision of raw materials, manufacturing, transportation, maintenance, and disposal and their related CO_2 emissions.

9.8.2
Biomass

The use of biomass is generally considered as a "carbon-neutral" process because the carbon dioxide released during the generation of energy is balanced by that absorbed by plants during their growth [65]. However, it is crucial to account for any other energy inputs that may affect this carbon-neutral balance on a case-by-

Table 9.4 Contribution to CFP in gCO_2/kWh primary energy [66].

Origin of biomass	Germany	Baltic	Sweden	Russia	Canada
Pelletizing	11	13	15	20	13
Local transport	1	1	2	2	2
Sea/River transport	4	6	5	7	15
River transport	2	2	2	2	2
Total	18	22	24	31	32

case basis, for example, any use of fertilizer, or energy consumed in vehicles when harvesting or transporting the biomass to its point of use.

There are three main groups of biomass: dependent resources (coproducts and waste generated from agricultural, industrial, and commercial processes), dedicated energy crops (short-rotation crops), and multifunctional crops (which can be used to create different types of energy). Domestic applications mainly involve pelletized biomass for use in CHP generation.

The transportation of large quantities of feedstock increases life-cycle CO_2 emissions, so biomass electricity generation is most suited to small-scale local generation facilities, or in the operation of CHP plants. Pelletizing the biomass considerably reduces volume and transportation needs and consequently is one of the solutions to increase the energy intensity of the feedstock. In the UK, the carbon footprint of biomass can vary widely depending on the kind of the feedstock: 25 gCO_2/kWh (for high-density wood gasification) to 93 gCO_2/kWh (combustion of low-density miscantus) [4].

The CFP of the biomass should take into account the energy requirements for wood pellet production. This electricity is subtracted from the green power produced by the combustion of the pellets as this consequently reduces the overall output and increases the CFP per kWh. Depending on the type of biomass used, the figures may be in the range of 50 to 300 kWh/t [66]. To preserve the green support of the wood pellets, biomass should be used for drying the pellets. Depending on the origin of the pellets, the transportation carbon emissions may include those from truck, train, sea vessels, and river boats.

In Belgium (Wallonia), Green Certificates are granted according to the CO_2 balance of the supply chain. The wood pellets have reference CFP from the LCA of 55 gCO_2/kWhp (in comparison with natural gas 251 gCO_2/kWhp and coal 396 gCO_2/kWhp). Ryckmans [66] shows the CO_2 balance in gCO_2/kWh primary energy of various pellet supply chains for Wallonia in Table 9.4.

9.8.3
Solar

Electricity and heat produced from photovoltaics have a far smaller impact on the environment than traditional methods of electrical and heat generation. During their operation, PV cells use no fuel other than sunlight, give off no atmospheric

Table 9.5 Third-generation solar cell analysis [58].

Capacity of the technology	5 kWp
Chain efficiency	30%
PV area	40 m^2
Amount of electrical energy delivered to the load by the PV system	4.465 MWh
Total costs of the system	€3862
Energy supply costs	866 €/MWh
A household unit: CO_2 emission reduction vs. the electricity from the UK grid	5 t/y

or water pollutants, and require no cooling water. Like all equipment, during manufacture energy is used and because of the current fuel mix, emissions are produced.

The silicon required for PV modules is extracted from quartz sand at high temperatures. This is the most energy intensive phase of PV module production, accounting for 60% of the total energy requirement. It has been estimated that for building integrated PV systems, the total life-cycle emissions of CO_2 are between 13 and 731 gCO$_2$/kWh produced (depending mainly on PV cell technology) [67]. Life-cycle CO_2 emissions for UK photovoltaic power systems are currently 58 gCO$_2$eq/kWh. These figures are very much dependent on the geographical location: in South European countries, the CFP can be about 35 gCO$_2$eq/kWh, due to more sunlight, longer operating hours, and higher energy output [4].

If the PV system replaces diesel generation, which produces about 700 gCO$_2$/kWh, then substantial CO_2 CFP reduction can be made. Disposal of PV cells generally after 30 years of service can present some waste disposal problems which slightly increase the overall CFP of the technology.

The EMINENT software tool [68] has been tested and applied to analyze the solar film photovoltaic cell potential for application in the UK. The analysis finds that first- and second-generation PV cells are not economically efficient and are not a sustainable option for mass scale application in the household sector. The analysis also shows that the application of third-generation solar cells in the household sector approaches the point when it can become economically viable and contribute to the reduction of the CO_2 footprint (Table 9.5). If economic incentives are applied at the national and local levels to stimulate the use of the solar technology, it can become commercially viable in the household sector in the UK. The details of the analysis are given in Klemeš et al. [58].

An example of an evaluation, which includes LCA, was an EC project developing a solar energy storage system (consisting of general solar heating and a thermochemical storage unit) detailed in Masruroh et al. [7]. In this LCA study, six options for the selection of appropriate salt/binder combinations were considered, three situations in which different energy requirements are to be fulfilled by using the SOLARSTORE system, and two possible market places. Thirty-six LCA case studies were performed to select the most appropriate reactive compounds for different situations and different market places. The phase-based breakdowns of the four LCA-based environmental impacts are shown quantified as CFP as well.

kg CO$_2$/y

g CO$_2$/kWh

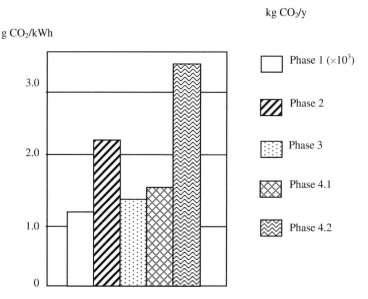

Fig. 9.14 The results of LCA-based CFP evaluation [69].

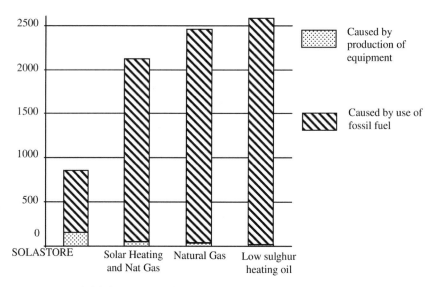

Fig. 9.15 Annual global environmental impact [7].

The phases in these figures are as follows: phase 1, raw material acquisition and components manufacturing; phase 2, transportation of raw materials to manufacturing site; phase 3, transportation of units from the manufacturing site to the assembling site; phase 4.1, transportation of the SOLARSTORE product to consumers in France; phase 4.2, transportation of the SOLARSTORE product to consumers in Spain (Fig. 9.14).

Although the SOLARSTORE system was shown to improve the efficiency of traditional solar heating systems, it cannot fulfill the total annual heating requirement in most locations around Europe. A backup boiler is used to accommodate the remaining heating requirement. In this case the use of conventional energy production, i.e., electricity/gas will produce certain emissions, which should be taken into consideration. Figure 9.15 shows the annum-based comparison of global warming potential impact for different heating systems.

9.8.4
Wind

Wind generated electricity has one of the lowest CFPs. Manufacturing and construction account for almost all of the carbon emissions, the rest being maintenance. In the UK, a typical wind generation CFP is about 4.64 gCO_2eq/kWh [4]. The availability should also be taken into consideration, and in most cases is rarely above 20–30% [69]. Backup systems increase the effective CFP.

9.8.4.1 Energy Efficiency/Saving
This is the most obvious and usually easiest way to reduce emissions and improve the CFP. It ranges from simple energy-saving measures based on observation or energy audits, and implementation of energy-saving products, such as insulation, windows, and energy-saving bulbs.

What has to be considered is that each of these energy-saving actions also produces some CFP [76]. More sophisticated methodologies such as heat integration [19] connected with decarbonization [62, 77] have been discussed previously in this chapter.

With the increase in the application of renewables and the variety of renewables available, these methodologies are becoming vital for single building as well. A recent example examines the integration potential of fuel cells and CHP and combined cycles [12].

Carbon sequestration is also an option to reduce the CO_2 released by combustion of carbon-based fuels. The economic implications have been studied [1, 59] and shown that these measures also have a CFP, which can be considerable.

9.8.5
Overall CFP Performance of the Technologies for Energy Use

CO_2 emissions are studied in terms of elimination, minimization, and capture and sequestration, for a number of technologies. Zero or low carbon technologies are in the forefront of the attention of scientists, engineers, the general public, and politicians. However, if we want to obtain a clear vision of how various technologies can affect the environment we need to take into account their CFP.

This means that life-cycle analysis should be carried out for the technologies suggested rather than their immediate effect on the environment assessed. The minimization of emissions and the reduction of the CFP are required to be looked

Table 9.6 CFPs of some building-related technologies.

Technology	CFP (gCO$_2$eq/kWh)	Factors
Coal large scale [4]	790–830	Power station type, quality of coal, distance from the source of coal, type of the mine, efficiency, emission treatment.
Gas large scale [4]	400–420	Power station type, distance from the source of gas, gas extraction, efficiency, emission treatment.
micro-CHP gas [70]	200–250	Materials used, type of CHP and its efficiency, type of fuel, distance from the manufacturer, maintenance.
Biomass [4]	25–93	Distance from the source of biomass, type of the transport used, type of biomass, type of combustions process, its efficiency, emission (e.g., NO$_x$) treatment.
Solar [4]	30–80 (backup not included)	Geographical location related to sun, distance from the manufacturer, type of solar cells and their efficiency, materials used, maintenance needed, backup.
Wind [4]	2–8 (backup not included)	Location related to wind, place of installation, distance to the manufacturer, maintenance needed related to the environment, backup.

at in a wider context which can more accurately determine the overall effectiveness of systems which can be related to buildings and building complexes. In these cases low-carbon technologies may not be as effective as first appears (Table 9.6).

9.9
A Selection of Industrial and Residential Sector Applications: Petrochemical, Hospital Complex, Food and Drink

9.9.1
Petrochemicals – Fluid Catalytic Cracking

The fluid catalytic cracking (FCC) unit is a major process in oil refineries, and over the years attempts have been made to improve yields and efficiencies in response to external drivers. The retrofit of the associated heat exchanger network (HEN) of the FCC can frequently result in improvements in the energy recovery, and consequently a reduction in the energy use or increasing throughput performance. In this example, the HEN of the FCC process consists of a main column and a gas concentration section [3].

The required data for the study, provided by a simulation of the process, were extracted from the existing network, which then establishes the heating and cooling requirements of the process. The stream data are made up of 23 hot streams and

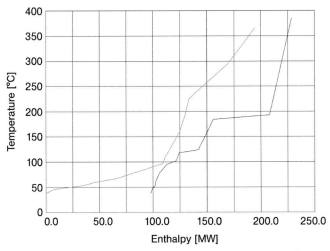

Fig. 9.16 Composite Curve of FCC process with optimum ΔT_{min} (after [3]).

11 cold streams. The associated cost and economic data required for the analysis were specified by the refinery owners. Incremental area efficiency is used for the targeting stage of the design, and the design is then carried out using the network pinch method, which consists of a diagnosis stage and an optimization stage. In the diagnosis stage a few promising designs were generated using a software package called SPRINT developed by UMIST (now The University of Manchester). SPRINT was also used to optimize the initial design, in order to trade off capital cost against energy savings. The design options were then compared and evaluated and the final retrofit design put forward for final inspection.

The existing ΔT_{min} of the process was established as 24 °C and the hot utility consumption of the process was 46.055 MW. The area efficiency of the existing design was 0.805. The targeting stage of the analysis made use of the incremental area efficiency which produced a minimum approach temperature of 11.5 °C. The potential for energy saving was then derived from the resultant composite curve.

The composite curve of the process is shown in Fig. 9.16. It can be seen that the composite curves are relatively wide apart except in the area around the pinch. The total annualized plot target for a grass root design was performed using SPRINT. In the case of the existing or retrofit design, the tradeoff between capital and energy was performed by taking the area and energy target from SPRINT and translating these into energy saving and investment. It was assumed that the utilities would be used in the same ratio as that in the existing design.

The capital cost was estimated assuming that the distribution of area would be the same as that for the existing network. The resulting optimum minimum approach temperature was found to be around 11.5 °C for incremental α and about 17.5 °C for constant α. The area efficiency, α, of the existing network was found to be 0.804, a value which indicated that the existing design was using the area reasonably efficiently. However, despite this, it was found that there still existed the potential for improvement. As the constant α targeting produced a conservative approach, an incremental α value of 1.0 was used to set the target for the retrofit. The target was set and the potential for energy savings was found to be about 12.117 MW.

Analysis of the existing design revealed that there were four processes-to-process heat exchangers that transferred heat across the pinch (from above to below the pinch). It was also found that there existed hot utility units which supplied heat to below the pinch and cold utility units that removed heat from above the pinch. These energy violations of the established pinch rules have produced the energy scope of the project.

The retrofit design was carried out using the network pinch method, which allowed the setting of an absolute limit on the energy recovery of the structure. The network pinch can be overcome by changes in the topology of the network which identifies opportunities to shift heat from below to above the network pinch. The minimum approach temperature for the network which was set by targeting, using the area efficiency concept, was used. The next stage consisted of testing a set of modifications which resulted in the increase of the energy recovery of the process. However, the increase in energy recovery was related to an increase in the heat exchange area. Consequently any benefit in energy cost reduction had to be balanced with the additional capital cost associated with the increase in the heat exchanger area. A number of promising design solutions were generated, which were then optimized in order to minimize the total cost.

The identified four designs all produced a payback period of less than 2 years. The final design chosen for the retrofit situation was that of option C (Fig. 9.17), which had the shortest payback period and the least additional area required.

In total four new heat exchangers were added in this design option and one existing exchanger (number 1 in Fig. 9.17) was removed as the duty on this exchanger approached zero. In reality this exchanger need not removed but used as one of the additional exchangers which were required by the changed design, by re-piping either the hot stream or the cold stream as in exchanger 42 or 40, respectively, in Fig. 9.17, further reducing the additional area required. The main features of the data used in the study and the design are given in Fig. 9.18. All the modifications carried out in the diagnosis stage have moved heat from below the network pinch to above the network pinch.

The final resultant design produced energy saving of 8.955 MW, which is about 74% of the potential of the design. The utility cost saving is $2,388,600/year, which represents an approximate 27% decrease in the utility bill. The modified heat exchanger network required an investment of $3,758,420, which produced a payback period of just below 19 months.

The study shows that a combination of targeting, using pinch technology, and the network pinch method can produce improvements in an existing HEN resulting in reduction in total costs of the network. The employed method can recognize bottlenecks in an existing system, and then provides a series of potential improvements by searching for modifications that are able to shift heat from below to above the network pinch. It was found that targeting for maximum energy saving at each potential modification tends to give good tradeoff between area and energy.

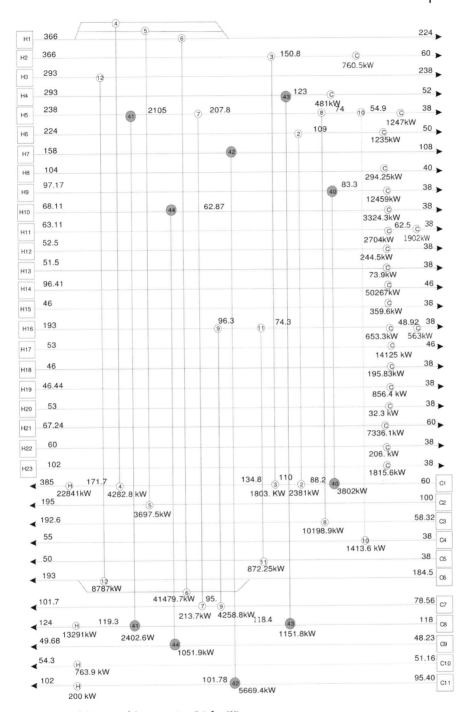

Fig. 9.17 Grid diagram of design option C (after [3]).

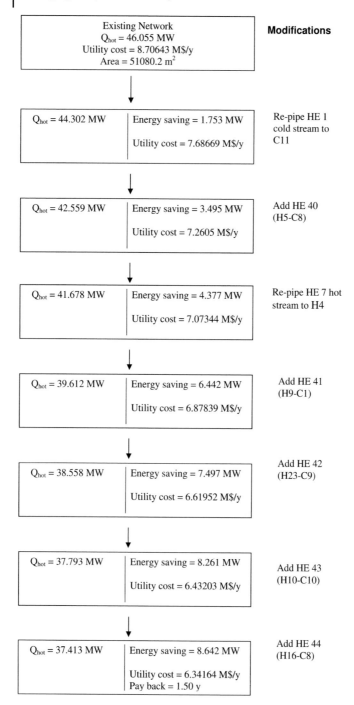

Fig. 9.18 Sequential search of modifications of the existing network for design option B (after [3]).

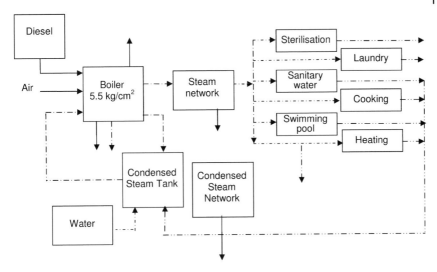

Fig. 9.19 Hospital heat flow block diagram (after [72]).

Although the authors showed the potential of the targeting and network pinch method in existing networks, they also stated that a number of other issues could have been investigated. These include the following:

- Further investigations of the resultant network for impact on the operational, control, and safety constraints. The results of the initial study would be used as a preliminary design for further detailed energy-related studies of the process.
- Inclusion of the piping cost, which can be estimated, in order to obtain an improved approximation of the total cost required to implement the modification. Additionally, the pressure drops in the network need to be further studied to evaluate impact on the pumping requirement of the process.
- Evaluation of the potential of heat transfer enhancement techniques which can be employed to reduce the additional area requirements.
- Evaluation of the potential of integration of other equipment such as the main column and using column pinch analysis to scope and screen beneficial improvements.

9.9.2
Hospital Complex

Herrera et al. [72] presented a study of a hospital complex which included an institute, a general hospital, a regional laundry center, a sports center, and some other public buildings. The use of diesel as a fuel represented 75% of its total energy consumption and 68% of its total energy cost which was US$ 396,131 in 1999.

The hospital complex process diagram is shown in Fig. 9.19. In the hospital complex, the heat demand is met by producing steam in boilers fuelled by a high-price diesel fuel. There is no heat recovery between the existing heat sources and heat

Table 9.7 The input data extracted from Fig. 9.19.

Streams	ΔH (kW)	CP (kW/°C)	T_{in} (°C)	T_{out} (°C)
Hot streams				
Soapy water (1)	23.7	0.53	85	40
Condensed steam (2)	96.32	2.41	80	40
Cold streams				
Laundry sanitary water (LS)	17.60	0.59	25	55
Laundry (L)	77.12	2.20	25	60
Boiler feed water (BF)	7.13	0.24	30	60
Sanitary water (SW)	77.12	2.20	25	60
Sterilization (S)	12.50	0.14	30	121
Swimming pool water (SP)	151.67	50.56	25	28
Cooking (CO)	59.63	0.85	30	100
Heating (H)	100.82	14.40	18	25
Bedpan washers (B)	4.94	0.05	21	121

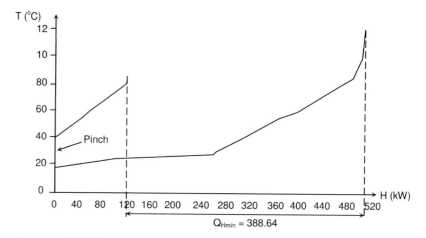

Fig. 9.20 CC for the hospital complex process streams (after [72]).

sinks. The hot streams were identified as the soiled soapy water from the laundry and the flow of condensed steam not recovered in the condensation network. The stream data are presented in Table 9.7.

It needs to be noted that the CP represents mass flow m multiplied by the specific heat capacity C_p and in this case are considered constants. As can be seen from the resultant composite curves (CC) (Fig. 9.20) the amount of external heating required, or the hot utility target, of this hospital complex is 388.64 kW. The grand composite curve (GCC) has been employed to determine the temperature levels of the utilities necessary to satisfy this requirement. This graphical method allows a more precise analysis to be undertaken in order to integrate the thermal utilities considered for the process heating and cooling requirements (Fig. 9.21). This GCC

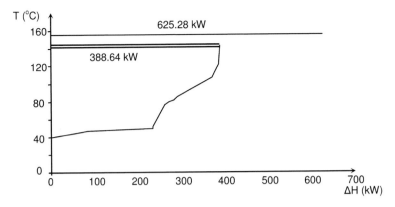

Fig. 9.21 GCC of the hospital (after [72]).

indicates that the required heating of complex should be 388.64 kW. This means a yearly energy requirement of 12.26 TJ.

The actual amount of heating provided is in fact 625.28 kW, which represents the heat services that are currently transferred to the complex. This results in a potential energy saving of 38% equivalent to saving 246,000 L of diesel/year (US$ 100,000). To reduce energy demands in the form of heating to the value targeted by the analysis, the heat-integration-based analysis suggests that four extra heat exchangers should be added to the network. Two in the laundry to cover part of the heat demand, a third in the machinery rooms which help to heat boiler feed water, and the final one in the condensation tank area that heats the sanitary water. Several other issues could also have been considered to further refine the analysis, such as fouling, pressure drop, and nonconstant heat demand.

9.9.3
Food and Drink

There have been many studies which have employed Pinch technology and its associated heat integration analysis in the food processing industry. Generally the food processing industry has a far different thermodynamic profile than that which occurs in the refining and petrochemical industry. The food processing industry is characterized by the relatively low temperatures of the process streams, normally in the range of 120–140 °C, a small number of hot streams, low boiling point elevation of food solutions, considerable deposition of scale in evaporator and recovery systems, and seasonal performance.

Many studies have also found that pinch technology/heat integration application is hampered by particular features of the food processing systems, such as direct steam heating, difficulties in cleaning heat exchanger surfaces and high utility temperatures.

Despite these drawbacks, the benefits that can be obtained by the application of the technologies and result in optimized heat systems and low energy consumption far outweigh the difficulties of performing these studies. Additionally, there are other advantages, in that technological improvements, such as reduced deposition of scale due to reduced utility temperatures, self-regulation of heat process, and reduced emissions can result.

The study of the production of refined sunflower oil, performed by Klemeš et al. [73], provides an example of the benefits that can be obtained by performing a pinch technology/heat integration analysis. The process was operated with a minimal temperature difference of 65 °C at the process pinch. The external heating required by the system was provided by two types of hot utilities: dautherm steam and water steam. The external cooling required was provided by two cold utilities: cooling water and ice water. The study resulted in the suggestion of increasing heat recovery, and reducing the minimum temperature difference to 8–14 °C. The increase in heat recovery provided by the reduction in the minimum driving force for the process produced an increase in the heat transfer area, but this was more than offset by the reduction in the hot and cold utility requirements. A further benefit of the analysis was the reduction in the number of utilities needed, by the elimination of water steam and cooling water, thereby considerably simplifying the overall design.

Further examples of the benefits of pinch technology/heat integration have been provided by Klemeš et al. [73]. In the case of a processing plant extracting raw sunflower oil, there existed the inappropriate placement of a one-stage evaporation system for separating the solvent, benzene. Because of the flammability of the benzene and the sharp increase in the boiling point elevation, the problem could not be eliminated by changes in the pressure. However, an appropriately placed indirect heat pump, pumping heat from the condensation temperature level to the utility temperature level in the evaporation system, partially solved the problem. In the case of a crystalline glucose production processing plant, the same authors found that vapors bleeding from a multiple-stage evaporation system for concentrating the water–glucose solution resulted in the unnecessary overexpenditure of utilities. This resulted from the pinch technology/heat integration analysis finding that the multiple-stage evaporation system is inappropriately placed across the process pinch.

There are also many other studies in the food processing industry that have resulted in the reduction of energy use and increased energy efficiency. In the production of tomato concentrate the main consumer of heat is a multiple-stage evaporation system. The appropriate placement of this evaporation system is difficult because of the very large differences in the heat consumption of the evaporation and recovery systems. Furthermore, the available temperature difference in the evaporator was found to be insufficient for a greater number of stages. However, increasing the two- and three-stage evaporators to four stages and by using steam ejecting compressors, the energy savings produced were beneficial when accounting for additional capital costs of the modifications.

A case study of a whisky distillery by Smith and Linnhoff [74] provides another example of how pinch technology/heat integration can reduce energy use and increase energy efficiency. It was found that steam was being used below the process pinch, and consequently produced an overall increase in utility usage. As the steam

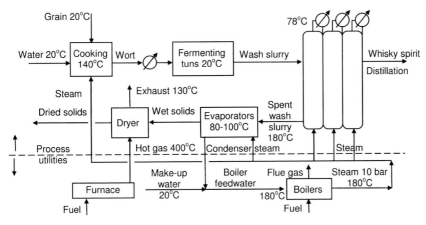

Fig. 9.22 The whisky distillery process, represented by a simple flow diagram (after [38]).

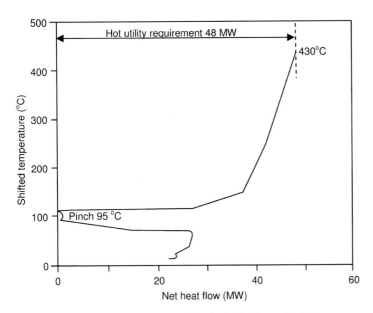

Fig. 9.23 The grand composite curve of the whisky distillery (after [38]).

use was related to the use of a heat pump, the steam used below the process pinch was eliminated by reducing the size of the heat pump. Although the steam now had to be used for process heating above the process pinch, the overall energy costs were reduced due to the reduction in the compressor duty.

A further study, involving a whisky distillery, has been provided by Kemp [38]. The flowsheet of the process (Fig. 9.22) shows the principal units of the processing systems including the cooking, fermentation, and distillation of the whisky, the processing of the spent grains, and the utility system providing the drying duty and steam requirements for the distillation. The hot utility requirement for the

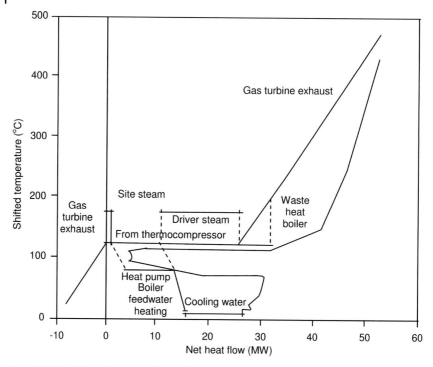

Fig. 9.24 The final configuration of the utility system matched against the grand composite (after [38]).

process is 48 MW and the pinch, as is at 95 °C, both of which can be clearly seen on the grand composite curve in Fig. 9.23. The hot utility requirements are mainly steam for the distillation system and hot air for the drying system.

Kemp [38] indicated that the form of the grand composite and the temperature of the pinch could be exploited for heat pumping, and also suggested that the process would benefit from the introduction of a combined heat and power scheme for improved process integration and energy efficiency. As the power demand of the site was 12 MW, there existed two possibilities for providing both the power and the heat demands from the same utility system. A gas turbine that produced 12 MW of power would also supply approximately 30 MW of high-grade heat from the exhaust. The second option involved the use of back-pressure steam turbines to produce the required power, but this required the production of approximately 100 MW, much larger than that was required. Additionally, the gas turbine exhaust would be sufficiently clean to be used for drying purposes.

The final configuration of the utility system matched against the grand composite is shown in Fig. 9.24. The heat was provided principally from the gas turbine exhaust and from the existing thermocompressors. The existing package boilers were used to provide steam for the thermocompressors. An increase in the efficiency of this part of the system was produced by using below pinch waste heat to

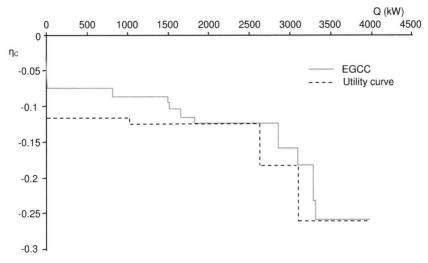

Fig. 9.25 The exergy grand composite curve (EGCC) for the plant (after [75]).

preheat the boiler feedwater. Waste heat boilers driven by the exhaust from the gas turbine provided additional steam.

A large number of processes in the food and drink industry make use of chilling and refrigeration systems. Pinch technology/heat integration has also been used to increase the efficiency of these systems. An example of a Swedish slaughtering and meat processing plant has been provided by Fritzson and Berntsson [75]. In order to simplify the analysis, the plant was initially divided into two sections. The first section involved the below ambient part of the process, where compression power is required to provide refrigerated cooling. The process above ambient temperature mainly involves the use of heat for the production of steam and hot water for cleaning purposes. Excess heat was also available from the flue gases in the slaughterhouse, and from two installed heat pumps.

The analysis of the below ambient temperature section makes use of the method proposed by Linnhoff and Dhole [46] for low-temperature process changes involving shaftwork targeting. Figure 9.25 shows the exergy grand composite curve (EGCC) for the plant. There is a large gap between the EGCC and the utility curve indicating a low efficiency in the use of available shaftwork. This is an area that could be potentially improved, and would result in a 15% reduction in energy demand.

Fritzson and Berntsson [75] started the reduction process by adjusting the loads on the subambient utilities by first maximizing the load on the highest temperature subambient utility (−10 °C). After this the load on each lower temperature utility was maximized (Fig. 9.26). The reduction process was started at the highest level (the highest temperature), as this utility can be provided at a lower cost than refrigeration at lower temperatures. The modified system was then modeled and simulated in HYSYS, which showed a 5% reduction in the shaftwork required.

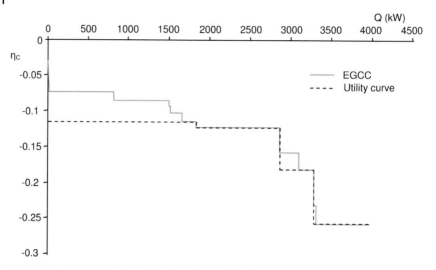

Fig. 9.26 Adjusted the loads on the subambient utilities by maximizing the load on the highest temperature subambient utility ($-10\,°C$) and then maximizing the load on each of the lower temperature utilities in turn (after [75]).

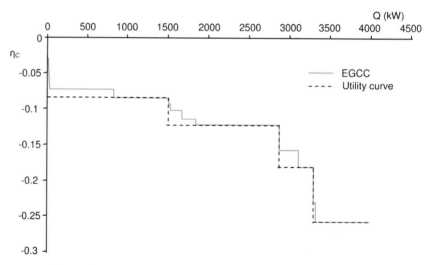

Fig. 9.27 The area between the EGCC and the utility curve is further decreased by changing the temperature of the first refrigeration level from $-10\,°C$ to $-3\,°C$ (after [75]).

The resultant EGCC diagram (Fig. 9.26), however, still revealed a relatively poor fit between the EGCC and the utility curve. To reduce the gap further it was suggested that the temperature of the highest temperature refrigeration level should be increased from $-10\,°C$ to $-3\,°C$, and then re-adjusting the loads as done before (Fig. 9.27). This configuration provided a 10% reduction in the shaftwork requirement. Other temperature changes were also suggested, which although reduced further the shaftwork requirements, were found to be less cost effective.

9.10
Conclusions

Increasing the energy efficiency of energy using processes is the most effective initial method of reducing cost and reducing the emissions which affect the stability of the world climatic system. Increasing energy efficiency is the cleanest method in the short term to produce green energy, i.e., that energy that produces the least amount of emissions. Unfortunately, in many situations, this is ignored as many in society are seduced by the more publicity exciting technologies that are available as renewable energy sources.

An effective measurement of the impact of different energy efficiency methods and energy reduction methods involves the use of the recently developed carbon footprint (CFP) which takes into account all carbon emissions over the whole life cycle of a process or product.

This chapter has been written with an intention and target of raising the awareness of the main methodologies which could contribute to the increase of energy efficiency in processing plants and reduce the environmental impact as measured by the carbon footprint.

There are many more examples that could have been discussed, but limited space makes it possible to tackle only some, but by the best opinion of the authors, the most important features from various fields including processing industries as well as buildings have been included.

A wide list of references has been provided to assist in the provision of more information, and if authors manage to raise the interest and awareness of potential readers this chapter has fulfilled its intentions.

References

1 KLEMEŠ, J., BULATOV, I., COCKERIL, T., Techno-economic modelling and cost functions of CO_2 capture processes. *Computers & Chemical Engineering* 31(5–6) (2007), pp. 445–455.

2 KLEMEŠ, J., FRIEDLER, F., Recent novel developments in heat integration – total site, trigeneration, utility systems and cost-effective decarbonisation: case studies waste thermal processing, pulp and paper and fuel cells. *Applied Thermal Engineering* 25(7) (2005), pp. 953–960.

3 AL-RIYAMI, B. A., KLEMEŠ, J., PERRY, S., Heat integration retrofit analysis of a heat exchanger network of a fluid catalytic

cracking plant. *Applied Thermal Engineering* 21 (2001), pp. 1449–1487.

4 PARLIAMENTARY OFFICE FOR SCIENCE AND TECHNOLOGY (POST). Carbon Footprint of Electricity Generation, 2006.

5 ALBRECHT, J., The future role of photovoltaics: a learning curve versus portfolio perspective. *Energy Policy* 35(4) (2007), pp. 2296–2304.

6 FIASCHI, D., CARTA, R., CO_2 abatement by co-firing of natural gas and biomass-derived gas in a gas turbine. *Energy* 32(4) (2007), pp. 549–567.

7 MASRUROH, N. A., LI, Bo, KLEMEŠ, J., Life cycle analysis of a solar thermal system

with thermochemical storage process. *Renewable Energy* 31(4) (**2006**), pp. 537–548.

8 SEMPRA ENERGY WEBSITE – ENERGY SAVING TIPS. www.sdge.com/business/ee_tips.shtml (accessed August 21, **2007**).

9 US DEPARTMENT OF HOUSING AND URBAN DEVELOPMENT. How to identify primary opportunities to save energy – energy saving techniques. www.hud.gov/offices/cpd/ affordablehousing/training/web/energy/ techniques/techniques.cfm (accessed September 21, **2007**).

10 INTERNATIONAL ENERGY AGENCY. www.iea.org/Textbase/subjectqueries/ index.asp (accessed October 10, **2007**).

11 AEA TECHNOLOGY. Study on Energy Management and Optimisation in Industry, 2000, was prepared by AEA Technology plc at the request of the Environment DirectorateGeneral of the European Commission. www.ec.europa.eu/environment/ippc/pdf/ energy.pdf (accessed August 23, **2007**).

12 VARBANOV, P., KLEMEŠ, J., SHAH, R. K., SHIHN, H., Power cycle integration and efficiency increase of molten carbonate fuel cell systems (research paper). *Transactions of the ASME Journal of Fuel Cell Science and Technology* 3 (**2006**), pp. 375–383.

13 KLEMEŠ, J., LUTCHA, J., VAŠEK, V., Recent extension and development of design integrated system – DIS. *Computers & Chemical Engineering* 3(4) (**1979**), pp. 357–361.

14 TRIANTAFYLLOU, C., SMITH, R., The design and optimisation of fully thermally coupled distillation columns. *Transactions of the Institute of Chemical Engineers* 70 (**1992**), pp. 118–132.

15 HERNÁNDEZ, S., JIMÉNEZ, A. Design of energy-efficient Petlyuk systems. *Computers & Chemical Engineering* 23(8), pp. 1005–1010.

16 KELLOGG, M. W., Kellogg divided wall column. Promotional brochure, **1998**.

17 EEO (ENERGY EFFICIENCY OFFICE). Opportunities for compact heat exchangers. Best Practice Programme R&D Profile 36, **1993**.

18 LINNHOFF, B., VREDEVELD, D. R., Pinch technology has come of age. *Chemical Engineering Progress* 33 (**1984**), p. 40.

19 KLEMEŠ, J., DHOLE, V. R., RAISSI, K., PERRY, S. J., PUIGJANER, L., Targeting and design methodology for reduction of fuel, power and CO$_2$ on total sites. *Applied Thermal Engineering* 17 (**1997**), pp. 993–1003.

20 SMITH, R., *Chemical Process Design*, McGraw–Hill, Inc, New York, **1995**, 460 pp.

21 NRCan (NATURAL RESOURCES CANADA). Success Stories. cetcvarennes.nrcan.gc.ca/en/indus/agroa_fd/ ip_pi/ap_pa.html (accessed October 1, 2007) **2003**.

22 U.S. DEPARTMENT OF ENERGY, INDUSTRIAL ASSESSMENT CENTERS, ENERGY EFFICIENCY AND RENEWABLE ENERGY, www.iac.rutgers.edu/database/ (accessed September 30, **2007**).

23 Carbon Trust. www.actionenergy.org.uk/action+energy/ (accessed July 6, **2007**).

24 UK Government's Energy Efficiency Best Practice programme (EEBPp) Benchmarking tool for industrial buildings – heating and internal lighting, May **2002**, 32 pp.

25 KLEMEŠ, J., Chemical process simulation programs – present state and future development of flowsheeting. *Collection of Czechoslovak Chemical Communications* 42(2) (**1977**), pp. 426–445.

26 HEYEN, G., KALITVENTZEFF, B., *Process Monitoring and Data Reconciliation. Computer Aided Process Engineering*, Puigjaner, L., Heyen, G. (Eds.), Wiley, New York, **2005**.

27 VALI III USER GUIDE, BELSIM S.A., Liege, Belgium, December **2003**.

28 LANGENSTEIN, M., JANSKY, J., LAIPPLE, B., Finding megawatts in nuclear power plants with process data reconciliation. *Proceedings of ICONE12, 12th International Conference on Nuclear Engineering*, April 25–29, **2004**, Arlington, VA, USA, ICONE12-49152. Available at www.btbjansky.com/pdf/Q_70_ICONEfinal-paper.pdf (accessed October 3, 2007).

29 MINET, F., GEORGES HEYEN, G., KALITVENTZEFF, B., Dynamic data reconciliation of regenerative heat exchangers coupled to a blast furnace. *11th European Symposium on Computer Aided Process Engineering*, **2001**, pp. 1053–1058.

30 Wikipedia, en.wikipedia.org/wiki/List_of_Chemical_Process_Simulators (accessed October 3, **2007**).

31 EUROSTAT. Eurostat Press Office. ec.europa.eu/eurostat [15/03/2007], **2006**.

32 DTI. UK Energy in Brief. www.dti.gov.uk/energy/statistics/publications/in-brief/page17222.html [15/03/2007], **2006**.

33 UK GOVERNMENT. Government Response to Kate Barker's Review of Housing Supply: The Supporting Analysis. Office of the Deputy Prime Minister. www.odpm.gov.uk [15/03/2007] **2005**.

34 CPI (CENTRE FOR PROCESS INTEGRATION). Heat integration and energy systems. MSc Course, School of Chemical Engineering and Analytical Science, The University of Manchester, UK, **2004/2005**.

35 LINNHOFF, B., TOWNSEND, D. W., BOLAND, D., HEWITT, G. F., THOMAS, B. E. A., GUY, A. R., MARSLAND, R. H., *User Guide on Process Integration for the Efficient Use of Energy*, IChemE, Rugby, UK, **1982**, last edition, **1994**.

36 SMITH, R., *Chemical Process Design and Integration*, Wiley, New York, **2005**, 685 pp.

37 SHENOY, U. V., *Heat Exchanger Network Synthesis, Process Optimisation by Energy and Resource Analysis*, Gulf Publishing Company, Houston, **1995**, 635 pp.

38 KEMP, I., *Pinch Analysis and Process Integration. A User Guide on Process Integration for Efficient Use of Energy*, Butterworth-Heinemann, Elsevier, **2007**.

39 KLEMEŠ, J., PERRY, S., Process optimisation to minimise energy use in food processing. In: *Handbook of Waste Management and Co-product Recovery in Food Processing*, Vol. 1, Waldron K. (Ed.), Woodhead Publishing, Cambridge, **2007**.

40 TAAL, M., BULATOV, I., KLEMEŠ, J., STEHLIK, P., Cost estimation and energy price forecast for economic evaluation of retrofit projects. *Applied Thermal Engineering* 23 (**2003**), pp. 1819–1835

41 DONNELLY, N., KLEMEŠ, J., PERRY, S., Impact of economic criteria and cost uncertainty on HEN network design and retrofit. *Proc of PRES'05*, Klemeš, J. (Ed.), AIDIC, **2005**, pp. 127–132.

42 ASANTE, N. D. K., ZHU, X. X., An automated and interactive approach for heat exchanger network retrofit. *Trans. IChemE* 75 (**1997**), pp. 349–360 (Part A).

43 URBANIEC, K., ZALEWSKI, P., KLEMEŠ, J., Application of process integration methods to retrofit design for Polish sugar factories. *Sugar Industry* 125(5) (**2000**), pp. 244–247.

44 SPRINT. *Process Integration Software*. Centre for Process Integration, CEAS, The University of Manchester, UK, **2007**.

45 STAR. *Process Integration Software*. Centre for Process Integration, CEAS, The University of Manchester, UK, **2007**.

46 LINNHOFF, B., DHOLE, V. R., Shaftwork targets for low temperature process design. *Chemical Engineering Science* 47 (**1992**), pp. 2081–2091.

47 DHOLE, V. R., Distillation column integration and overall design of subambient plants. PhD Thesis, UMIST, UK, **1991**.

48 LEE, G. C., Optimal design and analysis of refrigeration systems for low temperature processes. PhD Thesis, UMIST, UK, **2001**.

49 LEE, G.-C., SMITH, R., ZHU, X. X., Optimal synthesis of mixed refrigerant systems for low temperature processes. *Industrial & Engineering Chemistry Research* 41 (**2003**), p. 5016.

50 KIM, J.-K., KLEMEŠ, J., Sustainable energy integration of refrigeration and heat pumps systems. *PRES 2006*, Prague, G8.3 [1465], **2006**.

51 LINNHOFF, B., AHMAD, S., SUPERTARGETING: optimum synthesis of energy management systems. *Journal of Energy Research Technology* 111 (**1989**), p. 121.

52 MURPHY, L., WU, F. F., An open design approach for distributed energy management systems. *IEEE Transactions on Power Systems* 8 (**1993**), p. 3.

53 LEVERMORE, G. J., *Building Energy Management Systems*, E&FN SPON, **1992**, 239 pp.

54 TURNER, W. C. (Ed.), *Energy Management Reference Library*, Marcel Dekker, New York and Basel, **2003**, 1904 pp.

55 WORRELL, E., GALITSKY, CH., Energy efficiency improvement and cost saving opportunities for petroleum refineries. *An ENERGY STAR Guide for Energy and Plant Managers*. Ernest Orlando Lawrence Berkeley National Laboratory, U.S. Environmental Protection Agency, LBNL-56183, **2005**, 114 pp.

56 U.S. EPA's GUIDELINES FOR ENERGY MANAGEMENT. Available at www.energystar.gov (accessed Aug. 12, **2006**).

57 Rajan, G. G., *Practical Energy Efficiency Optimization*, PennWell Books, Tulsa, USA, **2006**, 401 pp.

58 Klemeš, J., Zhang, N., Bulatov, I., Jansen, P., Koppejan, J., Novel energy saving technologies assessment by EMINENT evaluation tool. *PRES'05, Giardini Naxos*, Klemeš, Jiri (Ed.). *Chemical Engineering Transactions* 7 (**2005**), pp. 163–167.

59 Klemeš, J., Bulatov, I., Koppejan, J., Friedler, F., Hetland, J., Novel energy saving technologies evaluation tool, T5-142. *17th European Symposium on Computer Aided Process Engineering – ESCAPE17*, Bucharest, Plesu, V., Agachi, P. S. (Eds.), Elsevier, Amsterdam, **2007**.

60 The Carbon Trust's Small-Scale CHP field trial update. CTC513 www.carbontrust.co.uk/Publications/publicationdetail.htm?productid=CTC513&metaNoCache=1 (accessed August 12, 2007), **2005**.

61 Ellis, M. W., *Fuel Cells for Building Applications*, ASHRAE, USA, **2002**.

62 Varbanov, P., Perry, S., Klemeš, J., Smith, R., Synthesis of industrial utility systems: cost-effective de-carbonisation. *Applied Thermal Engineering* 25 (**2005**), pp. 985–1001.

63 Alanne, K., Saari, A., Sustainable small-scale CHP technologies for buildings: the basis for multi-perspective decision-making. *Renewable and Sustainable Energy Reviews* 8 (**2004**), pp. 401–431.

64 Peacock, A. D., Newborough, M., Impact of micro-CHP systems on domestic sector CO2 emissions. *Applied Thermal Engineering* 25 (**2005**), pp. 2653–2676.

65 www.ccap.org/pdf/biopub.pdf (accessed August 14, **2007**).

66 Ryckmans, Y., Biomass sustainability certification in Belgium. *Presentation at IEA Bioenergy Task 40 and EUBIONET 2 Joint Workshop*, Rotterdam, The Netherlands, **2007**.

67 www.nei.org/index.asp?catnum=2&catid=260 (accessed Sept. 14, **2007**).

68 EMINENT project. www.eminentproject.com (accessed Sept. 20, **2007**).

69 www.ucte.org/pdf/Publications/2004/UCTE-position-on-wind-power.pdf (accessed March 12, **2007**).

70 www.cga.ca/publications/documents/Jan2506SRatAWMAmtgfinal.pdf (accessed March 12, **2007**).

71 Asante, N. D. K., Automated and interactive retrofit design of practical heat exchanger networks. PhD Thesis, UMIST, Manchester, UK, **1996**.

72 Herrera, A., Islas, J., Arriola, A., Pinch technology application in a hospital. *Applied Thermal Engineering* 23 (**2003**), pp. 127–139.

73 Klemeš, J., Kimenov, Nenov, Application of pinch-technology in food industry. *Proceedings of the 1st Conference on Process Integration, Modelling and Optimisation for Energy Saving and Pollution Reduction – PRES'98*, Prague, August **1998**, lecture F6.6.

74 Smith, R., Linnhoff, B., The design of separators in the context of overall processes. *Chemical Engineering Research and Design* 66 (**1988**), pp. 195–228.

75 Fritzson, A., Berntsson, T., Efficient energy use in a slaughter and meat processing plant – opportunities for process integration. *Journal of Food Engineering* 76 (**2006**), pp. 594–604.

76 Tan, R., Foo, D. C. Y., Pinch analysis approach to carbon-constrained energy sector planning. *Energy*, doi:10.1016/j.energy.2006.09.018, **2007**.

77 Perry, S., Klemeš, J., Bulatov, I., Integrating waste and renewable energy to reduce the carbon footprint of locally integrated energy sectors. *Energy*, doi:10.1016/j.energy.2008.03.008, **2007**.

78 Klemeš, J., Bulatov, I., *Novel Energy Saving Technologies and Their Assessment by EMINENT Evaluation Tool, Efficient Exploitation of Renewable Energy*, CAPE Forum, Maribor, **2007**.

10

Optimization of Structure and Operating Parameters of a Sequence of Distillation Columns for Thermal Separation of Hydrocarbon Mixtures

Mariusz Markowski and Krzysztof Urbaniec

Keywords

distillation columns, pinch technology, refrigeration subsystems, heat exchanger network (HEN), reboilers, Carnot refrigeration cycle

10.1
Introduction

The thermal separation of gas mixtures using a sequence of distillation columns coupled with a refrigeration plant is an energy intensive process. To reduce energy consumption, the following measures can be applied [1, 2]:

- optimization of the sequence of separation of components,
- use of side reboilers and condensers,
- selection of pressure levels in columns so that heat of condensation of their respective overhead vapors can be used in reboilers of other columns [3].

In the industrial practice, side reboilers and condensers are seldom applied and in addition, the prevailing tendency is to choose simple separation sequences. Assuming that the feed is a mixture of n components $A_1, A_2, \ldots, A_{n-1}, A_n$ ordered after increasing volatility, preferred sequences can be exemplified by the following two options of the composition of outgoing streams: (a) distillate A_1; retentate $A_2, \ldots, A_{n-1}, A_n$; (b) distillate $A_1, A_2, \ldots, A_{n-1}$; retentate A_n. Owing to simplification of this kind, the structure of column sequence can be relatively simple facilitating automatic control by not so complex circuitry. However, such an approach may lead to a unnecessary high energy consumption.

It is felt that the consequences, in terms of energy expenditure, of the industrial simplification of column sequences deserve more attention. A design approach based on optimization models would make substantial energy savings possible. Furthermore, advantage could be taken of a considerable energy-saving potential associated with the application of heat-integrated distillation columns [4–7] as already proved in pilot-plant scale [8, 9].

The present work is concerned with the design of energy-efficient gas separation systems comprising a sequence of conventional distillation columns and a refriger-

Process Systems Engineering: Vol. 5 Energy Systems Engineering
Edited by Michael C. Georgiadis, Eustathios S. Kikkinides and Efstratios N. Pistikopoulos
Copyright © 2008 WILEY-VCH Verlag GmbH & Co. KGaA, Weinheim
ISBN: 978-3-527-31694-6

ation subsystem. The application of concepts and tools derived from pinch technology is proposed. Using the notions of ideal column and its profile, thermal integration of the systems employing conventional columns is considered first. The reasoning is subsequently extended to systems employing heat-integrated columns.

10.2
Methodology of Thermal Integration of a System Comprising a Sequence of Distillation Columns and a Refrigeration Subsystem

Irrespective of the type of distillation columns, when operating a column sequence for thermal separation of gas mixtures, process streams must be cooled down, respectively heated up in an adjoined heat exchanger network (HEN). The heat deficit of heating operations is covered by heat supply from heat sources, and the heat surplus of cooling operations is discharged to heat sinks. Furthermore, the column sequence is coupled with a refrigeration subsystem employing a set of compressors. A considerable energy input, typically in the form of electric power, is required to drive the compressors.

A two-stage procedure schematically shown in Fig. 10.1 is proposed for designing an energy-efficient separation system [10]:

I. Preliminary determination of process parameters by thermodynamic inspection;
II. Selection of system structure from the set of feasible structures, sizing of equipment, and final adjustment of process parameters.

For known chemical composition of the feed-containing components A_1, A_2, ..., A_n (listed in the order of increasing volatility), stage I is initiated by determining the column sequence. In an ideal distillation column, the distribution of heat surplus or deficit in the rectifying section and the stripping section is governed by the assumption of thermodynamic equilibrium between liquid and vapor at any point along the column, implying that $y_i \rightarrow k_i \cdot x_i$, where x_i and y_i are molar fractions of the ith component in the liquid and vapor, respectively, and k_i is the equilibrium constant ($i = 1, 2, \ldots, n$).

To satisfy these conditions, a specific order of separation of the components is required. According to Gibbs phase rule for equilibrium systems, a two-phase mixture of n components has n degrees of freedom. Assuming that the pressure in the column is known and the remaining $n - 1$ state variables are

- in the rectifying section, the concentration of components $A_1, A_2, \ldots, A_{n-1}$,
- in the stripping section, the concentration of components $A_2, A_3, \ldots, A_n$,

it becomes possible to determine the concentration of components A_1 and A_n in the distillate and in the residue, respectively.

It can thus be concluded that if the feed is a mixture of n components as specified above, the energy-optimal order of separation of components is governed by the

(a)

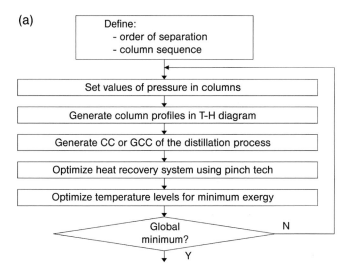

(b)

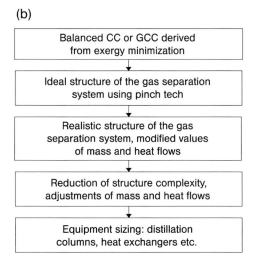

Fig. 10.1 Procedure of heat integration of a system for thermal separation of gas mixtures [10]: (a) stage I – selection of optimum process parameters (pressure, temperature, and mass flow values), (b) stage II – selection of system structure, adjustment of process parameters, and equipment sizing.

principle that the distillate should be a mixture of components $A_1, A_2, \ldots, A_{n-1}$, and the residue – components $A_2, A_3, \ldots, A_{n-1}, A_n$. By applying this principle to each separation stage, the separation sequence is determined and it becomes possible to identify the structure of connections, via incoming and outgoing streams, between distillation columns in the sequence.

As a criterion for the remaining part of decision making in stage I, minimum exergy flow resulting from the interaction between the column sequence and heat sources and sinks in the refrigeration subsystem is proposed. This makes it possible to determine process parameters taking the chemical composition of process streams into account but disregarding system structure and equipment sizing. By minimizing exergy flow, the lowest possible compressor shaftwork in the refrigeration subsystem is ensured. As main decision variables, the pressure levels in columns, the number of external heat sources and their respective temperature levels are regarded. In the case of conventional columns, additional decision variables are the temperature levels in condensers and reboilers. In the case of heat integrated columns, the pressure levels in both the rectifying sections and stripping sections should be regarded as decision variables.

In order to make the application of pinch technology concepts possible, it is proposed to consider the columns as ideal ones. An ideal column is a theoretical notion as it is based on the assumption of reversible separation process characterized by thermodynamic equilibrium along the column; this implying an infinite number of trays as well as infinitely many condensers and reboilers. The entropy increment corresponding to the separation of gas mixture in such an ideal column is equal to the entropy increase caused by mixing of the gaseous components. For an ideal column, the temperature–enthalpy diagram can be used for constructing so-called column profiles to visualize the energy potential of internal streams in both the rectifying section and stripping section of the column.

Having determined column profiles for all the columns in the sequence under consideration and knowing the parameters of process streams linking the subsystems of gas separation system, pinch technology rules can be used to plot, in the temperature–enthalpy diagram, composite curves, and/or grand composite curve for the ideal separation process. It becomes then possible to select temperature levels for heat exchange with external heat sources and sinks, this being equivalent to selecting pressure levels in the individual stages of the compression refrigeration subsystem.

Stage II is concerned with the selection of system structure possibly using an investment-related criterion like the number of equipment units or the total system cost comprising investment and operation costs. No specific criterion can be recommended for general use because in various situations encountered in the industrial practice, one should consider not only system complexity or cost but also operation safety, environmental impact, etc. At this stage, the application of pinch technology is straightforward. One can use the data determined in stage I for ideal columns operated in an ideal system structure, that is:

- theoretical values of mass flow of process media,
- pressure values of process media,
- temperature levels selected for condensers and reboilers,
- temperature levels selected for heat exchange with heat sources and sinks.

After constructing composite curves and grand composite curve, the ideal system structure is determined. Following the reduction of system complexity and the necessary adjustment of process parameters to ensure nonzero driving forces for heat and mass exchange, the values of vapor flow in the respective columns will typically be 5–10% greater than the theoretical ones.

10.3
Mathematical Model of an Ideal Distillation Column

The model of an ideal column is based on the following assumptions:

- the feed point corresponding to the point of thermodynamic equilibrium,
- the thermodynamic state of the feed stream is that of saturated liquid,
- infinitesimally small temperature differences between external heat sources or heat sinks and internal streams,
- thermodynamic equilibrium between liquid and vapor at any point along the column.

The column profiles in the temperature–enthalpy diagram (for both the stripping section and the rectifying section) can be generated using differential equations of mass, energy, and component balances.

Vapor and liquid streams at any point in the column are identified by: L and V – molar flowrates of liquid and vapor, T – temperature, x_i and y_i – molar fraction of the ith component in liquid and gas phase, h' and h'' – molar enthalpy of liquid and vapor.

The energy balance of an infinitesimal element of the column (Fig. 10.2):

$$L \cdot h' - V \cdot h'' + dQ_n - (L + dL) \cdot (h' + dh')$$

$$+ (V + dV) \cdot (h'' + dh'') = 0 \tag{10.1}$$

can be transformed, by neglecting lower order terms, into

$$dQ_n = L \cdot dh' + h' \cdot dL - V \cdot dh'' - h'' \cdot dV \tag{10.2}$$

From the mass balance of the same element:

$$L - V - (L + dL) + (V + dV) = 0 \tag{10.3}$$

it follows that

$$dL = dV \tag{10.4}$$

The mass balance of the ith component in the same element

$$L \cdot x_i - V \cdot y_i - (L + dL) \cdot (x_i + dx_i) + (V + dV) \cdot (y_i + dy_i) = 0 \tag{10.5}$$

can be transformed, by neglecting lower order terms, into

$$L \cdot dx_i + x_i \cdot dL - V \cdot dy_i - y_i \cdot dV = 0 \tag{10.6}$$

The equation expressing thermodynamic equilibrium between liquid and vapor

$$y_i = k_i \cdot x_i \tag{10.7}$$

where k_i – equilibrium constant of the ith component, can be differentiated to yield

$$dy_i = k_i \cdot dx_i + x_i \cdot dk_i \tag{10.8}$$

The equation system (10.4), (10.6), (10.7), (10.8) can be transformed to yield

$$dx_i = \frac{x_i \cdot (k_i \cdot dL - dL + V \cdot dk_i)}{L - V \cdot k_i} \tag{10.9}$$

$$dy_i = \frac{y_i \cdot (dL - dL/k_i + L \cdot dk_i/k_i^2)}{L/k_i - V} \tag{10.10}$$

Knowing that

$$\sum_{i=1}^{n} dx_i = 0 \tag{10.11}$$

and

$$\sum_{i=1}^{n} dy_i = 0 \tag{10.12}$$

and using Eqs. (10.9) and (10.11), we obtain

$$\sum_{i=1}^{n} \frac{x_i \cdot (k_i \cdot dL - dL + V \cdot dk_i)}{L - V \cdot k_i} = 0 \tag{10.13}$$

In a similar way, using Eqs. (10.10) and (10.12), we obtain

$$\sum_{i=1}^{n} \frac{y_i \cdot (dL - dL/k_i + L \cdot dk_i/k_i^2)}{L/k_i - V} = 0 \tag{10.14}$$

By applying Eqs. (10.13) and (10.14) to the stripping section and the rectifying section of the distillation column, the column profiles can be determined.

An ideal distillation column operated at thermodynamic equilibrium is a purely theoretical concept as it would require an infinite number of trays as well as infinite numbers of reboilers and condensers. It can however be illustrated by column profiles in the temperature–enthalpy diagram as shown in Fig. 10.3(a). On the basis of ideal column profiles, the profiles of a real column can be obtained by reducing, to finite values, the numbers of reboilers and condensers.

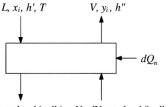

L, x_i, h', T ⟶ V, y_i, h''

⟵ dQ_n

$L+dL, x_i+dx_i, h'+dh'$ $V+dV, y_i+dy_i, h''+dh'', T+dT$

Fig. 10.2 Inputs and outputs of an infinitesimal element of the column.

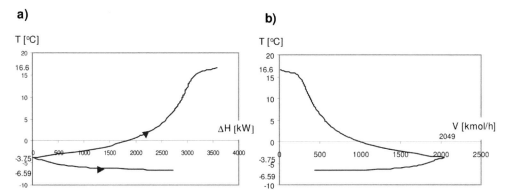

Fig. 10.3 Sample column profiles: (a) heat surplus/deficit in the rectifying section and stripping section, (b) minimum flowrate of vapors. Parameter values correspond to column C-4 in Fig. 10.7.

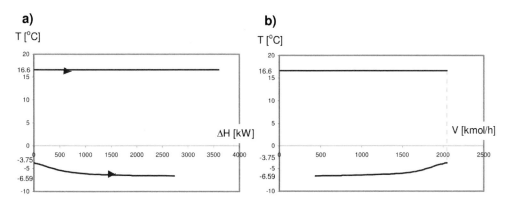

Fig. 10.4 Transformation of the theoretical profiles shown in Fig. 10.3 for the case of a single reboiler and infinitely many condensers.

In Fig. 10.4, an example of reduction of the number of reboilers is presented. The minimum heat flow to be supplied to the reboiler in the rectifying section is represented by the horizontal line in Fig. 10.4(a), while the minimum mass flow of vapor at reboiler outlet is represented by the horizontal line in Fig. 10.4(b). In a

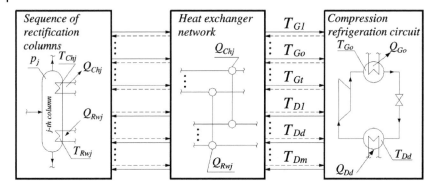

Fig. 10.5 Scheme of interactions between subsystems of the gas separation system employing conventional distillation columns.

real column incorporating the rectifying section that corresponds with the profiles shown in Fig. 10.4, to ensure proper values of the driving forces of heat and mass exchange, the real values of heat and mass flows must be ~5–10% greater than their values indicated in the diagrams.

It should be stressed that the described transformation of column profiles makes it possible to exactly determine, at a definite temperature level, the heat surplus or heat deficit for binary mixtures only. Where the number of components is greater than two, the assumption of constant condensation/evaporation temperature is no longer valid and the resulting uncertainty margin is several percent.

10.4
Thermal Integration Procedure for a System Employing Conventional Distillation Columns

10.4.1
Theoretical Model

The gas separation system can be decomposed into three subsystems schematically shown in Fig. 10.5:

- sequence of conventional distillation columns,
- heat exchanger network,
- compression refrigeration circuit.

As the above subsystems are coupled by multiple process streams, the compressor shaftwork depends on the following factors:

- order of separation of components in the column sequence,
- pressure values in the individual columns,

- number of temperature levels and the corresponding temperature values in re-boilers and condensers,
- number of temperature levels and the corresponding temperature values in the refrigeration circuit,
- structure of the heat exchanger network.

As a first step, the order of separation of components of the feed stream is determined according to the procedure outlined in Section 10.2. This makes it possible to identify the sequence of distillation columns and reduce the optimization task to the problem of minimizing compressor shaftwork, which depends on the HEN structure and the following variables:

- pressure levels in the individual columns,
- number of temperature levels and the corresponding temperature values in the reboilers and condensers,
- number of temperature levels and the corresponding temperature values in the refrigeration circuit.

The knowledge of the separation sequence makes it also possible to determine ideal column profiles, generate composite curves characterizing the ideal separation process and identifying the position of pinch point.

10.4.2
Thermal Integration

Instead of directly dealing with compressor shaftwork, one can consider the objective function expressing the exergy increment, that is, difference between exergy stream of low-temperature heat sources (operated at temperatures T_{Dd}, $d = 1, 2, \ldots, m$, lower than temperature at pinch point) and exergy stream of high-temperature heat sources (operated at temperatures T_{Go}, $o = 1, 2, \ldots, t$, higher than temperature at pinch point). The minimum of this function corresponds to the minimum of compressor shaftwork in the refrigeration subsystem. The objective function can be expressed as

$$FCK = \sum_{d=1}^{m} Q_{Dd}\left(\frac{T_a}{T_{Dd}} - 1\right) - \sum_{o=1}^{t} Q_{Go}\left(\frac{T_a}{T_{Go}} - 1\right) \tag{10.15}$$

where Q_{Dd}, Q_{Go} – heat flow from dth low temperature, respectively, oth high-temperature source, T_a – ambient temperature.

Heat flows Q_{Dd} and Q_{Go} depend on pressure levels in the columns and temperature levels in the reboilers and condensers. This can be symbolically expressed as

$$Q_{Dd} = f_d(T_{Chj}, T_{Rwj}, p_j, T_{D1}, T_{D2}, \ldots, T_{Dm});$$

$$Q_{Go} = g_o(T_{Chj}, T_{Rwj}, p_j, T_{G1}, T_{G2}, \ldots, T_{Gt})$$

where $j = 1 \ldots e$; $h = 1 \ldots u$; $w = 1 \ldots \tau$; e – number of columns; u – number of temperature levels in condensers, and τ – number of temperature levels in reboilers.

The above symbolic equations illustrate the complex nature of relationships involved in the objective function expressed by Eq. (10.15) in the general case of arbitrary number of temperature levels in condensers and reboilers. However, it is reasonable to expect that under industrial conditions, the number of condensers and reboilers is minimized. At known composite curves characterizing the separation sequence (these curves are obtained by combining all the column profiles), the minimum number of temperature levels for condenser and reboiler operation can be graphically determined as illustrated in Fig. 10.6.

The graphical technique is based on two simplifying assumptions:

- in each condenser or reboiler, the condensation or evaporation temperature is constant,
- enthalpy streams described by the composite curves are not affected by the reduction of the number of condensers and reboilers.

Figure 10.6(a) illustrates the principle of determining suitable temperature levels for condensers and reboilers, respectively, in an arbitrarily selected part of the separation sequence. However, assuming the ideal case of heat transfer at zero temperature difference, condensation and evaporation are performed at the same temperature level; this is the underlying principle of graphical construction shown in Fig. 10.6(b), where the enthalpy range of heat regeneration is divided into the minimum number of parts together with their respective temperature levels for condensers and reboilers. If this principle is accepted, then the minimum number of temperature levels for condensers and reboilers depends on the minimum temperature approach at pinch point $\Delta T_{\min \text{PT}}$. Consequently, at known temperature of each level, heat flows from all the external sources also depend on $\Delta T_{\min \text{PT}}$ and therefore the number of variables affecting the heat flows is reduced:

$$Q_{Dd} = f_d'(\Delta T_{\min \text{PT}}, p_j, T_{D1}, T_{D2}, \ldots, T_{Dm})$$

$$Q_{Go} = g_o'(\Delta T_{\min \text{PT}}, p_j, T_{G1}, T_{G2}, \ldots, T_{Gt})$$

As a consequence of the above transformation, the objective function depends on the minimum temperature approach, which can be regarded as a parameter of the minimization problem.

The minimization procedure is started by setting initial values of temperature levels in the refrigeration subsystem. An iteration loop is then entered by setting pressure values in the individual columns, constructing composite curves for the separation sequence (the minimum temperature approach at pinch point $\Delta T_{\min \text{PT}}$ is set at a value characteristic of industrial column sequences, typically 4–10 K), determining the minimum number of condensers and reboilers and so selecting temperature levels in the refrigeration circuit that the value of exergy increment is locally minimized. In the consecutive iterations, the pressure values in the individ-

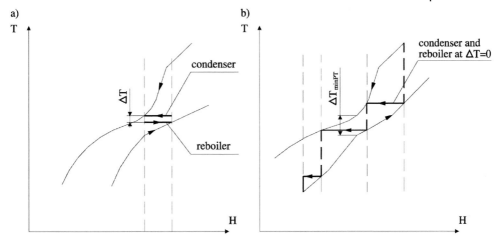

Fig. 10.6 Minimization of the number of temperature levels for condensers and reboilers in a column sequence using composite curves derived from ideal column profiles. (a) determination of suitable temperature levels for condensers and reboilers in an arbitrarily selected part of the separation sequence; (b) division of the range of heat regeneration into minimum number of parts with their respective temperature levels for condensers and reboilers.

ual columns are adjusted and the selection of temperature levels is repeated until the global minimum of the exergy increment is attained. Finally, composite curves or grand composite curve characterizing the entire gas separation system are constructed and details of the optimum structure of HEN are determined using pinch technology.

10.4.3
Example

Let us consider the separation of a four-component hydrocarbon mixture according to the following specification [11]:

- mass fractions: methane: 0.18; ethene: 0.54; ethane: 0.12; propene: 0.16,
- feed flowrate 20 000 kg/h,
- feed temperature 25 °C,
- ambient temperature 25 °C.

In order to facilitate the separation of components, the maximum allowable pressure should be much lower than the critical pressure of any component, so its value is assumed at 3500 kPa. The ethene–propene refrigeration plant is assumed to have three temperature levels at the cold source (three heat sinks) and one temperature level at the hot source.

The main features of the separation system are determined according to the two-stage procedure presented in Chapter 5. The structure of the sequence of ideal distillation columns is shown in Fig. 10.7. Using equilibrium constants of the components calculated from Peng–Robinson equation, ideal column profiles are deter-

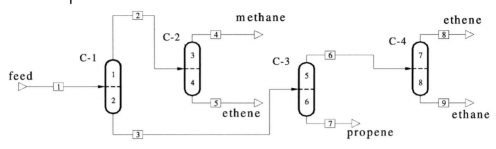

Fig. 10.7 Structure of the sequence of ideal distillation columns.

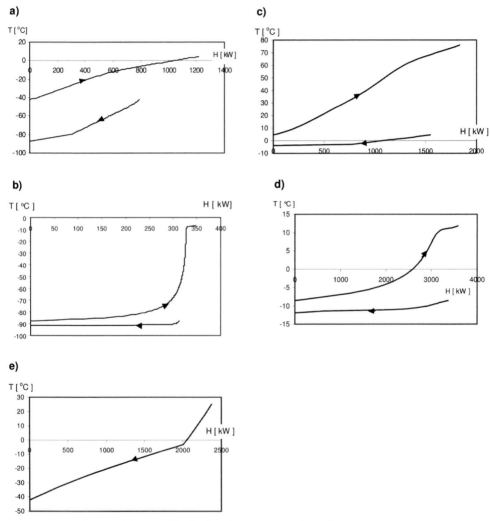

Fig. 10.8 Column profiles for the column sequence shown in Fig. 10.7: (a) column C-1, (b) column C-2, (c) column C-3, (d) column C-4, (e) feed curve [11].

Table 10.1 Temperature levels and heat flows in the refrigeration plant [11].

Temperature [°C]	Heat flow [kW]
+15	+2380
−15	−3313
−45	−1538
−94.5	−788

Table 10.2 Parameters of process streams indicated in Figure 10.7 [11].

Stream No.	Temperature [°C]	Pressure [kPa]	Flowrate [kg/h]	Mass fraction of component	
1	−42.25	3500	20 000	Methane	0.18
				Ethene	0.54
				Ethane	0.12
				Propene	0.16
2	−87.7	3500	4124.9	Methane	0.87
				Ethene	0.13
3	4.65	3500	15 875	Ethene	0.65
				Ethane	0.15
				Propene	0.2
4	−91.5	3500	3597.6	Methane	∼1
5	−6.84	3500	527.3	Ethene	∼1
6	−3.75	3500	12 677	Ethene	0.81
				Ethane	0.19
7	75.92	3500	3197.9	Propene	∼1
8	−11.92	3150	10 289	Ethene	∼1
9	11.85	3150	2388.1	Ethane	∼1

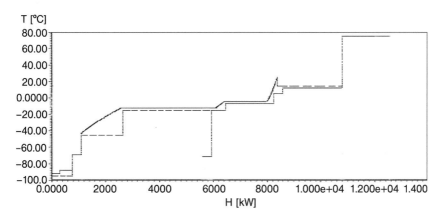

Fig. 10.9 Composite curves for the sequence of ideal distillation columns plotted together with the temperature levels of the refrigeration circuit corresponding to the minimum exergy increment.

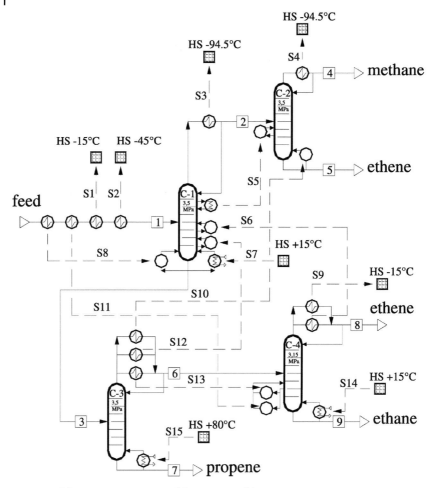

Fig. 10.10 Schematic representation of the structure of the sequence of conventional columns coupled with the HEN and heat sources/sinks in the refrigeration subsystem. The numbers in column icons correspond to the numbers given in Fig. 10.7; HS – heat sink (−) or source (+); parameters of energy streams are listed in Table 10.3.

mined as shown in Fig. 10.8. To enable correct positioning of composite curves characterizing the ideal separation sequence, the minimum temperature approach at pinch point is assumed $\Delta T_{\min PT} = 5$ K and the temperature of heat supply from an external source is assumed at 80 °C. After carrying out calculations to minimize the objective function expressed by Eq. (10.15), the optimum temperature levels are shown in Table 10.1 and the optimum parameters of process streams in Table 10.2. Composite curves based on the optimization results are shown in Fig. 10.9 and the realistic structure of column sequence together with the associated HEN in Fig. 10.10.

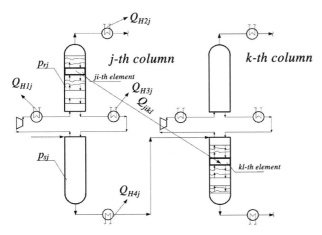

Fig. 10.11 Example of thermal coupling between elements of heat-integrated columns.

10.5
Thermal Integration Procedure for a System Employing Heat-Integrated Distillation Columns

10.5.1
Adjustment of the Theoretical Model

Each heat-integrated column is equipped with a compressor to increase the pressure of vapors leaving the stripping section. A higher pressure and, consequently, a higher temperature in the corresponding part of the column profile can be maintained in the rectifying section making it possible to exchange heat with the stripping section [6, 8]. Assuming that both the sections are divided into series of subsections (elements), thermal coupling in accordance with the principles of pinch technology can be proposed between subsections of different columns [10]. In general, as can be seen in Fig. 10.11, thermal coupling employs transferring heat Q_{jikl} from the ith element of the rectifying section of the jth column to the lth element of the stripping section of the kth column. As a result, a thermally integrated sequence of HIDiC columns is generated.

In order to minimize exergy losses resulting from the heat transfer, each column is equipped with the following heat exchangers as schematically shown in Fig. 10.11:

- vapor cooler to desuperheat compressed vapor (cooler duty Q_{H1j}),
- distillate vapor condenser/cooler to condense and cool down the distillate to the saturation temperature at the inlet of the subsequent column (condenser duty Q_{H2j}),
- cooler to cool down the liquid leaving the rectifying section to the saturation temperature at the inlet of stripping section (cooler duty Q_{H3j}),

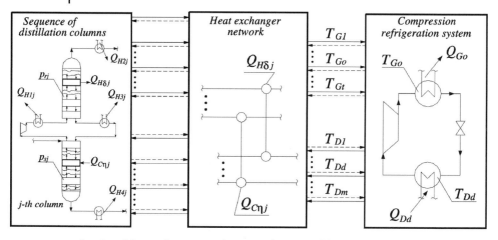

Fig. 10.12 Scheme of interactions between subsystems of the gas separation system employing heat-integrated distillation columns.

- cooler (heater) to cool down (heat up) the depleted liquid to the saturation temperature at the inlet of the subsequent column (cooler/heater duty Q_{H4j}).

It is assumed that the process streams on which the above heat exchangers are installed can interact, at zero temperature difference, with the refrigeration circuit and the elements of stripping sections of the columns in the sequence under consideration. The assumption of theoretical heat exchange at zero temperature difference makes it possible to focus on the thermodynamic aspects of the problem aiming at a preliminary selection of values of process parameters (temperature, pressure, mass flowrate, etc.). At this preliminary stage, the aspects of column and heat exchanger sizing are disregarded. Equipment sizing is carried out in the final step of the design procedure when the preliminary values of process parameters have been adjusted to arrive at realistic requirements on columns and heat exchangers.

The structure of the HEN can be determined using pinch technology. It is schematically shown in Fig. 10.12 that the gas separation system can be decomposed into three subsystems: the sequence of distillation columns, the heat exchanger network, and the compression refrigeration circuit.

As these subsystems are coupled by multiple process streams, the compressor shaftwork depends on the following factors:

- order of component separation in the column sequence,
- structure of the column sequence and HEN (determined on the basis of GCC),
- pressure values in the stripping and rectifying sections of each column,
- number of temperature levels and the corresponding temperature values in the refrigeration circuit.

10.5.2
Thermal Integration

Having defined the sequence of distillation columns, the compressor shaftwork depends on the HEN structure and the following variables:

- pressure values in the stripping and rectifying sections of each column,
- number of temperature levels and the corresponding temperature values in the refrigeration circuit.

To minimize the compressor shaftwork, it is sufficient to minimize the objective function expressed by Eq. (10.15). The heat flows Q_{Dd}, Q_{Go} appearing in the formula (10.15) depend on pressure levels p_{rj} in the rectifying section and p_{sj} in the stripping section of each column and temperature levels T_{Dd}, T_{Go} of heat sources, and are determined numerically. Their dependence on the decision variables can be indicated by the symbolic equations [10]:

$$Q_{Dd} = f_d(p_{rj}, p_{sj}, T_{D1}, T_{D2}, \ldots, T_{Dm})$$

$$Q_{Go} = g_o(p_{rj}, p_{sj}, T_{G1}, T_{G2}, \ldots, T_{Gt}) \quad (j = 1, \ldots, e)$$

where $j = 1, \ldots, e$ and the symbols are consistent with those used in Eq. (10.15) and Fig. 10.12.

The minimization procedure is started by setting initial values of pressure in each column, constructing GCC and selecting temperature levels in the refrigeration circuit so as to locally minimize the objective function. In the consecutive iterations, the pressure values in the individual columns are adjusted and the selection of temperature levels is continued until the global minimum of the objective function is attained. Simultaneously, the structure of the HEN is determined using pinch technology.

10.5.3
Example

Let us consider the separation, in a sequence of HIDiC columns, of a four-component hydrocarbon mixture according to the specification listed in Section 10.4.3. The refrigeration subsystem is assumed to have four temperature levels at the cold source (four heat sinks). The temperature of heat supply from an external source is assumed at 60 °C.

As the chemical composition of the feed is the same as in the previous example, the optimum order of component separation is also the same and therefore the scheme of column sequence does not differ from that shown in Fig. 10.7. After carrying out iterative calculations according to the procedure schematically shown in Fig. 10.1(a), the minimum value of objective function expressed by Eq. (10.15) is found at 610 kW and the values of process parameters in the column sequence are determined. The results of pressure calculation are following:

Table 10.3 Parameters of energy streams indicated in Figure 10.10.

Stream No.	Heat duty [kW]	Stream No.	Heat duty [kW]
S1	475	S9	2838
S2	1538	S10	50
S3	476	S11	150
S4	312	S12	270
S5	300	S13	1230
S6	530	S14	2160
S7	220	S15	1838
S8	200		

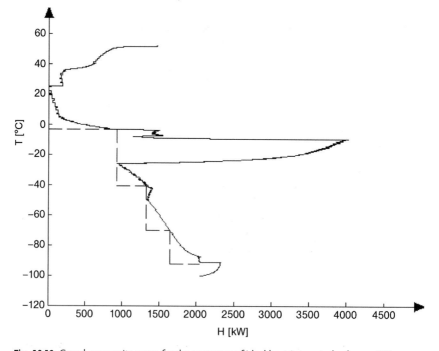

Fig. 10.13 Grand composite curve for the sequence of ideal heat-integrated columns [10].

○ for the rectifying sections of columns C-1 to C-4,

$$p_r = [3450, 3500, 3500, 3250] \text{ kPa};$$

○ and for the corresponding stripping sections,

$$p_s = [3000, 2320, 2200, 2040] \text{ kPa}.$$

The results of calculations based on idealized models are converted into engineering design according to the procedure schematically shown in Fig. 10.1(b). As

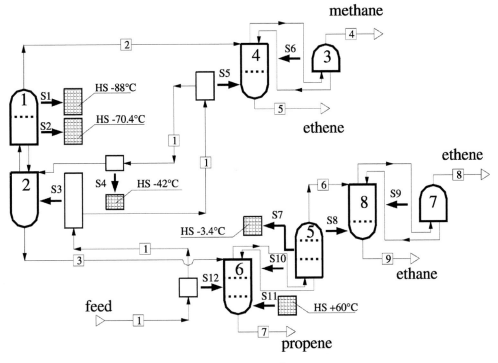

Fig. 10.14 Schematic representation of the realistic structure of
the sequence of heat-integrated columns coupled with the HEN
and heat sources/sinks in the refrigeration subsystem. The
numbers in column icons correspond to the numbers given in
Fig. 10.7 [10]; HS – heat sink (−) or source (+); parameters of
energy streams are listed in Table 10.4.

the initial step, the grand composite curve is constructed as shown in Fig. 10.13.
This makes it possible to determine the optimum values of temperature, also
shown in Fig. 10.13, at its four levels in the refrigeration plant. In Table 10.5, the
values of temperature and heat flows at low-temperature sources in the refrigeration circuit are compared with those previously obtained for the sequence of conventional distillation columns (see also Table 10.1).

After selecting system structure, the values of process parameters including temperature and pressure levels, and mass flows are changed. The resulting values of
heat flows and compressor shaftwork are larger than those corresponding to the
theoretical optimum. As the next step, system complexity is reduced taking engineering constraints into account. The realistic structure of the sequence of HIDiC
columns and its interfaces to other subsystems (heat sinks corresponding to the
cold source of the refrigeration system, and heat sources) is shown in Fig. 10.14.
The implementation of gas separation process in the realistic system structure ne-
cessitates adjustments in mass flows; the grand composite curve is changed to the
form presented in Fig. 10.15 and the adjusted minimum value of objective function

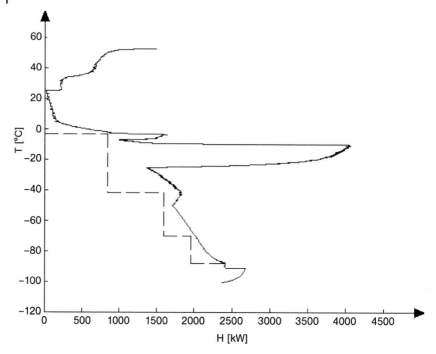

Fig. 10.15 Grand composite curve corresponding to the realistic column sequence presented in Fig. 10.14.

Table 10.4 Parameters of energy streams indicated in Figure 10.14.

Stream No.	Heat duty [kW]	Stream No.	Heat duty [kW]
S1	430	S7	840
S2	360	S8	540
S3	1317	S9	3230
S4	760	S10	220
S5	70	S11	1520
S6	340	S12	230

(10.15) is 741 kW. Bearing in mind that the objective function has a physical meaning of the work of Carnot refrigeration cycle, the increase in function value, relative to its theoretical minimum by a factor of 1.21, indicates that the compressor shaftwork in the realistic system is more than 20% larger than compressor shaftwork at its theoretical minimum.

Table 10.5 Temperature levels and heat flows at hot and cold sources of the refrigeration subsystem.

Sequence of conventional columns		Sequence of HIDiC columns	
Temperature [°C]	Heat flow [kW]	Temperature [°C]	Heat flow [kW]
+15	+2380	−3.4	−924
−15	−3313	−40.7	−402
−45	−1538	−70.4	−322
−94.5	−788	−88	−405

10.6
Conclusions

The proposed methodology of thermal integration of a system comprising a sequence of distillation columns and a refrigeration subsystem can be regarded as an engineering tool to facilitate energy efficient application of sequences of distillation columns. The methodology works well with examples of systems for separation of hydrocarbon mixtures.

The results of examples solved illustrate the energy-saving potential of heat-integrated distillation columns. If applied instead of conventional columns in a system comprising also a HEN and a refrigeration subsystem, a sequence of heat-integrated columns makes it possible to significantly reduce the compressor shaft-work needed for gas separation. In Table 10.5, the optimized parameters of refrigeration cycle coupled with a sequence of conventional distillation columns are compared with those of refrigeration cycle coupled with heat-integrated columns. The value of objective function according to Eq. (10.15) at its theoretical minimum, that is, prior to defining realistic system structure, is 1431 kW in the former case, and 610 kW in the latter case – a reduction by 57% that can be attributed to heat-integrated columns. These are theoretical figures to be interpreted on the basis of Carnot refrigeration cycle and in addition, energy needed for driving compressors in the heat-integrated columns is disregarded. Consequently, the corresponding reduction of compressor shaftwork in a real system will probably be less significant, but a reduction of 25–30% seems possible.

Symbols

h' molar enthalpy of liquid (J/mol)
h'' molar enthalpy of vapor (J/mol)
H enthalpy flow (W)
k_i equilibrium constant of the ith component [−]
L flowrate of liquid (mol/s)
p_j pressure in the jth column (Pa)
p_{rj} pressure in the rectifying section of the jth column (Pa)
p_{sj} pressure in the stripping section of the jth column (Pa)
Q_{Chj} heat flow in the hth condenser of the jth column (W)

Q_{Dd} heat flow in the dth low-temperature source of the refrigeration circuit (W)
Q_{Go} heat flow in the oth high-temperature source of the refrigeration circuit (W)
Q_n surplus or deficit of the heat flow (W)
Q_{Rwj} heat flow in the wth reboiler of the jth column (W)
T temperature (°C)
T_a ambient temperature (K)
T_{Chj} temperature in the hth condenser of the jth column (K)
T_{Dd} temperature of the dth low temperature of the refrigeration circuit (K)
T_{Go} temperature of the oth high-temperature source of the refrigeration circuit (K)
T_{Rwj} temperature in the wth reboiler of the jth column (K)
V flowrate of vapor (mol/s)
x_i molar fraction of the ith component in liquid [−]
y_i molar fraction of the ith component in vapor [−]
$\Delta T_{\min PT}$ minimum temperature difference at pinch point (K)

Indices

e number of columns
m number of temperature levels for low-temperature heat sources in the refrigeration circuit
n number of components in the gas mixture
t number of temperature levels for high-temperature heat sources in the refrigeration circuit
u number of condensers
w number of reboilers

References

1 SMITH, R., *Chemical Process Design*, McGraw-Hill, New York, **1995**.
2 ULLMANN'S ENCYCLOPEDIA OF INDUSTRIAL CHEMISTRY, Vol. B3, Unit operations II, 5th Edition, VCH, New York, **1988**, pp. 4–57.
3 KING, C. J., *Separation Processes*, 2nd Edition, McGraw-Hill, New York, **1980**.
4 IWAKABE, K., NAKAIWA, M., HUANG, K., NAKANISHI, T., ZHU, Y., ROSJORDE, A., OHMORI, T., ENDO, A., YAMAMOTO, T., PRES **2004**.
5 MAH, R. S. H., NICHOLAS, J. J., WODNIK, R. B., *AIChE* 23 (**1977**) 651.
6 OLUJIC, Z., GADALLA, M., SUN, L., DE RIJKE, A., JANSENS, P. J., PRES 2004, paper No. 676 (**2004**).
7 SUN, L., DE RIJKE, A., OLUJIC, Z., JANSENS, P. J., in: *The 5th International Conference on Process Intensification for the Chemical Industry*, Gough, M. (ed.), Maastricht, The Netherlands, **2003**, p. 151.
8 NAKAIWA, M., HUANG, K., ENDO, A., OHMORI, T., AKIYA, T., TAKAMATSU, T., *Trans. IChemE, Part A, Chem. Eng. Res. Des.* 81 (**2003**) 162.
9 OLUJIC, Z., FAKHRI, F., DE RIJKE, A., DE GRAAUW, J., JANSENS, P., *J. Chem. Technol. Biotechnol.* 78 (**2003**) 241.
10 MARKOWSKI, M., TRAFCZYŃSKI, M., URBANIEC, K., *Appl. Therm. Eng.* 27 (7) (**2007**) 1198.
11 MARKOWSKI, M., URBANIEC, K., BUDEK, A., GRABARCZYK, R., PRES 2004, paper No. 879 (**2004**).
12 FONYO, Z., *Int. Chem. Eng.* 14 (**1974**) 18.

Index

21st Century Energy Plant   4

a
alternative technologies
– key parameters   16
– nomenclature   16

b
biomass   279f.
black-oil model   163
branch-and-bound algorithm   45ff.
– decision variables   142

c
capacity expansion planning   216, 219ff.
– capacity expansion model   219ff.
– chance constraints concept   219
– decomposition algorithm   226ff.
– probabilistic model   223ff.
– stochastic programming   217ff.
carbon dioxide capture and sequestration, CCS   5
carbon footprint, CFP   251, 274ff.
– LCA-based evaluation   281ff.
– technology performance   283f.
Carnot refrigeration cycle   129, 220
CO_2 capture   231ff.
– flowsheet   233f.
– optimization   238ff.
– process model   234
– Ryan–Holmes process   232
– separation techniques   232
CO_2 emissions   39, 57, 59ff., 251, 274ff.
coal gasification   127
cold gas cleanup, CGCU   21
combined heat and power, CHP   276ff.
– biomass   279f.
– fuel cells   278f.
– micro-CHP   277, 279
– reciprocating engine   277

– solar   280ff.
– Sterling engine   277f.
– wind energy   283
compact heat exchangers   253
composite curves, CC   77f., 204, 209ff., 219f., 259ff.
– balanced composite curve, BCC   261
– balanced grand composite curve, BGCC   261
– exergy grand composite curve, EGCC   295f.
– grand composite curve, GCC   261ff.
compressed hydrogen, CGH2   90f.
– compression   91ff.
– operation issues   91ff.
– storage   90ff.
compression refrigeration system   264
conditional value-at-risk (CVaR) constraint   219, 223
configuration design   18ff., *see also* process design
cryogenic tank   96f.

d
Darcy's law   89
decomposition algorithm   226ff.
– Benders decomposition   227
– L-shaped method   227f.
distillation columns   201ff.
– column profiles   204
– column sequence   201
– composite curves   204, 209ff., 219f.
– conventional column model   208ff.
– heat-integrated column model   215ff.
– ideal column model   204ff.
– thermal integration   209f., 217
distribution networks   39, 55f.
district energy system design   39ff.

– composite curves 77f.
– costing functions 57ff.
– data preprocessing 61f.
– empirical constraints 60f.
– energy balances 48ff.
– energy consumption profile 42, 64f.
– mass balances 55
– master optimization 46ff.
– optimization phase 44ff., 70ff.
– postprocessing phase 44, 81
– resolution strategy 45
– slave optimization 47f.
– structuring phase 42ff.
– test case 63ff.
district heating systems 39
– distribution networks 39
dynamic performance simulation 203ff.

e
economic efficiency 32
electric power industry 215ff.
– capacity expansion problem 216ff.
– management problems 216
EMINENT software tool 268, 272ff., 281
energy conversion technologies 42ff., 56f.
– centralized 43
– decentralized 43, 69
– test case design 63ff.
energy efficiency 249ff., 253ff.
– audit 254f.
– cogeneration 253
– heat integration 250, 257ff.
– improvement factors 253f.
– pinch technology 252
energy efficiency optimization 265ff.
– management program 266
energy management systems, EMS 266f.
energy-saving analysis 255ff.
– case studies 284ff., 289ff.
– process data reconciliation 255
– software tools 256f.
energy-saving technologies 251ff., 267ff.
– early stage technologies, ESTs 268ff.
– EMINENT software tool 268, 272ff., 281
– energy supply algorithm 268ff.

f
fifth-order model 201f.
flowsheet 233ff.
– design 233ff., 242ff.
– optimization 239ff.
flowsheeting simulation 251, 255f.

fluid catalytic cracking, FCC 284ff.
food processing industry 291ff.
fuel-cell technology 95, 130ff., 252, 278f.
– high-temperature fuel cell 130
– molten carbonate fuel cell, MCFC 130
– solid oxide fuel cell (SOFC) combined cycle 130ff.

g
gas separation 231ff.
– costs 235
– optimization 238
– subsystems 216
– techniques 232
gas turbine 23, 28
gas-lift well 160ff., 175, 186ff.
gas-oil ratio, GOR 160
– incremental gas-oil ratio, IGOR 161
General Algebraic Modeling System, GAMS 32
genetic algorithm, GA 45ff., 161
global warming potential, GWP 144
greenhouse gas, GHG 1, 125
– emissions 144ff.

h
heat exchanger network, HEN 202, 284ff.
heat integration 250, 257ff.
– case studies 284ff.
– composite cold stream 260
– composite curves, CC 259ff.
– composite hot stream 260
– compression refrigeration system 264
– exergetic efficiency 263f.
– flow diagram 263
heat recovery steam generator, HRSG 20ff., 29
high-temperature fuel cell 130
hot gas cleanup, HGCU 21
hydrogen adsorbents 99ff.
– carbon-based adsorbent 108ff.
– metal hydrides 101f.
– modeling equations 103ff.
– pore size distribution 109
hydrogen compressors 90ff.
– diaphragm compressors 92f.
– piston compressors 92f.
hydrogen infrastructure 138ff.
– pathway options 139
– planning 140ff.
hydrogen infrastructure strategic planning model 141ff.

– Chinese case study   147ff.
– MILP modeling   141, 144
– superstructure representation   142
hydrogen production technologies   125ff.
– partial oxidation   126ff.
– steam reforming   126ff.
hydrogen storage systems   96ff.
– compressed hydrogen storage   90ff.
– dynamic optimization framework
  114ff.
– liquid hydrogen storage   93ff.
– "solid" hydrogen storage   98ff.,
  108ff.
– underground storage   97ff.
hydrogen types   94
– ortho–para conversion   95
hydrogen-electricity polygeneration systems
  127ff.
– capital investment   280
– case study   146ff.
– cost of electricity, COE   280ff.
– economic analysis   130ff.
– hydrogen cost   133ff.
– hydrogen production system   127ff.
– IGCC-based   127, 129ff.
– infrastructure   138ff.
– SOFC combined cycle-based   130ff.

i
incremental gas-oil ratio, IGOR   161
induction generator   200ff.
– fifth-order model   201f.
– reduced-order model   202f.
Integrated Gasification Combined Cycle,
  IGCC   2ff., 130
– cogeneration system   129
Intergovernmental Panel for Climate
  Change, IPCC   144
investment planning, *see* strategic
  investment planning

j
Joule–Thompson cycle   94

l
life-cycle assessment, LCA   251
linear programming, LP   141, 161
linearization algorithm   13f.
liquid hydrogen, LH2   93ff., 98f.
– boil-off   97f.
– liquefaction   94ff.
– operation issues   94ff.
– ortho–para conversion   95

m
metal hydrides   101f.
– hydrogen storage modeling   102ff.
– hydrogen storage optimization
  112ff.
metal-organic frameworks, MOFs   99ff.
methanol pathway   146ff.
methanol synthesis   21, 26ff., 34
MIP modeling techniques   8ff.
– linearization   13f.
– model variables   12
– objective function   11
mixed-integer linear programming, MILP
  141, 226ff.
– hydrogen infrastructure model
  141ff.
mixed-integer nonlinear programming,
  MINLP   18ff., 161ff., 45
mixed-integer programming, MIP   8ff.
– net present value, NPV   8f.
– superstructure   9f.
multilayer insulation, MLI   96f.
multiobjective optimization problem   44ff.,
  60ff.
– evolutionary optimizer   142
multiparametric controller, MPC   196,
  207ff.
– explicit MPC   208ff.
– optimization problem   209f.

n
natural gas   231
– autothermal reforming, ATR   127ff.
– CO_2 separation techniques   232
– partial oxidation, POX   127ff.
– steam methane reforming, SMR
  127ff.
net present value, NPV   8f., 11, 144
NODAL analysis   160
nonlinear programming, NLP   116

o
offshore oilfield   159
oil production optimization   159ff.
– accuracy   59
– examples   183ff.
– heuristic methods   160
– IGOR heuristic rule   161, 187
– mathematical programming
  techniques   161ff.
– NODAL analysis   160
– oil flow rate   170
– well management routines   160
open-loop control problem   208

operating costs, OC   235, 244ff.
ordinary differential equation, ODE   166f.

p

Pareto optimal frontier   44, 72ff., 142ff., 145
 – configuration types   72
photovoltaic, PV   280ff.
pinch technology   204, 250, 252, 258, *see also* heat integration
pipeline network simulators   160ff.
 – four-well dry gas model   61ff.
 – gas-lift well   186ff.
 – naturally flowing well   183ff.
 – treelike structure pipeline network 188f.
 – two-well network   163ff.
polygeneration energy systems   1ff., 15ff., 40ff., 127ff.
 – advantages   4
 – capital investment   280
 – characters   18
 – Chinese development plan   7f.
 – design optimization   18ff.
 – development plans   4ff.
 – district energy systems   39ff.
 – features   9
 – hydrogen production   127ff.
 – Integrated Gasification Combined Cycle, IGCC   2ff.
 – models   1, 8ff.
polygeneration plant
 – investment planning   15ff.
 – superstructure representation   20
primary energy technologies   1f.
process design   19ff.
 – case study   31ff.
 – economic parameter   29f., 33
 – mathematical model   23ff.
 – physical representation   24ff.
 – sensitivity analysis   34ff.
 – superstructure representation   20ff.
process integration technology   257ff., *see also* heat integration
 – software   261
process model development   234ff.
 – optimization   241ff.
 – overall installation factor, OIF   235
 – SAFT–VR equation of state   235ff.
production optimization   160ff., 170ff., *see also* oil production optimization
 – mathematical models   163ff.
 – optimization formulation   179ff.
 – separability programming   170ff.
 – solution procedure   177ff.

 – stability condition   175
production wells   159ff.
 – gas-lift well   160ff., 175, 186ff.
 – gas-to-gas lift well   160ff.
 – naturally flowing well   183ff.
 – well oil rate   172ff.
pseudoglobal sensitivity analysis   34ff.

r

reboilers   206ff., 210f.
reciprocating engine   277
reduced-order model   202f.
refrigeration subsystem   202ff., 210
renewable energy sources   252
Ryan–Holmes process   232

s

SAFT–VR equation of state   235ff.
scenario tree   226
sensitivity analysis   34ff., 159
 – model parameters   35
 – pseudoglobal sensitivity   34
separability programming   170ff.
sequential linear programming, SLP   161, 177ff.
 – procedure   177f.
sequential quadratic programming, SQP 161
solar energy   280ff.
 – case studies   281
"solid" hydrogen storage   98ff.
 – absorption   98ff.
 – adsorbents   99ff.
 – adsorption   98
 – dynamic optimization framework 114ff.
 – kinetics   104ff.
 – metal hydrides   101f.
 – model solution   108ff.
 – operation issues   102f.
 – process modeling   103ff.
 – tank optimization   112ff.
solid oxide fuel cell, SOFC   127ff.
SOFC combined cycle   127ff.
steam reforming   126
Stirling engine   277f.
stochastic programming, SP   217ff.
 – modeling framework   218f.
strategic investment planning   8ff., 215ff.
 – case study   15ff.
 – levels   10
 – MIP algorithm   8ff.
 – model assumptions   11
 – model variables   12

successive linear programming, SLP   161,
    176ff.
superstructure representation   9f.
syngas production   3
syngas purification   21ff.
    – physical representation   24ff.

t

technology database   43ff.
Texaco gasification technology   31, 127f.
thermal efficiency   249f.
thermoeconomic simulation model   42
total capital investment, TCI   235, 243ff.

u

underground hydrogen storage   97ff.
    – reservoirs   97
    – operation issues   89ff.
    – sealing effect   98

v

Vision 21 plan   4ff.
    – supporting technologies   6

w

wall distillation technology   253

water-gas shift reaction, WGS   127
well management routines   160
    – levels   160
well models   164ff.
    – choke model   165, 169
    – flow line model   165
    – manifold model   165
    – Peaceman's well model   164
    – surface flow line model   165
    – tubing model   165
    – well oil rate   172
"well-to-wheel" (WTW) life cycle   125
wind conversion system   195f.
wind energy   195, 283
wind turbines   195ff.
    – aerodynamics modeling   197ff.
    – control techniques   195
    – drive train   197, 199f.
    – dynamic behavior   203ff.
    – efficiency   204
    – induction generator   200f.
    – pitch angle control   204
    – power control   207ff.
    – simulations   205ff.
    – system modeling   196ff.